Science and Ecosystem Management in the National Parks

Science and Ecosystem Management in the National Parks

William L. Halvorson and Gary E. Davis, Editors

The University of Arizona Press Tucson

The University of Arizona Press

∞ This book is printed on acid-free, archival-quality paper.
Manufactured in the United States of America

01 00 99 98 97 96 6 5 4 3 2 1

Library of Congress Cataloging-in-Publication Data

Science and ecosystem management in the national parks / William L. Halvorson and Gary E. Davis, editors ; foreword, Paul G. Risser.
p. cm.
Includes bibliographical references and index.
ISBN 0-8165-1566-2 (cloth : alk. paper)
1. National parks and reserves—United States—Management. 2. Conservation of natural resources—Government policy—United States. 3. Ecosystem management—Government policy—United States. 4. Natural resources—United States—Management. 5. National parks and reserves—Research—United States. 6. United States. National Park Service—Management. I. Halvorson, William L. (William Lee), 1943– . II. Davis, Gary E.
SB482.A4S35 1996
333.78'3'0973—dc20 95-32530

British Library Cataloguing-in-Publication Data
A catalogue record for this book is available from the British Library.

Contents

Foreword

The U.S. national park system comprises the most widely recognized scenic areas in all of America. Virtually everyone knows, at least by reputation, of Yellowstone, the Grand Canyon, the Everglades, or the Great Smokies. These places are valued as treasures of wilderness of spectacular grandeur or as examples of awesome biological or geological processes. They offer recreational opportunities as well as places for inspiration, reflection, and mental and emotional renewal for millions of people each year.

Beyond all this, the national parks are a source of comforting constancy, i.e., an emotional anchor. Despite changes in economic status, political upheaval, social injustices, or disasters, the national parks are always available to serve as actual or potential refuges. The parks are traditionally "American," are always welcoming, and serve as symbols of all that we value.

This constancy, this anchor in turbulent times, however, is threatened. The national park system is at risk from both internal and external forces. These threats, which are well documented in this book, are caused by excessive use of fragile habitats, overuse and consumption of natural resources, air and water pollution, and invading species.

Sometimes threats to the unique and precious features of the parks are caused by improper management decisions. Such decisions are not the result of irresponsibility, but rather due to a lack of crucial information. For example, park administrators may not understand the natural processes well enough to make informed decisions to protect the saguaro cactus in Saguaro National Monument or to manage the wolves and moose on Isle Royale. In some areas, the population status of individual or interacting species may not be sufficiently known, as was the case with species invasions into Hawaii's national parks and with the fisheries in Yellowstone National Park. In other areas, management decisions may face public opposition, such as the use of fire to manage the sequoia trees in Sequoia and Kings Canyon National Parks.

Managing the national parks is complicated and difficult. Ecosystems are highly complex, with many interactions among the living and nonliving components. In addition, science has demonstrated that these ecosystems have considerable natural variation. Management decisions must take into account the dynamic nature of plant and animal populations as well as the ramifications of such decisions. As was demonstrated repeatedly, decisions made to protect one resource may, and did, jeopardize another.

It would be possible theoretically to manage these complex ecosystems by simply using them until disaster, whether natural or man-made, strikes, resulting in the loss of vital characteristics and landmarks. Some changes, however, are irreversible, such as loss of species, and some recovery processes are exceedingly slow. A resource damaged by human activities may also respond in subtle ways and may already be severely jeopardized before the damage can be recognized by casual observation. A far better approach, therefore, would be to regularly monitor the essential characteristics of each park so that any signs of ecosystem degradation could be readily identified and remedied. Park administrators also need to understand the structure and function of these ecosystems well enough to predict how the park will respond to internal and external threats and to various management decisions.

The only way for park administrators to evaluate the status of the natural resources in the national parks and to know how to manage these resources is to develop a systematic monitoring program and to conduct research on the most important processes that control these ecosystems. Such monitoring and research programs yield information that can be used to protect the parks and to guide management decisions that will preserve the natural resources. Research can also provide data concerning park usage and maintenance levels to determine which areas can withstand additional usage, thereby providing added enjoyment for the public. Research can also assist in identifying areas that require some usage constraints to ensure their survival for the enjoyment of future generations.

This book uses a series of 12 extremely interesting case studies to demonstrate effectively the power of science in guiding the management of our national parks. Each of these cases is fascinating in itself; collectively, they offer a powerful argument for expanding and strengthening the research program of the National Park Service. Indeed, one cannot read this book without reaching the inescapable conclusion that past research has dramatically saved natural resources in some parks. As a nation, we are being irresponsible to ourselves and to future generations unless we develop a stronger and more comprehensive research program from which to manage our national parks.

Past generations have enjoyed the constancy, comfort, and pleasures of the national parks. Today, those parks are facing well-documented threats. If we are to preserve these complex and irreplaceable national treasures, they must be managed strategically and scientifically.

Paul G. Risser
President, Miami University, Oxford, Ohio

Abbreviations

AQD	Air Quality Division
AQRV	air-quality-related values
BART	best available retrofit technology
CAWCD	Central Arizona Water Conservation District
CPSU	Cooperative Park Studies Unit
CPV	canine parvo-virus disease
CRF	Cave Research Foundation
DDT	dichloro-diphenyl-trichloro-ethane
DMB	differential mass balance
EOF	empirical orthogonal function
EPA	Environmental Protection Agency
ESA	Ecological Society of America
GAO	General Accounting Office
GIS	Geographic Information System
KWIS	Karst Water Instrumentation System
LVVWD	Las Vegas Valley Water District
MCNPA	Mammoth Cave National Park Association

MSY	maximum sustainable yield
NBS	National Biological Service
NGS	Navajo Generating Station
NIH	National Institutes of Health
NPC	National Park Concessions, Inc.
NPS	National Park Service
NS	nonsignificance
RMD	Rincon Mountain District (Saguaro National Monument)
SCENES	Subregional Cooperative Electric Utility, NPS, EPA, and Department of Defense Study
SNM	Saguaro National Monument
SRP	Salt River Project
T&E	threatened and endangered (species)
T&EMP	Threatened and Endangered Plant-Monitoring Program
TMBR	tracer mass balance regression
TMD	Tucson Mountain District (Saguaro National Monument)
USDA	U.S. Department of Agriculture
USDI	U.S. Department of the Interior
USFS	U.S. Forest Service
USFWS	U.S. Fish and Wildlife Service
USGS	U.S. Geological Survey
WHITEX	Winter Haze Intensive Tracer Experiment

1

Historical Perspective

1

Long-term Research in National Parks: From Beliefs to Knowledge

Gary E. Davis and William L. Halvorson

Management of natural resources in national parks has struggled to evolve from belief-based advocacy to knowledge-based consensus in the late twentieth century. This evolution paralleled a shift from the view of park ecosystems as static, isolated, independent landscapes to the scientific understanding that parks are dynamic, integrated with larger landscapes, and affected by human activities far and near. This understanding brought a recognition that parks require active, iterative, experimental, and adaptive management (Leopold et al. 1963; Robbins et al. 1963; Orians et al. 1986). Early on, park managers believed that fire and predators were "bad," so they suppressed fire and removed wolves and coyotes from national parks to protect the forests and "good" game animals, such as elk and deer (Sumner 1983). Based on this belief and deductive reasoning alone, it was logical to remove "bad" influences from the parks.

These attitudes began to change as early as 1933, when George Wright and others challenged traditional National Park Service (NPS) belief-based concepts by recognizing, and publicizing in Fauna Series 1, the legitimate roles of native tree-boring insects and naturally occurring fire in parks (Wright et al. 1933). In 1940 Adolph Murie reinforced this ecological view of parks. He also pressed to uphold an NPS policy of protecting predators, contrary to administrative sentiment to continue coyote "control" in Yellowstone National Park (Murie 1940; Sumner 1983). In spite of these early efforts, little progress was made to move away from preconceived notions of "right" and "wrong" biota in parks and to embrace the emerging science of ecology. In the 1960s NPS was still spraying native forest insects to protect desirable trees, "controlling" native plant populations, and planting fish in naturally fishless lakes to provide recreational fishing opportunities. Simultaneously, heated discussions about the appropriateness of active management versus natural regulation in parks embroiled the agency. Pressed to adopt a scientific approach to park management by the conservation and scientific communities through numerous advisory boards, surveys, and blue-ribbon panels (11 in 30 years from 1963 to 1992; see chapter 2 of this volume), NPS still struggles to break from the past. The old ways linger on. Many policymakers and park administrators still rely on their beliefs and opinions to protect our common heritage, rather than on scientific inquiry and verifiable knowledge (Layden and Manfredo 1994).

Modern ecology is slowly moving park management beyond belief-based treatments of symptoms to science-based ecosystem management of causes. We envision that in the long-term, management decisions based on adequate understanding of ecosystem function will prevail over belief-based advocacy. This collection of case studies provides tests of this hypothesis from many ecosystems and sociopolitical situations. Our vision is that the knowledge of ecosystem structure and function from sustained research, such as that described in these case studies, will help administrators to accept the paradigm shift from belief to science-based natural-resource management in national parks and elsewhere.

Resource Trends

Current national attitudes toward natural resources are driving alarming trends in biological systems, both in and out of national parks. Oddly, the nation's approach to natural-resource-based economic development has not changed much from the time when the American frontier closed in the nineteenth century. We continue to consume resources as if they were inexhaustible, as if we could still go over the next mountain range to find more when we run out. Some people are still shocked to learn that we live in a world with finite resources. Nevertheless, the time has passed when we could afford to act as if technology would always bail us out, no matter how much we degraded the environment. The rapidly increasing human population in the United States continues to demand more from its finite resource base. Unsustainable consumption of "renewable" resources; habitat fragmentation; human alterations of air, water, and soil; and the spread of alien species all require attention to avert economic, social, and environmental catastrophe.

Unsustainable consumption of "renewable" resources drives populations and communities to failure and reduces economic productivity. Serial depletion of coastal fishery stocks and the harvest of ancient forests are prime examples. In southern California, the diving fishery exhausted stocks of five abalone species one after another from 1950 to 1980, shifted to red sea urchins in the mid 1970s, expanded into northern California in the late 1980s when southern California urchin stocks declined, and began developing new markets for purple urchins in the early 1990s (Dugan and Davis 1993). This pattern of unsustainable biotic resource exploitation is common worldwide (U.S. Department of Commerce 1992).

Most land-use practices fragment habitats and thereby erode society's productive resource base when populations and communities collapse for lack of appropriate space, i.e., critical habitat (Robinson et al. 1992). Coastal development threatens migratory birds and coastal fisheries with the loss of marshes and estuaries. Loss of large, wide-ranging predators alters community structure and function, thereby accelerating loss of biodiversity (Wilson 1988).

Human alterations of air, water, and soil drive ecosystems toward unstable and less productive states: pollution-simplified systems, reduced productivity of contaminated wildlife, groundwater extraction, and surface-water diversion are all common.

The spread of alien species causes loss of natural biodiversity and disrupts ecosystem structure and function. The virtual extinction of native birds on Guam caused by introduced brown tree snakes provides a sobering example of the serious ecological impacts of alien species. Alien species are wreaking havoc on native flora and fauna in many areas of the world. In addition to tropical rainforests, two of the most notable examples of ecosystems threatened by alien species are national parks in Hawaii and southern Florida.

Evaluation of NPS Science

In 1980 NPS tried to identify, catalog, and describe threats to park resources in a "State of the Parks" report. Sadly, NPS has very little reliable information about the nature and condition of natural resources under its stewardship (Stohlgren and Quinn 1992). The 1980 report listed 4,300 threats, but could describe the nature and extent of only 25% of them. In 1985 Congressman Bruce Vento, Chairman of the House Subcommittee on National Parks and Recreation, asked the U.S. General Accounting Office (GAO) to review actions taken by NPS to address threats to resources of the national parks. The GAO review, issued in 1987, found that limited progress had been made in documenting and mitigating threats to the parks. Although the NPS budget for natural resources more than doubled between 1980 and 1984, it was still less than 1% of the NPS operating budget and inadequate to resolve the threats.

The GAO report put the 4,300 park threats in six categories: ecosystem integrity (aesthetic degradation), polluted air, altered water quality or quantity, resource consumption, alien species invasions, and visitor impacts (including park operations). The report made two recommendations: (1) that NPS maintain up-to-date resource management plans and use that information to identify and rank resource management needs and to prepare annual budget requests, and (2) that long-term resource-monitoring programs be initiated. This volume grew out of the latter recommendation.

NPS Strategy for Sustained Studies

In 1987 NPS had little experience in long-term resource-monitoring programs. Indeed, very few agencies or academic institutions had much experience with long-term ecological research then (Likens 1989; Risser 1991). A recent review of the journal *Ecology* (Tilman 1989) revealed that nearly 70% of the papers in the journal over the last decade that reported field manipulations were based on 2 years or less of study. More than 90% lasted fewer than 5 years, and almost none of the studies longer than 10 years were based on direct observational data. Limited experience with designing monitoring programs in Channel Islands National Park in California and Shenandoah National Park in Virginia and with research programs in Everglades, Great

Smoky Mountains, and Yellowstone National Parks indicated that a minimal monitoring program would cost about $200 million per year, nearly 20% of the agency's operating budget (Davis 1993).

Some park managers have attempted to adopt an ecosystem management approach. Faced with having to reallocate significant internal resources by increasing science expenditures from 1% to 20% of their budgets to meet the challenges of science-based management, they have largely given up in frustration (Haskell 1993). With few models to follow and a serious need for more certainty about costs, value, and approaches to resource monitoring, NPS devised a three-pronged strategy for developing long-term resource-monitoring programs based on the following: (1) a review of sustained research efforts in parks, (2) completion of resource inventories and conceptual models of all 252 park units with significant natural resources, and (3) development and evaluation of 10 prototype monitoring programs. This book is a product of the review of sustained research.

Purpose of the Review

In 1990 we asked scientists and senior park managers across the country to nominate specific examples of natural resource issues that could be used to evaluate the role and value of sustained scientific investigations in resolving critical management questions in national parks. They nominated more than 100 candidates. We convened a panel of 21 senior scientists and park managers to review and discuss the candidates and to select a representative sample of issues and examples for case studies. We identified the case studies presented here (Fig. 1.1) as the best examples for evaluating applications of sustained research to the six management issues identified in the 1987 GAO report.

This book reviews the history of NPS-guided research in the parks. It provides vivid examples of the high cost and futility of managing park ecosystems by belief-based advocacy and without adequate knowledge of system structure or function. The case studies also show how scientific research can effectively reduce this cost and teach park managers how to better protect and preserve national parks.

By presenting these 12 case studies, we hope to demonstrate to natural resource policymakers and administrators that decisions based on sound sci-

Figure 1.1 Case-study parks used in this volume to discuss long-term monitoring of natural resources.

entific information are more enduring and cost-effective than decisions derived from opinion-based beliefs. We want to provide scientists with models of study designs, analyses, and applications of results that have been effective in communicating resource threats and potential remedial actions. We also wish to alert the scientific community that the scientific value of national parks as natural areas is in serious jeopardy. Finally, we want to reinforce the paradigm that ecosystem-level resource management in national parks is essentially a series of large-scale, long-term experiments; that external influences alter national park resources despite protection within park boundaries; and that sustained studies or monitoring are required to adequately understand and protect the dynamics and diversity of park ecosystems.

The book begins with a historical overview to explore why research has not been an important aspect of NPS management in the United States. The first case studies show the need for a long-term view to gain an appropriate appreciation and understanding of park ecosystems. The next case studies explore the issue of park integration with the rest of the landscape, to help readers recognize that simply putting a fence around a management unit does not keep out the problems. The last case studies look at the issue of protection versus use of natural resources and at attempts to find an appropriate balance. The final section of the book summarizes findings of the case studies and analyzes the lessons one can learn from them.

Acknowledgments

We wish to thank Bruce Kilgore, Dennis Fenn, Gary Williams, and Gene Hester, whose encouragement and support made this project possible. We are also indebted to the hundreds of NPS scientists and managers who nominated case studies for our consideration. We especially appreciate the work of James Carrico, Robert Chandler, Jay Goldsmith, Paul Haertel, Ron Hiebert, Gary Johnston, Randy Jones, Lloyd Loope, Cliff Martinka, David Mihalec, William Paleck, Stanley Ponce, C. Mack Shaver, Michael Soukup, Brian Underwood, Jan van Wagtendonk, John Varley, Ro Wauer, and Richard Whitman, who helped us select the case studies used in this book. Gloria Maender provided editorial assistance. Gary Davis would especially like to thank William Robertson, Jr. for his inspiration and encouragement to attempt long-term studies in national parks.

Literature Cited

Davis, G. E. 1993. Design elements of monitoring programs: the necessary ingredients for success. Environmental Monitoring and Assessment 26:99–105.

Dugan, J. E., and G. E. Davis. 1993. Applications of marine refugia to coastal fisheries management. Canadian Journal of Fisheries and Aquatic Sciences 50: 2029–2042.

Haskell, D. A. 1993. Is the U.S. National Park Service ready for science? The George Wright Forum 10(4):99–104.

Layden, P. C., and M. J. Manfredo. 1994. National park conditions: a survey of superintendents. National Parks and Conservation Association, Washington, D. C. 18 p. + 15 apps.

Leopold, A. S., S. A. Cain, C. M. Cottam, I. M. Gabrielsen, and T. L. Kimball. 1963. Report of the advisory board on wildlife management. National Parks Magazine, Insert 4-63, April:I–VI.

Likens, G. E., ed. 1989. Long-term studies in ecology, approaches and alternatives. Springer-Verlag, New York. 214 p.

Murie, A. 1940. Ecology of the coyote in the Yellowstone. National Park Service Fauna Series 4. x + 206 p.

Orians, G. H., J. Buckley, W. Clark, M. E. Gilpin, C. F. Jordan, J. T. Lehman, R. M. May, G. A. Robilliard, and D. S. Simberloff. 1986. Ecological knowledge and environmental problem solving, concepts and case studies. National Academy Press, Washington, D.C. 388 p.

Risser, P. G., ed. 1991. Long-term ecological research: an international perspective. John Wiley & Sons, New York. 294 p.

Robbins, W. J., E. A. Ackerman, M. Bates, S. A. Cain, F. F. Darling, J. H. Fogg, Jr., T. Gill, J. M. Gillson, E. R. Hall, and C. L. Hubbs. 1963. A report by the advisory committee to the National Park Service on research. National Academy of Science, National Research Council, Washington, D.C. 75 p.

Robinson, G. R., R. D. Holt, M. S. Gaines, S. P. Hamburg, M. L. Johnson, H. S. Fitch, and E. A. Martinko. 1992. Diverse and contrasting effects of habitat fragmentation. Science 257:524–526.

Stohlgren, T. J., and J. F. Quinn. 1992. An assessment of biotic inventories in western U.S. national parks. Natural Areas Journal 12(3):145–154.

Sumner, L. 1983. Biological research and management in the National Park Service: a history. The George Wright Forum 3(4):3–27.

Tilman, D. 1989. Ecological experimentation: strengths and conceptual problems. P. 136–157 in G. E. Likens, ed. Long-term Studies in Ecology, Approaches and Alternatives. Springer-Verlag, New York. 214 p.

U.S. Department of Commerce. 1992. Our living oceans, report on the status of living marine resources, 1992. National Oceanic and Atmospheric Administration Technical Memorandum NMFS-F/SPO-2. 148 p.

Wilson, E. O., ed. 1988. Biodiversity. National Academy Press, Washington, D.C. 521 p.

Wright, G. M., J. S. Dixon, and B. H. Thompson. 1933. A preliminary survey of faunal relations in national parks. National Park Service Fauna Series 1. iv + 157 p.

2

Management in National Parks: From Scenery to Science

Ervin H. Zube

There is an old, trite, but nevertheless true saying that is particularly meaningful to this book: "You can't manage what you don't know." For more years than any current manager can remember, National Park Service (NPS) managers have been trying to manage our national parks without knowing all the physical, biological, and cultural pieces; their interrelationships; and the management implications for individual pieces and for the ecosystems and landscapes of which they are a part. This volume represents an important benchmark in helping to rectify that glaring knowledge gap. It identifies some of the pieces of selected park ecosystems, begins to reveal relationships among them, and contributes to our understanding of ecosystem function.

It has been convincingly argued that the lack of a long-term science program in NPS can be traced to early Congressional and management emphases on tourists and scenery (Sellars 1989, 1992). The legislation that created Yellowstone National Park in 1872, the nation's first national park, referred to the area as "a pleasuring ground," and the legislation creating NPS in 1916 stated that its mission was "to conserve the scenery and the natural and historic objects and the wild life therein." This people- and scenery-centered emphasis has also been linked to a nineteenth-century search for national identity and pride (Runte 1979). Lacking cultural resources comparable to those found in Europe, the great scenic landscapes of the West provided a national resource that could be pointed to with pride and claimed as better than anything of its kind in Europe.

Early NPS directors were challenged to attract tourists and build a constituency that would provide support for this new, growing, and relatively unknown agency. To that end, scenery was marketed. And, it was marketed with the eager cooperation of railroad companies that not only provided access to many of the parks in the pre-automobile era, but also provided luxury hotels and other creature comforts in parks for affluent American tourists. Travel to the West was not for those of modest means. Resources were managed to provide these affluent tourists with neat, tidy, landscape scenes and panoramas and with opportunities to see wildlife close-at-hand without fear of predators intruding on the scene, a practice that has been described as "facade management" (Sellars 1993).

Because of such an emphasis on visitor accommodation and scenery management, it is not surprising that scientific research has had a checkered history within NPS. The first scientific studies were undertaken in 1929 by George Wright, a naturalist in Yosemite National Park who, using his own funds, conducted a study of wildlife management issues in the national parks. This initiative led to the formation in 1932 of a Wildlife Division, under Wright's leadership, which grew rapidly to a staff of 27. Following his untimely death in 1936 and with the growing economic effects of the Great Depression, the division was reduced to a staff of three, who were transferred in 1939 to the Bureau of Biological Survey, an agency now known as the U.S. Fish and Wildlife Service (Wright 1992). Thus was born a short-lived research program and a pattern of research emergence and recession that would be repeated in the future.

Several general histories of NPS, a biography, and three autobiographies of prominent NPS directors give further clues about management emphasis on scenery and visitor accommodation as well as the difficulties encountered in trying to change that emphasis, both within NPS and Congress.

Ise (1961), in his policy history of natural and archaeological parks, presented an interesting view on pre- and post-World War II scientific research. He noted that in 1941, 97 college and university groups conducted research in the parks, suggesting that most research at that time was by academic scientists. He also noted that research expanded in 1956 when 160 projects were undertaken by state universities, other federal agencies, and NPS staff.

Perhaps of most interest today are his comments about special park problems of that time, including loss of species, with special reference to predators, overpopulation of some species, introduction of exotic flora and fauna, fishing in the parks, wilderness preservation, conflicting adjacent land uses, and potential impacts of pesticides from adjacent lands. This could be easily interpreted as a priority list for resource monitoring and research. Many of these problems resurfaced approximately 20 years later in the report, *State of the Parks: A 1980 Report to the Congress,* prepared by NPS at the request of Congress.

Everhart's quasi-official history of NPS, published in 1972, made no mention of research or science, nor of the brief but important efforts of George Wright. That omission is interesting, particularly in light of the developments in research that emerged in the 1960s under the combined leadership of Stewart Udall, Secretary of the Interior, and George Hartzog, Director of NPS. The 1983 edition of Everhart's book does include several pages on research, but they reflect a distinct disciplinary bias, albeit with brief mention of Udall's efforts to address wildlife management issues. Everhart marked the start of NPS research as associated with the establishment in 1930 of the George Washington Birthplace National Memorial. He suggested that NPS finally recognized the need for research when the agency was engaged in historic restoration projects. Referring to President F. D. Roosevelt's 1933 transfer by Executive Order of 57 historical areas from other agencies to NPS, he noted that NPS assembled "a competent research staff after acquiring the national historical parks" (p. 53). Everhart suggested that "The Park Service has always had more success in getting funds from Congress for historical and archaeological research than for the necessary scientific studies in natural areas" (p. 53).

The biography of Stephen Mather, the first NPS director (Shankland 1970), and the autobiographies of Horace Albright as told to Robert Cahn (Albright and Cahn 1985), Conrad Wirth (1980), and George B. Hartzog, Jr. (1988) provide additional insights into the perceived importance of research by four dominant individuals who occupied the top administrative position in NPS. Mather served as director from 1917 to 1928, Albright from 1929 to 1933, Wirth from 1951 to 1964, and Hartzog from 1964 to 1972. Collectively they served as director for approximately one-half of the life of NPS. They also occupied the director's position during two particularly critical eras:

Mather and Albright during the birth and early years of the agency, and Wirth and Hartzog during the boom times of post-World War II. The post-war years were a time of significant increases in park visitation and a booming economy, as well as a time when the United States became an automobile-oriented society. The automobile made national parks accessible to many for whom they had previously been physically and economically out-of-reach.

The administrative direction that Mather took was largely spelled out in a 13 May 1918 letter that he received from then Secretary of the Interior Franklin K. Lane. It was, however, Albright, Mather's associate, who drafted the policy that Lane sent under his signature to Mather. Nevertheless, it is likely that Mather played a strong role in defining the content before Lane received the document from Albright (Ise 1961; Albright and Cahn 1985). Echoing the words of the 1916 act that created NPS, the policy emphasized maintaining the parks in "unimpaired form for future generations as well as those of our time." However, it also set forth a number of more specific policy mandates, including (1) allowing the grazing of cattle but not sheep; (2) harmonizing roads with the landscape; (3) encouraging all outdoor sports; (4) eliminating all private in-holdings; (5) providing both low-priced and high-class accommodations; (6) accommodating the auto; (7) working with railroads, chambers of commerce, tourist bureaus, and auto-highway associations to advertise travel to the parks; (8) cooperating with municipal, county, and state parks and with the Canadian Park Service; and (9) seeking new parks that offer "scenery of supreme and distinctive quality or some natural features so extraordinary or unique as to be of national interest and importance" (Ise 1961, p. 194–195). Clearly, the dominant policy emphasis was on tourism and scenery management.

According to Sellars (1992), Mather and Albright, the first two directors, were both actively involved in getting the 1916 act creating NPS passed in Congress, and they believed they understood the intent of Congress. Their management emphases on promoting the parks, accommodating tourists, and managing scenery were, in their minds, what Congress intended. These management emphases, plus establishing an effective management agency and expanding the system, are confirmed in Shankland's (1970) biography of Mather. Little could Mather have imagined the magnitude of park visitation one-half century later, the threatening developments on park borders, or the negative effects of technological innovations emerging during his lifetime—

innovations such as the automobile. Mather considered the automobile to be very important to the future of NPS, "as a source of abundant strength to the national parks" (Shankland 1970, p. 147). His strong endorsement of the car and his membership in early automobile organizations led to the marketing of national parks via automobile interests as well as railroad interests, with the former gradually gaining prominence.

Mather's successor, Horace Albright, followed in his mentor's footsteps, notably in building the institution, improving and expanding visitor facilities, and expanding the system, particularly in the eastern United States. He also emphasized developing a systematic approach to park planning and development. In 1930 he created, in the Washington, D.C. office, the Branch of Research and Education headed by an assistant director. The primary research emphasis of this branch, however, appears to have been on generating information for park interpretive programs and not for park management. At different times during its existence, this branch consisted of various subunits, including public relations, photography, field education, history, museums, naturalist, and wildlife, none of which suggest an emphasis on scientific research. The branch existed until 1941. Ten years later, under the directorship of Conrad Wirth, research reappeared in the table of organization for the Washington, D.C. office in the form of an assistant director for Research and Interpretation (Olsen 1985). The parallel with the office that Albright established is striking.

Wirth's term as director can, perhaps, be best described as one that emphasized growth and development. The keystone of his tenure was Mission 66, which was visitor oriented. He saw as his challenge the need to update park infrastructure, including visitor accommodations. The goal was to have the parks restored to top condition following years of neglect attributed to the Depression and World War II. The target date for this restoration was 1966, the 50th anniversary of the establishment of NPS. There were, and still are, strong and mixed opinions about this program because of its dominant emphasis on development rather than protection and management of resources. The key point for this introduction, however, is that research was not a major factor in the Wirth administration, as is indicated by the absence of any significant mention of either research or resource management in his autobiography.

George Hartzog became director of NPS at a critical time in its history: the environmental movement was beginning to surface; dissatisfaction with the perceived development orientation of the Wirth administration emerged among some NPS advocates and critics; and a new environmentally oriented Secretary of the Interior, Stewart Udall, came into office—all calling for change.

Udall, responding to criticisms of NPS for failure to responsibly address wildlife management problems in the parks—notably the population explosions of selected species due to elimination of natural predators—called for a review of wildlife management policies and practices by a special commission headed by A. Starker Leopold. In addition, Udall requested a National Academy of Science review of the NPS research program. Both review groups issued their reports in 1963, and both reports were waiting for Hartzog when he assumed the directorship of NPS in 1964.

Leopold et al. (1963) commented on the dangerous policy implications of managing without information. They concluded that a greatly expanded research program related to the management of plants and animals was needed and that the skills and knowledge to undertake such a program did not exist within NPS. Having a broader mandate, the National Academy of Science report noted, "An examination of natural history research in the NPS shows that it has been only incipient, consisting of many reports, numerous recommendations, vacillations in policy and little action. . . . [T]he Committee is not convinced that the policies of the NPS have been such that the potential contribution of research and a research staff to the solution of the problems of the national parks is recognized and appreciated" (Robbins et al. 1963, p. x, xi). It also noted "a failure to recognize the distinctions between research and administrative decision making" (Robbins et al. 1963, p. 31). The report concluded that NPS research lacked continuity, coordination, and depth. A substantial list of recommendations was made, including the following:

1. define the purpose or objectives of each park;

2. inventory and map the natural resources of each park;

3. create a permanent, independent, and identifiable research unit;

4. appoint an assistant director for research, who would report to the director;

5. make research mission oriented;

6. establish research centers in individual parks when justified by the park and the importance of the research;

7. provide additional substantial support for research.

Shortly after Hartzog took office, Udall directed that the Leopold report be implemented. At the time the report was issued, the annual natural-science budget in NPS was $28,000 (Hartzog 1988, p. 88). Recognizing the need for additional funds to undertake the expanded research program, Hartzog met with the chairman of the Interior Department Appropriations Committee in the House of Representatives to discuss the matter. In Hartzog's words, "Flat-out, he bombed the idea." He quoted the chairman's response, "Research is a function of the NIH (National Institutes of Health), not the Park Service" (1988, p. 103). Hartzog then financed research under the guise of resource studies, with money from a park-management reserve fund, confirming Everhart's observation about the difficulty of getting money from Congress for scientific research.

Under Hartzog, the science program was expanded and the Office of Chief Scientist, with administrative responsibility for all natural science research, was established. Echoing an earlier emergence-recession cycle, this was to be a short-lived arrangement. In 1971, under a new director, the Washington office was reorganized, central oversight and coordination were eliminated, and research administration was decentralized to the then eight (now 10) regional offices, each with a chief scientist (National Research Council 1992). This pattern still prevails at the time of this writing.

Four years after the Leopold and Robbins reports were issued, Darling and Eichorn (1967) reported on their review of NPS policies for the Conservation Foundation. Although research was not a specific target of this review, the report is replete with indications of the need for research to guide park development and management, including decisions on development sites, wildlife management, and control of exotic species.

Five years later, in 1972, marking the centennial celebration of Yellowstone National Park, the Conservation Foundation issued another report, *National Parks for the Future* (Dennis 1972), which provided a substantial review of NPS research needs. Among those identified were (1) ecosystem and sociopsychological carrying capacity, (2) basic resource inventories, (3) ecosystem studies of principal plant-animal communities, and (4) studies of rare species. The report described the substantial need for an NPS research program and called for the development of a major program with adequate funding (p. 17–18).

Shortly after publication of the report, the Office of Chief Scientist, with reporting responsibility to the NPS director, was created in Washington, D.C. That position existed from 1973 to 1976, when it became three steps removed, down the table of organization, from the director's office. It occupied that position until 1978, when it was moved up one level to that of assistant director for Science and Technology.

The most recent study of research in NPS was conducted in 1992 (National Research Council 1992). It reviewed several reports on this topic. The Leopold and Robbins reports of 1963 were used as a baseline. The other reports included three internal NPS reports (Allen and Leopold 1977; National Park Service 1980; Castleberry 1987), two by the National Parks and Conservation Association (1988, 1989), and one by the Conservation Foundation (1979). The NPS research program has been reviewed 11 times in 30 years. Eight of these reviews occurred in 12 years between 1977 and 1989. In addition, a short-lived National Research Council project was initiated in 1980 at the request of Congress as a follow-up to the NPS *State of the Parks: A 1980 Report to Congress*. The study was terminated after one meeting of the appointed panel. It was apparently viewed as unimportant, and funds for the study were deleted from the NPS budget.

The following are among the recommendations made by most of these recent reviews of the science program:

1. Congress should enact a formal mandate for an NPS research program;

2. NPS should establish a research arm that is independent of other activities, such as management or interpretation;

3. a separate budget line should be established along with a major expansion of the research program;

4. the research program should serve short-term, applied-resource-management information needs and long-term, basic-science interests to support both management programs and strong science;

5. coordination should be established with other academic and agency research programs.

Numerous other recommendations have also been advanced. Nevertheless, the reviews strongly agree that NPS research has not received the required budgetary support nor held a strong position within the central administration or among park administrators. What little research has been done has traditionally been focused primarily on day-to-day management problems at the expense of longer-term research issues. Several of the recent reports have also reflected the growing awareness of national parks as potential global-monitoring stations, have recognized their value as relatively undisturbed ecosystems, and have recommended greater use of them for investigating questions about regional and global change.

Research in the parks to date, sporadic as it appears to have been, was driven primarily by management issues and concerns. Furthermore, it has had an uncertain history and a conspicuous absence of strong direction. Relatively little research has been directed to basic understanding of ecosystem function and relationships among ecosystems. Research has too frequently been directed to fighting management brush fires and not to understanding broader ecosystem and landscape issues. The latter require gaining the scientific understanding that is essential for developing management plans that reflect the mandate of 1916: "to conserve the scenery and the natural and historic objects and the wild life therein." Congress did not say "save part of the scenery" or "save some of the wild life," nor did Congress mandate managing the parks as stage sets or, as Sellars (1993) described it, as facades. Without such caveats, it is not only reasonable but also prudent and responsible to interpret the congressional mandate as "to conserve the scenery, including all the elements thereof." These are also the elements of ecosystems. That mandate can only be accomplished through science.

This book is one of relatively few benchmarks along the journey to a different and (from a long-term perspective) more important and relevant research orientation for NPS. The chapters that follow address a broad range of topics that are of interest to both management and science. They range from scenery management at Grand Canyon to ecological functioning—e.g., interactions between moose and wolf populations on Isle Royale, control of exotic species in Hawaiian national parks, simulation of natural fire in the national parks of the Sierra Nevada, and study of rare species and habitats in Indiana Dunes and Death Valley—to impacts of an expanding urban area on Saguaro National Monument. It is fascinating to note that the chapters address some of the same issues Mather and Albright addressed nearly three-quarters of a century ago—visitor and scenery management and predator control—but from profoundly different perspectives. Those early NPS directors were not concerned about too many visitors, but rather too many predators. The tide has turned; park managers now worry about too many visitors and too few predators. The case studies in this volume also reflect the concerns that Ise (1961) noted more than 30 years ago and that were identified in the *State of the Parks* report of 1980—concerns such as exotic species, wilderness, and adjacent land uses. Time has passed, the concerns have not, and the needs for research, both for on-the-ground management and for long-term understanding, have only become more abundant and imperative.

This volume gives testimony that the talents and skills required to successfully undertake the necessary research exist in NPS, albeit in inadequate numbers. It also demonstrates that many of the important questions of today are the same ones that have been asked by nearly a dozen review commissions, committees, and panels over the last quarter-century. Furthermore, those questions relate to the structure of science in NPS, to funding, and to building a cadre of managers and scientists who see their responsibilities, both short-term and long-term, as interrelated with common goals.

Editors' Note

This chapter was written before the November 1993 reorganization of the U.S. Department of the Interior (USDI), when all of the research biologists of NPS were transferred—along with the research biologists of all the other USDI agencies—into the National Biological Survey. This reorganization of the USDI is still in turmoil as of this writing in 1995.

In January 1995, the new agency was renamed the National Biological *Service* under pressure that Congress was going to eliminate the National Biological *Survey*. During this same time, NPS was undergoing a massive restructuring and reduction in force. In May 1995, this restructuring eliminated the 10 regional offices and, in their stead, created 7 field directorates and 16 system support offices. This administrative turmoil has meant that NPS research is in an even more desperate state than at the writing of all of the chapters in this book.

Literature Cited

Albright, H. M., and R. Cahn. 1985. The birth of the National Park Service. Howe Brothers, Salt Lake City. 340 p.

Allen, D. L., and A. S. Leopold. 1977. A review and recommendations relative to the NPS science program. Memorandum to director. National Park Service, Washington, D.C. 15 p.

Castleberry, D. 1987. Report on the workshop of National Park Service regional chief scientists, Omaha, Nebraska, 3–5 December 1986. National Park Service, Washington, D.C. 10 p.

Conservation Foundation. 1979. Federal resource lands and their neighbors. Washington, D.C. 98 p.

Darling, F. F., and N. D. Eichorn. 1967. Man and nature in the national parks. Conservation Foundation, Washington, D.C. 86 p.

Dennis, R. T. 1972. National parks for the future. Conservation Foundation, Washington, D.C. 254 p.

Everhart, W. C. 1972. The National Park Service. Praeger Publishers, New York. 276 p.

———. 1983. The National Park Service. Westview Press, Boulder, Colorado. 197 p.

Hartzog, G. B., Jr. 1988. Battling for the national parks. Moyer Bell Ltd., Mt. Kisco, New York. 284 p.

Ise, J. 1961. Our national park policy: a critical history. Johns Hopkins University Press, Baltimore. 701 p.

Leopold, A. S., S. A. Cain, C. M. Cottam, I. M. Gabrielson, and T. L. Kimball. 1963. Report of the advisory board on wildlife management. National Parks Magazine, Insert 4-63, April:I–VI.

National Parks and Conservation Association. 1988. Research in the parks: an assessment of needs. Volume 2, Investing in park futures: a blueprint for tomorrow. Washington, D.C. 322 p.

———. 1989. National parks: from vignettes to a global view. Commission on Research and Resource Management Policy in the National Park System. Washington, D.C. 13 p.

National Park Service. 1980. State of the parks: A 1980 report to Congress. Washington, D.C. 36 p.

National Research Council, Committee on Improving the Science and Technology of the National Park Service. 1992. Science and the national parks. National Academy Press, Washington D.C. 122 p.

Olsen, R. 1985. Administrative history: organizational structures of the National Park Service 1917 to 1985. National Park Service, Washington, D.C. 136 p.

Robbins, W. J., E. A. Ackerman, M. Bates, S. A. Cain, F. F. Darling, J. M. Fogg, Jr., T. Gill, J. M. Gillson, E. R. Hall, and C. L. Hubbs. 1963. A report by the advisory committee to the National Park Service on research. National Academy of Science, National Research Council, Washington, D.C. 75 p.

Runte, A. 1979. National parks: the American experience. University of Nebraska Press, Lincoln. 240 p.

Sellars, R. W. 1989. Scenic or scenery. Wilderness 52(185):29–33.

———. 1992. The roots of national park management. Journal of Forestry 90(1): 16–19.

———. 1993. Manipulating nature's paradise. Montana: The Magazine of Western History 43(2):2–13.

Shankland, R. 1970. Steve Mather of the national parks. Alfred A. Knopf, New York. 370 p. + xxii index.

Wirth, C. L. 1980. Parks, politics, and the people. University of Oklahoma Press, Norman. 397 p.

Wright, R. G. 1992. Wildlife research and management in the national parks. University of Illinois Press, Urbana. 224 p.

2

Long-term Versus Short-term Views

3

Fire Research and Management in the Sierra Nevada National Parks

David J. Parsons and Jan W. van Wagtendonk

Fire, both lightning- and human-caused, is important to the natural ecosystems of California's Sierra Nevada (Kilgore 1973; Vankat 1977; van Wagtendonk 1986). Paleoecology has shown the close ties between fire, climate, and vegetation change in Sierra ecosystems over at least the past 10,000 years (Davis et al. 1985; R. S. Anderson 1990), but these relationships were disrupted by fire suppression during the past century. Recent research has led scientists to understand that long-term preservation of Sierra Nevada ecosystems in a close-to-natural state largely depends on land managers restoring fire to a semblance of its natural role (Parsons 1981).

The 1890 establishment of the first national parks in the Sierra Nevada marked the beginning of an era of fire suppression, a dramatic shift from millennia of frequent fires ignited by lightning and Native Americans. Fire suppression led to increased fuel loads as well as subtle shifts in species composition and forest age structure. Yet, other than scattered voices of concern, nearly 75 years passed before the National Park Service (NPS) acknowledged that fire suppression was changing park resources (Leopold et al. 1963). The first systematic studies of fire ecology and the effects of fire suppression in the Sierra Nevada parks began shortly after the Leopold report. In documenting the important role of fire in reducing fuel hazards and maintaining natural communities, these studies provided the basis for a prescribed fire management program (Kilgore 1973). The program has been gradually refined through experience and improved understanding of the history and effects of fires of variable intensity and frequency on different vegetation types (Bonnicksen and Stone 1982a; Bancroft et al. 1985; Parsons 1990a). This paper reviews the history of fire research and its contributions to the development of a management program designed to restore natural fire to the national parks of the Sierra Nevada.

The Sierra Nevada Parks

Sequoia, General Grant (later enlarged to become the present Kings Canyon), and Yosemite National Parks were created by Congress in 1890 as the country's second, third, and fourth national parks, respectively. The expansion of Sequoia National Park to include the Kern River drainage in 1926 and the creation of Kings Canyon National Park in 1940 greatly enlarged the total area of the southern parks. Today, these three parks (Fig. 3.1) include more than 586,815 ha, the great majority of which is managed as designated or proposed wilderness. They include much of the southern and central Sierra Nevada and are largely surrounded by U.S. Forest Service wilderness. Because of the extreme elevation gradient represented in the parks (from approximately 426.5 m to more than 4,387.2 m), the area contains a wide variety of plant communities for which fire is important: oak woodland, chaparral, and conifer forests (Rundel et al. 1977).

The Pre-park Era (Pre-1890)

Studies of fire scars in tree rings of the giant sequoia (*Sequoiadendron giganteum*) document a highly variable fire frequency, partially determined by climate, during the several-thousand-year period before European settlement (Swetnam 1993). Frequent lightning ignitions, coupled with burning by Native Americans (Roper Wickstrom 1987), reduced fuels, stimulated regeneration, and maintained a generally open environment (Vankat 1977). Intensive sheep and cattle grazing, including the trampling and consumption of plant material, and fires set by sheepherders as they left the mountains each fall, further opened the forests in the 1860s and 1870s (Dudley 1896). John Muir's (1909) detailed description of an 1875 fire in a giant sequoia grove emphasizes the importance of fire even in that period of apparently reduced fuel accumulation.

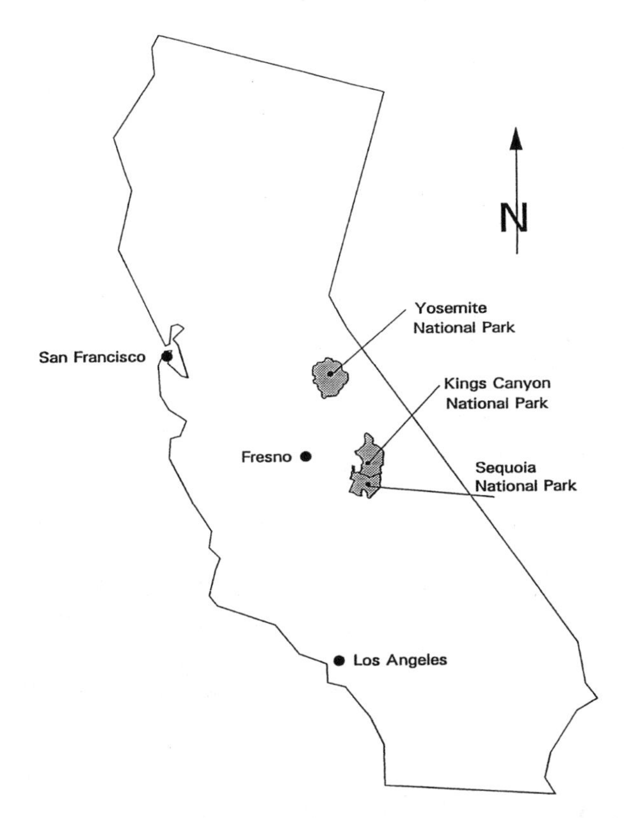

Figure 3.1 Sequoia, Kings Canyon, and Yosemite National Parks in the southern and central Sierra Nevada.

By the time Yosemite Valley and the Mariposa Big Tree Grove were set aside as a public trust in 1864, burning by Native Americans had essentially ended (M. K. Anderson 1991). By 1888, however, the alarm was being sounded about increased fuel hazards in Yosemite Valley (Gibbens and Heady 1964).

The Early Park Era (1890–1960)

The establishment of the three national parks in 1890 brought on a policy of total fire control. The legislative acts establishing Sequoia, General Grant, and Yosemite National Parks specifically stated that they would be managed "to provide for preservation from injury of all timber, mineral deposits, natural curiosities, or wonders . . . and their retention in their natural condition."

Suppressing all fires was considered essential to "preserve" the natural vegetation for which the parks had been largely created. In the early years, the U.S. Forest Service conducted firefighting in the parks (Pyne 1982). These efforts included the cutting of fire breaks and the removal of downed timber from sequoia groves (Vankat 1977). Suppression was less than complete in more remote areas because of limited funding and personnel and poor access (Vankat 1978).

Although several early superintendents at Sequoia National Park noted the growing fire hazard from increased fuel loading (Kilgore 1970), the first formal attention to the effects of fire suppression was given by Meinecke (1926) and Adams (1925). They highlighted the need to establish a reforestation program and the usefulness of "light burning" of the forest floor. Adams, however, also concluded that it would be too costly to carry out the needed burning program. In the 1940s Ernst (1949) raised concern over increased conifer invasions in the absence of fire in Yosemite meadows. Meanwhile, observations by naturalists (MacFarland 1949; Stagner 1952) and fire history research (Show and Kotok 1924; Reynolds 1959) documented the historical importance of fire in the conifer forests of the southern and central Sierra Nevada.

The Modern Era (1960–Present)

The evolution of NPS natural-resource policy since the 1960s has emphasized the restoration and preservation of natural ecosystems and ecological processes, including the restoration of fire as a native ecosystem element (Graber 1985; Parsons et al. 1986). Today, NPS management policies recognize the important role of fire in many park ecosystems and identify prescribed fire as a preferred means to achieve resource management objectives. This includes the use of fire to reduce unnatural fuel loads and to affect species composition. Policies also encourage research and monitoring to support management decisions. The 1984 California Wilderness Act further supports NPS policies by specifying that wilderness will be maintained in its natural condition.

In the Sierra Nevada parks, the decision to restore fire through prescribed burning and natural ignitions is based on data documenting the natural role of fire in the region. During the development of fire programs, park managers have had difficulties in translating scientific data for management actions and dealt with controversies over the definition of park purposes and objectives, including the meaning of "natural" (Bonnicksen and Stone 1982b, 1985; Lemons 1987; Parsons 1990a).

History of Fire Research and Management

The natural fire-management programs in Yosemite, Sequoia, and Kings Canyon National Parks have been built on an ever-improving understanding of fire history, fire ecology, and the effects of varying fire frequencies and intensities on vegetation structure and composition. Scientific research, coupled with increasingly sophisticated monitoring, has provided a basis for refining program goals, objectives, and methodologies (Bancroft et al. 1985; Parsons 1990b).

Montane Coniferous Forest

The Experimental Stage Publication of the findings of a special report to the Secretary of the Interior (Leopold et al. 1963) first drew attention to the need to reintroduce fire to the mixed conifer forests of the Sierra Nevada. The report specifically recommended controlled burning as the only way to successfully reduce the "dog-hair thicket of young pines, white fir, incense cedar, and mature brush—a direct function of over protection from natural ground fires" (p. 34). At about the same time, Biswell and others were providing convincing arguments for the importance of fire in the ponderosa pine (*Pinus ponderosa*) and giant sequoia forests of the Sierra Nevada (Biswell 1959; Hartesveldt 1964). Together with studies documenting the extent of fuel buildup in the absence of fire (Agee 1968), such information provided the basis for the first experimental prescribed burns. Richard Hartesveldt and Thomas Harvey planned and carried out these burns between 1964 and 1966 in the Redwood Mountain Grove of Kings Canyon National Park.

Detailed studies before and after burning documented significant fuel reduction and the dependence of giant sequoia on fire for regeneration (Harvey et al. 1980). Although most of the seedlings died within a few years, both survival and growth rates were highest where the fires burned hottest (Harvey and Shellhammer 1991). In addition to confirming the value of prescribed burning, these findings led to the conclusion that "fire as a tool probably should not be applied evenly in a short period of time throughout a large area" but rather that "prescription fires should be applied in a patchy manner" (Harvey et al. 1980).

Bolstered by the success of the first experimental burns, park scientists recommended restoring fire to park ecosystems. In 1969 plots in the Redwood Mountain Grove were burned. Studies carried out in conjunction with these burns documented burning conditions, heat release, and effects on vegetation and fuels (Kilgore 1972). Other studies explored the effects of burning and physical fuel manipulation in giant-sequoia and adjacent mixed-conifer forest communities on fuel conditions (Kilgore and Sando 1975), physical and hydrological properties of forest floor and soil (Agee 1973), and giant sequoia germination and seedling survival (Stark 1968). These studies proved the feasibility of prescribed burning and confirmed the important ecological role of fire in sequoia–mixed-conifer forests. What they failed to do was to provide criteria for determining the conditions under which fires should be ignited to accomplish varying objectives or criteria for evaluating the relative success of individual burns in terms of the naturalness of the fire effects. Concurrent studies in Yosemite documented the effects of varying fire prescriptions in ponderosa pine and incense cedar forests (van Wagtendonk 1974).

A Management Program By 1972 all three parks had established prescribed burning as a long-term management program. Through 1993, 275 prescribed burns covering 24,289 ha were carried out (Table 3.1). Most of these burns were in the mixed-conifer forest zone, including giant sequoia forests.

In the 1970s, burn objectives for the mixed conifer forest focused largely on fuel reduction and understory removal, with little attention given to preserving the patchiness of the forest (Bancroft et al. 1985). Research provided an improved understanding of the relationships between fire behavior and fuel accumulation (Parsons 1978) and forest structure (van Wagtendonk 1983) and showed that burning conditions had an important effect on ecosystem properties. Data on fuel characteristics (van Wagtendonk and Sydoriak 1985) provided the type of quantitative information necessary to refine burning prescriptions to achieve desired objectives. Unfortunately, park management quickly developed a general perception that sufficient information existed from those early studies to carry out an effective prescribed fire program in the sequoia–mixed-conifer forest. This perception blocked

Table 3.1. Number of prescribed burns and prescribed natural fires and total acres and hectares burned by each in Sequoia, Kings Canyon, and Yosemite National Parks, 1968–1993.

	No. Burns	Total Acres	Total Hectares
PRESCRIBED BURNS			
Sequoia/Kings Canyon	157	29,640	12,000
Yosemite	118	30,355	12,289
Subtotal	275	59,995	24,289
PRESCRIBED NATURAL FIRES			
Sequoia/Kings Canyon	426	36,773	14,888
Yosemite	463	49,668	20,109
Subtotal	889	86,441	34,997
TOTAL	1,164	146,436	59,286

funding required for research needed to further refine the sequoia fire program. Fire research then focused on other vegetation types during much of the 1970s and 1980s.

Also during the 1980s, the parks began to monitor fire behavior and fire effects (Ewell and Nichols 1985). The value of the monitoring programs in evaluating structural and compositional changes due to prescribed burning will be realized only after decades of documenting these changes.

Park managers and the public alike seriously questioned the expansion of prescribed burning. Was it needed? What were the effects of prescribed fire in the sequoia–mixed-conifer forest? Arriving at a clearly defined goal for the fire program was very difficult. A dearth of research left the park scientists and managers without a clear understanding of the natural forest structure and the magnitude of change due to fire suppression or prescribed burns. Bonnicksen and Stone (1982a) used aggregation theory to document an increase in the abundance of pole-size and mature trees and a decrease in sapling- and seedling-sized trees over presettlement conditions. Studies of fire history (Kilgore and Taylor 1979) and forest structure (Parsons and DeBenedetti 1979) provided further evidence of changes in fire regime and forest characteristics from presettlement times. Research documented that presettlement fire-return intervals averaged between 9 and 14 years for the Redwood Mountain Grove (Kilgore and Taylor 1979). This interval is far more frequent than even the most optimistic prescribed-burning scenarios can hope to accomplish because of funding and personnel constraints.

Program Challenges and Changes A debate over fire program goals soon surfaced. Should parks try to preserve the ecosystem structure that was in place when the first European settlers arrived? Should parks try to preserve ecological processes under which the communities evolved? Should parks maintain a hands-off approach to management? Is some combination of these appropriate? Bonnicksen and Stone (1982b) argued that before natural processes can operate, the forest structure must be restored to its presettlement state. NPS scientists (Parsons et al. 1986) responded that (1) fire suppression has increased fuel loads and shifted the relative age structure but not the species composition, and (2) these forests have long adapted to wide variations in climatic and fire regimes. The scientists believed that reversal of the effects of fire suppression should not require such a heavy-handed approach. They argued instead that fire should simply be restored to its natural role, burning at similar intervals, intensities, and seasons. This should bring about ecological effects similar to those that occurred in the past or that would have occurred today had humans not intervened. Management intervention, they felt, should be limited to cases where anthropogenic factors have had irreversible impacts or where life, property, or featured resources need protection (Parsons et al. 1986). Vale (1987) and Lemons (1987) give a good presentation of both sides of this debate over structural versus process goals.

Stephenson (1987) criticized the use of aggregation theory to reconstruct past forest structure because of uncertainties in many assumptions that Bonnicksen and Stone used to reach their conclusions. He specifically cautioned against applying findings based on such assumptions to management and policy decisions. Christensen (1988) argued that efforts to restore a specific primeval structure are neither practical nor desirable in an environment driven by disturbance, where a wide range of forest structures exist at any given time. He claimed that even though fire suppression may have shifted the frequency distribution of landscape states, any existing state probably also existed at some time in the past. Because the distribution of such states has continually shifted over time as climate and fire regime varied, Christensen suggested that a goal of restoring a particular frequency distribution of landscape states is neither necessary nor desirable. To answer these concerns, additional studies are needed to document the variability in prehistoric fire regime and vegetation and to build predictive forest models.

Bonnicksen and Stone (1985) expressed additional concerns about the appropriateness of Native American burning as part of the "natural" fire regime and the importance of establishing quantitative standards against which program success could be measured. Bonnicksen (1989a) claimed that NPS policy should focus on the abstract ideal of "letting nature take its course." Letting nature take its course may be one appropriate management goal; however, in many cases direct intervention is essential. This is especially true in the mixed conifer forest of the Sierra Nevada, where some fires will always need to be suppressed, requiring management intervention to replace them. The goal of natural regulation or unimpeded process described by Bonnicksen (1989b) is simply inappropriate in such cases.

In the 1980s an increased awareness of the variability of natural fire regimes and their role in controlling ecosystem dynamics helped to modify management practices. Spot fires, backing fires, and night burning replaced strip headfires in an attempt to increase the heterogeneity of fire behavior and effects. Objectives for individual burns included (1) breaking up homogeneous fuel accumulations to approximate the conditions thought to be within the range of natural variability, and (2) restoring fuel and forest characteristics so natural fires could be permitted to burn or prescribed fires could mimic the effects of natural fires (Bancroft et al. 1985).

Although the 1980s saw an increased use of scientific information in developing management programs, funding for fire-related research was scarce. Senior management in the Sierra Nevada parks felt that they were far ahead of managers in other areas in their understanding of fire ecology and in their implementation of fire-management programs. The limited funding focused largely on predictive modeling of fire behavior (van Wagtendonk and Botti 1984) and fire effects on fuels (van Wagtendonk 1985).

As more area was burned, the questions for which there were no research dollars continued to mount. The managers questioned whether, when, how often, and how hot to burn. They lacked full understanding of natural fire regimes and the effects of variable fire frequencies and intensities on ecosystem properties. They needed to quantitatively define when fuel buildups became "unnatural" (Parsons et al. 1986). They were bombarded by visitors who were unhappy about the visual impacts of prescribed burns in sequoia groves (Cotton and McBride 1987). Even worse, they did not have enough information to know if their own program goals were being met. To deal with these problems, the managers decided in the spring of 1985 to carry out an external review of the fire-management program, including its supporting scientific justification.

Program Review A panel of outside experts with experience in fire ecology, fire management, sequoia–mixed-conifer forest ecology, and visual resource values was asked to review the history, current status, and scientific basis for the sequoia–mixed-conifer fire-management program in the Sierra Nevada parks (Parsons 1990b). The panel, which met several times throughout 1986, presented a final report to the NPS director in February 1987 (Christensen et al. 1987). The panel recognized the importance of restoring fire to the sequoia groves and emphasized the need to understand the dynamic nature of mixed conifer ecosystems, in which change may be the only real constant. It recommended that two types of prescribed burns be recognized: "restoration fires" conducted to manipulate unnatural fuel conditions and "simulated natural fires" conducted to simulate or maintain the primeval fire regime (Christensen et al. 1987). In addition, the panel recommended that park managers use pre- and postburn fuel manipulation to reduce scorch and bark char in areas where scene management was of primary concern. The panel recognized the difficulty in fully restoring natural processes, but supported it as an overall goal in areas other than those with overriding special scenic or cultural values. The panel specifically recommended that increased emphasis be given to the scientific study of fire history, demography and life history, fuel dynamics, computer simulation models, visitor response, and fire effects.

Park managers quickly executed the panel's recommendations. The recommendation that individual burns be classified as either restoration or simulated natural fires helped clarify objectives for both the overall program and specific burns. As recommended, fire-effects monitoring programs were improved and standardized for all burns. Recognition of the need for additional information helped in the acquisition of funding to support several new research initiatives (Parsons 1990b). The park delimited a "special management area" concept in which burn blocks in heavily used areas would be carefully planned to present a mixture of burned and unburned areas that show a range of fire effects and to minimize scorching of bark on large trees (Dawson and Greco 1991).

Recent Research After receiving the panel's recommendations, park managers undertook several research projects, including expanded studies of vegetation and fire history, fire and forest dynamics (including understory vegetation), forest pathogens, fire scar enlargement, fuel dynamics, soil and cambium temperatures during prescribed burns, and visitor perception of the fire-management program (see Parsons 1990b for review of these studies). Some results of these studies have particular management implications: e.g., pollen, plant macrofossils, and charcoal documented significant changes in forest composition and fire occurrence over the past 10,000 years (R. S. Anderson 1990). These findings are important because they confirm the dynamic nature of the mixed-conifer forest community. It is an ever-changing community, varying with shifts in climate and fire, rather than a static entity. The implications for managing a continually changing ecosystem are complex (Christensen 1991).

Tree-ring-based fire chronologies from giant sequoia (Swetnam 1993) provide a detailed reconstruction of past fire occurrence, including variations in fire frequency and intensity as the climate changed. For example, fires in the Giant Forest of Sequoia National Park were several times more frequent during the period between A.D. 900 and 1300 than in following centuries (the Little Ice Age), when global temperatures were as much as 1°C lower (Swetnam 1993). Besides emphasizing the close ties among climate, fire, and vegetation, such data provide both a target and a basis for assessing the practicality of restoring natural fire regimes.

Forest age-structure studies indicate that occasional intense fires may be a critical factor in perpetuating the groupings of specimen trees so characteristic of giant sequoia groves (Stephenson et al. 1991). These findings indicate that management may need to change future fire prescriptions to allow fires to burn hot enough to favor sequoia.

Since 1985 Yosemite park managers have used geographic information system (GIS) technology to assist fire management and research (van Wagtendonk 1991a). Historic records have been digitized and compared with other themes to understand trends between fire occurrence and topography, vegetation, fuels, and location. The GIS has also been used to develop a fuel model map that can be used, together with data on lightning strikes, topography, and climate, to help predict fire occurrence and growth (van Wagtendonk 1991a,b). As understanding of ecosystem processes and fire effects are incorporated into such systems, their predictive capabilities will be of increasing value to scientists and decision makers.

Studies of visitor perception of the fire-management programs at Sequoia, Kings Canyon, and Yosemite National Parks confirm a high level of awareness of the purposes of the program, as well as an interest in experiencing a "natural" environment (Quinn 1988).

Today, prescribed burning in the mixed conifer zone continues as a management program struggling to find ways to incorporate as much of the new scientific data as possible. As the primary goal of the prescribed burning program moves from reducing fuel to mimicking natural conditions, increased attention will have to be given to simulating the variability of natural fire regimes, including the variable effects on ecosystem properties.

Prescribed Natural Fire

The restoration of natural fire to park ecosystems has been most successful in the higher elevations. Recognizing that short growing seasons and slow growth rates had minimized the effects of fire suppression, managers of the Sierra Nevada parks implemented the first natural-fire management programs in the United States. In 1968 two lightning fires were permitted to burn in the subalpine forest of the Middle Fork of the Kings River in Kings Canyon National Park (Kilgore and Briggs 1972). During the same year, a 324-ha prescribed burn was carried out in a nearby red-fir forest in an effort to better understand fire behavior and fire effects (Kilgore 1971). The success of these experiments soon led to the expansion of the natural fire zone in Sequoia and Kings Canyon National Parks to include most areas above 2,400 m and some as low as 1,980 m. Today, the natural fire zone includes 286,000 ha, or 82% of the two parks. Areas within this zone are selected on the basis of vegetation, fire history, fuel, topography, and the presence or absence of natural or human-created barriers. As in Yosemite, natural fires in Sequoia and Kings Canyon that are permitted to burn must meet definite prescribed conditions. The criteria are comprehensive, including fire weather and behavior and the availability of park staff to monitor, manage, and potentially suppress the fires. This is why such fires are now called "prescribed natural fires" as opposed to manager-ignited "prescribed burns."

The first natural-fire management zone in Yosemite was established in 1972. The zone was more than doubled in size to 188,450 ha the following year, when 27 fires burned a total of 14 ha. The Starr King fire, started by lightning on 4 August 1974, provided the first real test of the Yosemite program. After a slow start, the fire eventually burned 1,586 ha before being declared out on 31 October. Control action was taken on one side to keep the fire within the natural-fire-zone boundary. On-site monitoring documented rates of spread, intensities, fuel consumption, and fire effects thought to be typical of presettlement fire regimes (van Wagtendonk 1978). Although some concern was expressed regarding smoke drifting into Yosemite Valley, little public criticism was heard (van Wagtendonk 1978).

By 1993, 889 natural fires had burned a total of 34,997 ha in the three Sierra Nevada parks (Table 3.1). More than three-fourths of these fires never exceeded 0.1 ha in size, although several burned thousands of hectares. Typically, these fires were of low intensity, spreading in runs driven by winds or topography where the sparse fuels permitted such runs. Natural fires are usually confined to the ground, burning downed fuel and brush. Crown fires are rare but sometimes occur in lodgepole pine (*Pinus contorta*) or red fir (*Abies magnifica*) forests. Many fires burn from midsummer until they are extinguished by fall or winter snows. The importance of fire in determining subalpine-forest stand structure and composition has been documented in several recent studies (Parker 1988; Keifer 1991). Although the role of fire in subalpine meadows is less well understood, the documentation of at least one such fire (DeBenedetti and Parsons 1979) provides a basis for determining their natural role.

The repeated pattern of fires in the Illilouette Creek basin of Yosemite has been of special interest. In 1978, 4 years after the Starr King fire, two different fires (222 and 129 ha) burned to within 2 km of the Starr King burn before going out. In 1980 the 607-ha Illilouette fire was stopped when it burned against the previous fires (a portion of the Starr King fire area was reburned in a low-intensity fire). In 1981 the 898-ha Buena Vista fire burned out along a 2.4-km front with the Illilouette fire, again reburning about 32 ha of the Starr King burn. Since 1981, 24 additional fires have burned in the

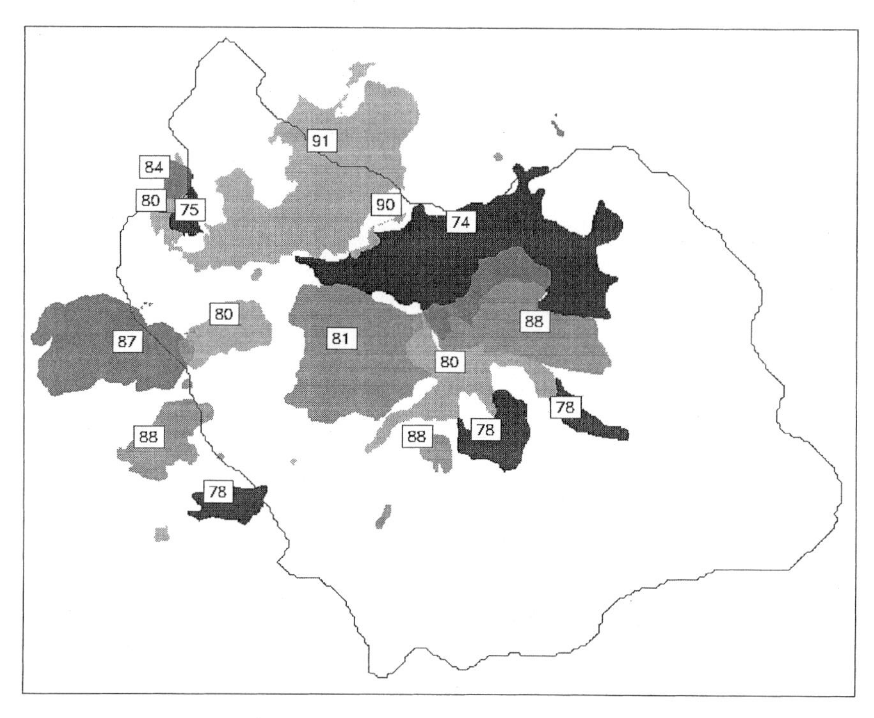

Figure 3.2 Major prescribed natural fires that burned within Illilouette Creek basin of Yosemite National Park, 1974–1991. Two-digit number refers to year of fire.

drainage, including two of more than 404 ha in size. The "jigsaw" pattern of burns that has developed in this area limits the size of additional fires (van Wagtendonk 1985). Fires in 1988 and 1991 largely filled in the missing pieces in the "jigsaw" pattern (Fig. 3.2). Thus, between 1974 and 1991, 61 fires, 12 of them more than 404 ha in size, burned a total of 5,048 ha in the Illilouette Creek basin. A similar pattern has occurred in the Roaring River drainage in Kings Canyon National Park, where repeated fires since 1972 have created a mosaic pattern of burned and unburned areas (Greenlee et al. 1980).

Documentation on the fire behavior and effects of the numerous prescribed natural fires that have occurred in recent years has provided a strong basis for expanding that program to include most of the higher elevations of all three Sierra Nevada parks. The successful expansion of the program to include more of the middle-elevation, mixed-conifer forest zone will depend largely on the success of integrating research and program results with management reality.

Challenges to Policy and Programs

As the restoration of fire to the Sierra Nevada parks has evolved from an experimental concept to a fully functioning management program, park scientists and managers have faced challenges at almost every step of the way. Major obstacles have included (1) developing criteria to guide decisions on where, when, how often, and how hot to burn; (2) weighing structural versus process goals and aesthetic versus ecological priorities; (3) defining the term "natural" (e.g., should Native American activities be included); (4) establishing standards by which to evaluate program success; (5) using the findings of research and monitoring to improve management programs; and (6) developing effective mechanisms to communicate program goals and accomplishments (Bancroft et al. 1985; Parsons 1990a).

At times, incomplete data have required that decisions be based on judgment, using the best information available. Ideally, as new information is acquired, programs are reevaluated and the necessary adjustments are made. Unfortunately, incomplete data, the lack of qualified staff, the lack of time, or differences in interpretation have occasionally hindered the coevolution of science and management.

Debates over the true purpose of national parks (to provide for public enjoyment or to preserve natural ecosystems), including the value of structural goals relative to process goals, have become acrimonious at times (Bonnicksen 1989a,b; Parsons 1990b; Stephenson et al. 1990). Perhaps the most difficult debate to resolve has been the argument that the visual appearance or aesthetics of sequoia groves is as important as the natural functioning of the forest (Cotton and McBride 1987). Considering visual or scenic values in what have been largely ecologically driven management plans requires reassessment of objectives and priorities (Cotton and McBride 1987; Dawson and Greco 1991). These are largely policy-level decisions that must be resolved before a consensus can be expected.

Perhaps the issue with the greatest potential to disrupt the park fire-management programs is that of smoke production and its relation to air-quality standards. Smoke produced from prescribed fires can cause discomfort to park visitors, obscure vistas, and contribute to California's overall air-pollution problem. At the time of this writing, the state of California is proposing that prescribed burns meet state particulate air-quality standards, a decision that could prohibit many future burns.

Another issue that must be considered in prescribed fire programs is that of cost-effectiveness. In addition to the direct costs of program operations, cost comparisons must include the costs of collecting preburn site information, research, and postfire monitoring. Prescribed natural fires can be particularly expensive because of the costs of monitoring long-burning fires in remote locations. In many areas, however, the cost per hectare of fire suppression will exceed that of prescribed fire (H. T. Nichols, pers. comm., 1992). Regardless, the intangible benefits from prescribed fire—the restoration of natural processes to a wilderness, including the excitement and educational opportunities to observe and learn about natural ecosystems—must ultimately be considered the most important benefit of such programs (H. T. Nichols, pers. comm., 1992).

The value of prescribed fire programs as educational opportunities to introduce park visitors to the intricacies of natural ecosystem processes has been one of the most exciting benefits to the Sierra Nevada parks. Although some critics have claimed park interpretive programs are biased and dogmatic (McBride 1993), the general belief of surveyed visitors is that the opportunity to experience and learn about "natural" settings is extremely important (Quinn 1988). Several recent articles have emphasized the opportunities presented by fire research and management to inform the public about the traditional role of fire and the importance of its reintroduction if park managers are to assure the long-term preservation of Sierra Nevada forests (Tweed 1987; Kunzig 1989).

Conclusions

Early fire research in the Sierra Nevada documented the effects of fire suppression on increasing fuel hazards, suppressing giant sequoia reproduction, and shifting successional patterns. These findings led to the initiation of a prescribed burning program designed to reduce fuels and stimulate reproduction. In the higher elevations, where suppression had had a minimal effect, lightning fires were permitted to burn, presumably simulating natural ignitions. Challenges to objectives and implementation strategies stimulated additional research, focused largely on fire history and the effects of variable fire intensity and frequency on vegetation structure and ecosystem processes. Other studies focused on refining burning prescriptions, developing predictive models of fire behavior and spread, improving monitoring protocols, and understanding effects on visual resources (see Parsons 1990b). Management responses included refining burning and monitoring techniques and more clearly articulating program objectives (Bancroft et al. 1985).

Park managers now recognize that prescribed fire can be used to create a wide range of forest structure and function. However, if goals and objectives are not clearly defined, there is no way to measure program success (Bonnicksen and Stone 1985). In the Sierra Nevada parks, the goal of the fire-management program is to preserve, or where necessary restore, fire as a natural ecosystem process. This goal will be accomplished when fires of similar size burn at similar intervals, intensities, and seasons, and thus with similar ecological effects, as fires that occurred in the past or would have occurred today had humans not intervened. This goal is accomplished by permitting natural fires to burn where the effects of suppression have been minimal and by using prescribed burns where the effects of suppression must be reversed before natural ignitions can be allowed. The goal is idealistic and will likely never be fully met. Wherever natural fires cannot be permitted to burn, management ignitions will continue to be required. Hands-on management, based on judgment and the best available information, will continue to be an integral part of the fire-management program.

Standards to evaluate program success must be based on an understanding of the variability and effects of natural fire regimes, including the relationships among fire, climate, and vegetation. Whether such standards are based on vegetation structure, fuels, fire behavior, or fire effects (Bonnicksen and Stone 1985; Parsons 1990a), they will require detailed information on past and present ecosystem properties.

Yet all the information possible will not be enough if ways are not found to greatly accelerate current rates of burning. Fire-history studies documenting the importance of frequent fire in the sequoia–mixed-conifer forest have not been adequately incorporated into management programs. Limitations on funding, together with political, visual, and air-quality concerns, have made it impossible to carry out the number and size of fires necessary to restore natural-fire return intervals.

Similar concerns surround the need for additional research. Increasing impacts of air pollution on Sierra Nevada forests (Duriscoe and Stolte 1989), together with predictions of future global climatic change, including changes in disturbance regimes, portend significant concerns for the future of prescribed fire programs. Shifts in species distributions, together with increased mortality of foliage or whole trees, would have unforeseen consequences on fire intensity and frequency. The potential impacts of such changes on what we think of as "natural" communities, including the role of fire, are largely unknown. The one sure fact is the need for improved understanding of interactions among climate, fire, and biota, including the ability to predict the effects of different climatic or management scenarios.

The accuracy with which park managers can predict the behavior, spread, and effects of individual fires will be increasingly critical to decisions on when and where to burn. Models to predict fuel accumulation and consumption, fire spread, smoke production, and the effects of individual fires or fire regimes on species distributions and abundances are needed to provide much of this capability (van Wagtendonk 1985).

Staff members in the Sierra Nevada parks have recently undertaken a lead role in developing an interdisciplinary research program to understand and predict the effects of climatic change on species and community distribution and on fire occurrence and effects in the southern and central Sierra Nevada (Stephenson and Parsons 1993). They hope that this program will begin to provide the predictive tools necessary to prepare managers for the critical resource issues of the twenty-first century.

Literature Cited

Adams, C. C. 1925. Ecological conditions in national forests and national parks. Scientific Monthly 20:561–593.

Agee, J. K. 1968. Fuel conditions in a giant sequoia grove and surrounding plant communities. Unpubl. M.S. thesis, University of California, Berkeley. 55 p.

———. 1973. Prescribed fire effects on physical and hydrologic properties of mixed-conifer forest floor and soil. Water Resources Center Contribution Report 143. University of California, Davis. 57 p.

Anderson, M. K. 1991. California Indian horticulture: management and use of redbud by southern Sierra Miwok. Journal of Ethnobiology 11:145–157.

Anderson, R. S. 1990. Holocene forest development and paleoclimates within the central Sierra Nevada, California. Journal of Ecology 78:470–489.

Bancroft, L., T. Nichols, D. Parsons, D. Graber, B. Evison, and J. van Wagtendonk. 1985. Evolution of the natural fire management program at Sequoia and Kings Canyon National Parks. P. 174–180 in J. E. Lotan, B. M. Kilgore, W. C. Fischer, and R. W. Mutch, tech. coords. Proceedings—Symposium and Workshop on Wilderness Fire. USDA, Forest Service General Technical Report INT-182.

Biswell, H. H. 1959. Man and fire in ponderosa pine in the Sierra Nevada of California. Sierra Club Bulletin 44(7):44–53.

Bonnicksen, T. M. 1989a. Fire gods and federal policy. American Forests 95(7&8): 14–16, 66–68.

———. 1989b. Nature vs. man(agement). Journal of Forestry 87(12):41–43.

Bonnicksen, T. M., and E. C. Stone. 1982a. Reconstruction of a presettlement giant sequoia-mixed conifer forest community using the aggregation approach. Ecology 63:1134–1148.

———. 1982b. Managing vegetation within U.S. national parks: a policy analysis. Environmental Management 6:101–102, 109–122.

———. 1985. Restoring naturalness to national parks. Environmental Management 9:479–486.

Christensen, N. L. 1988. Succession and natural disturbance: paradigms, problems, and preservation of natural ecosystems. P. 62–86 in J. K. Agee and D. R. Johnson, eds. Ecosystem Management for Parks and Wilderness. University of Washington Press, Seattle.

———. 1991. Variable fire regimes on complex landscapes: ecological consequences, policy implications, and management strategies. P. ix–xiii in S. C. Nodvin and T. A. Waldrop, eds. Fire and the Environment: Ecological and Cultural Perspectives. USDA, Forest Service General Technical Report SE-69.

Christensen, N. L., L. Cotton, T. Harvey, R. Martin, J. McBride, P. Rundel, and R. Wakimoto. 1987. Review of fire management program for sequoia-mixed conifer forests of Yosemite, Sequoia and Kings Canyon National Parks. Final report to National Park Service, Washington, D.C. 37 p.

Cotton, L., and J. R. McBride. 1987. Visual impacts of prescribed burning on mixed conifer and giant sequoia forests. P. 32–37 in J. B. Davis and R. E. Martin, tech. coords. Proceedings of the Symposium on Wildland Fire 2000. USDA, Forest Service General Technical Report PSW-101.

Davis, O. K., R. S. Anderson, P. Fall, M. K. O'Rourke, and R. S. Thompson. 1985. Palynological evidence for early Holocene aridity in the southern Sierra Nevada, California. Quaternary Research 24:322–332.

Dawson, K. J., and S. E. Greco. 1991. Prescribed fire and visual resources in Sequoia National Park. P. 192–201 in S. C. Nodvin and T. A. Waldrop, eds. Fire and the Environment: Ecological and Cultural Perspectives. USDA, Forest Service General Technical Report SE-69.

DeBenedetti, S. H., and D. J. Parsons. 1979. Natural fire in subalpine meadows: a case description from the Sierra Nevada. Journal of Forestry 77:477–479.

Dudley, W. R. 1896. Forest reservations: with a report on the Sierra Reservation, California. Sierra Club Bulletin 1(7):254–267.

Duriscoe, D. M., and K. W. Stolte. 1989. Photochemical oxidant injury to ponderosa pine (*Pinus ponderosa* Laws.) and Jeffrey pine (*Pinus jeffreyi* Grev. and Balf.) in the national parks of the Sierra Nevada of California. P. 261–278 in R. K. Olson and A. S. Lefohn, eds. Effects of Air Pollution on Western Forests. Air & Waste Management Association, Pittsburgh.

Ernst, E. 1949. Vanishing meadows in Yosemite Valley. Yosemite Nature Notes 38: 34–40.

Ewell, D. M., and H. T. Nichols. 1985. Prescribed fire monitoring in Sequoia and Kings Canyon National Parks. P. 327–330 in J. E. Lotan, B. M. Kilgore, W. C. Fischer, and R. W. Mutch, tech. coords. Proceedings—Symposium and Workshop on Wilderness Fire. USDA, Forest Service General Technical Report INT-182.

Gibbens, R. P., and H. F. Heady. 1964. The influence of modern man on the vegetation of Yosemite Valley. Agricultural Experiment Station Manual 36. University of California, Berkeley. 44 p.

Graber, D. M. 1985. Coevolution of National Park Service fire policy and the role of national parks. P. 345–349 in J. E. Lotan, B. M. Kilgore, W. C. Fischer, and R. W. Mutch, tech. coords. Proceedings—Symposium and Workshop on Wilderness Fire. USDA, Forest Service General Technical Report INT-182.

Greenlee, J. M., J. Villeponteaux, and E. A. Sheekey. 1980. Natural fire in the Sierra Nevada, California. P. 293–312 in Volume 10. Proceedings of the 2nd Conference on Scientific Research in the National Parks. National Park Service, Washington, D.C.

Hartesveldt, R. J. 1964. Fire ecology of the giant sequoias: controlled fire may be one solution to survival of the species. Natural History Magazine 73:12–19.

Harvey, H. T., and H. S. Shellhammer. 1991. Survivorship and growth of giant sequoia (*Sequoiadendron giganteum* [Lindl.] Buchh.) seedlings after fire. Madrono 38:14–20.

Harvey, H. T., H. S. Shellhammer, and R. E. Stecker. 1980. Giant sequoia ecology: fire and reproduction. National Park Service Scientific Monograph Series 12. 182 p.

Keifer, M. 1991. Age structure and fire disturbance in the southern Sierra Nevada subalpine forest. Unpubl. M.S. thesis, University of Arizona, Tucson. 111 p.

Kilgore, B. M. 1970. Restoring fire to the sequoias. National Parks and Conservation Magazine 44(277):16–22.

———. 1971. The role of fire in managing red fir forests. Transactions of the North American Wildlife and Natural Resources Conference 36:405–416.

———. 1972. Impact of prescribed burning on a sequoia-mixed conifer forest. P. 345–375 in Proceedings of the 12th Tall Timbers Fire Ecology Conference. Tall Timbers Research Station, Tallahassee, Florida.

———. 1973. The ecological role of fire in Sierran conifer forests: its application to national park management. Journal of Quaternary Research 3:496–513.

Kilgore, B. M., and G. S. Briggs. 1972. Restoring fire to high elevation forests in California. Journal of Forestry 70:266–271.

Kilgore, B. M., and R. W. Sando. 1975. Crown-fire potential in a sequoia forest after prescribed burning. Forest Science 21:83–87.

Kilgore, B. M., and D. Taylor. 1979. Fire history of a sequoia-mixed conifer forest. Ecology 60:129–142.

Kunzig, R. 1989. These woods are made for burning. Discover 10:86–95.

Lemons, J. 1987. United States' national park management: values, policy, and possible hints for others. Environmental Conservation 14:329–340.

Leopold, A. S., S. A. Cain, C. M. Cottam, J. M. Gabrielson, and T. L. Kimball. 1963. Wildlife management in the national parks. American Forests 69:32–35, 61–63.

MacFarland, J. W. 1949. A guide to the giant sequoia of Yosemite National Park. Yosemite Nature Notes 28:42–91.

McBride, J. R. 1993. Managing national parks. Renewable Resources Journal 11(1): 24–25.

Meinecke, E. 1926. Memorandum on the effects of tourist traffic on plant life, particularly big trees, Sequoia National Park, California. Unpubl. report in park archives. 18 p.

Muir, J. 1909. Our national parks. Houghton, Mifflin and Co., Cambridge, Massachusetts. 382 p.

Parker, A. J. 1988. Stand structure in subalpine forests of Yosemite National Park, California. Forest Science 34:1047–1058.

Parsons, D. J. 1978. Fire and fuel accumulation in a giant sequoia forest. Journal of Forestry 76:104–105.

———. 1981. The role of fire management in maintaining natural ecosystems. P. 469–488 in H. A. Mooney, T. M. Bonnicksen, N. L. Christensen, J. E. Lotan, and W. A. Reiners, tech. coords. USDA, Forest Service General Technical Report WO-26.

———. 1990a. Restoring fire to the Sierra Nevada mixed conifer forest: reconciling science, policy and practicality. P. 271–279 in H. G. Hughes and T. M. Bonnicksen, eds. Restoration '89: the New Management Challenge, Proceedings of the 1st Annual Meeting of the Society for Ecological Restoration. Society for Ecological Restoration, Madison, Wisconsin.

———. 1990b. The giant sequoia fire controversy: the role of science in natural ecosystem management. P. 257–268 in C. van Riper III, T. J. Stohlgren, S. D. Veirs, Jr., and S. C. Hillyer, eds. Examples of Resource Inventory and Monitoring in National Parks of California. National Park Service Transactions and Proceedings Series 8.

Parsons, D. J., and S. H. DeBenedetti. 1979. Impact of fire suppression on a mixed-conifer forest. Forest Ecology and Management 2:21–33.

Parsons, D. J., D. M. Graber, J. K. Agee, and J. W. van Wagtendonk. 1986. Natural fire management in national parks. Environmental Management 10:21–24.

Pyne, S. J. 1982. Fire in America: a cultural history of wildland and rural fire. Princeton University Press, Princeton, New Jersey. 654 p.

Quinn, J. A. 1988. Visitor perception of NPS fire management in Sequoia and Kings Canyon National Parks: results of a survey conducted summer 1987. National Park Service CPSU/UCD Technical Report 35. University of California, Davis. 35 p.

Reynolds, R. D. 1959. Effect of natural fires and aboriginal burning upon the forests of the central Sierra Nevada. Unpubl. M.A. thesis, University of California, Berkeley. 262 p.

Roper Wickstrom, C. K. 1987. Issues concerning Native American use of fire: a literature review. Yosemite National Park Publications in Anthropology 6. National Park Service, Washington, D.C. 68 p.

Rundel, P. W., D. J. Parsons, and D. T. Gordon. 1977. Montane and subalpine vegetation of the Sierra Nevada and Cascade Ranges. P. 559–599 in M. Barbour and J. Major, eds. Terrestrial Vegetation of California. John Wiley & Sons, New York.

Show, S. B., and E. I. Kotok. 1924. The role of fire in the California pine forests. U.S. Department of Agriculture Bulletin 1294. 80 p.

Stagner, H. R. 1952. The giants of Sequoia and Kings Canyon. Commercial Printing Co., Visalia, California. 32 p.

Stark, N. 1968. The environmental tolerance of the seedling stage of *Sequoiadendron giganteum*. American Midland Naturalist 80:84–95.

Stephenson, N. L. 1987. The use of tree aggregations in forest ecology and management. Environmental Management 11:1–5.

Stephenson, N. L., and D. J. Parsons. 1993. Implementing a research program to predict the effects of climatic change in the Sierra Nevada. P. 93–109 in S. D. Veirs, Jr., T. J. Stohlgren, and C. Schonewald-Cox, eds. Proceedings of the 4th Biennial Conference on Research in the National Parks of California. National Park Service Transactions and Proceedings Series 9.

Stephenson, N. L., D. J. Parsons, and H. T. Nichols. 1990. Replies from the fire gods. American Forests 96(3&4):35, 70.

Stephenson, N. L., D. J. Parsons, and T. W. Swetnam. 1991. Restoring natural fire to the sequoia-mixed conifer forest: should intense fire play a role? P. 321–337 in Proceedings of the 17th Tall Timbers Fire Ecology Conference. Tall Timbers Research Station, Tallahassee, Florida.

Swetnam, T. W. 1993. Fire history and climate change in giant sequoia groves. Science 262:885–889.

Tweed, W. 1987. Born of fire. National Parks 61(1–2):24–27, 45. Washington, D.C.

Vale, T. R. 1987. Vegetation change and park purposes in the high elevations of Yosemite National Park, California. Annals of the Association of American Geographers 77:1–18.

Vankat, J. L. 1977. Fire and man in Sequoia National Park. Annals of the Association of American Geographers 67:17–27.

———. 1978. Vegetation change in Sequoia National Park, California. Journal of Biogeography 5:377–402.

van Wagtendonk, J. W. 1974. Refined burning prescriptions for Yosemite National Park. National Park Service Occasional Paper 2. 21 p.

———. 1978. Wilderness fire management in Yosemite National Park. P. 324–335 in E. A. Schofield, ed. EARTHCARE: Global Protection of Natural Areas, Proceedings of the 14th Biennial Wilderness Conference. Westview Press, Boulder, Colorado.

———. 1983. Prescribed fire effects on forest understory mortality. P. 136–138 in Proceedings of the 7th Conference on Fire and Forest Meteorology. American Meteorological Society, Boston.

———. 1985. Fire suppression effects on fuels and succession in short-fire-interval wilderness ecosystems. P. 119–126 in J. E. Lotan, B. M. Kilgore, W. C. Fischer, and R. W. Mutch, tech. coords. Proceedings—Symposium and Workshop on Wilderness Fire. USDA, Forest Service General Technical Report INT-182.

———. 1986. The role of fire in the Yosemite Wilderness. P. 2–9 in R. C. Lucas, comp. Proceedings—National Wilderness Research Conference: Current Research. USDA, Forest Service General Technical Report INT-212.

———. 1991a. GIS applications in fire management and research. P. 212–214 in S. C. Nodvin and T. A. Waldrop, eds. Fire and the Environment: Ecological and Cultural Perspectives. USDA, Forest Service General Technical Report SE-69.

———. 1991b. Spatial analysis of lightning strikes in Yosemite National Park. P. 605–611 in Proceedings of the 11th Conference on Fire and Forest Meteorology. American Meteorological Society, Boston.

van Wagtendonk, J. W., and S. J. Botti. 1984. Modeling behavior of prescribed fires in Yosemite National Park. Journal of Forestry 82:479–483.

van Wagtendonk, J. W., and W. M. Sydoriak. 1985. Correlation of woody and duff fuel moisture contents. P. 186–191 in Proceedings of the 8th Conference on Fire and Forest Meteorology. American Meteorological Society, Boston.

4

Yellowstone Lake and Its Cutthroat Trout

John D. Varley and Paul Schullery

In 1868 an imaginative journalist named Legh Freeman, writing in the *Frontier Index,* described Yellowstone Lake as "the largest and strangest mountain lake in the world. It being 60 by 25 miles in size and surrounded by all manner of large game, including an occasional white buffalo, that is seen to rush down the perpetual snow peaks that tower above, and plunge up to its sides into the water. It is filled with fish half as large as a man, some of which have a mouth and horns and skin like a catfish and legs like a lizzard [*sic*]" (Varley and Schullery 1983, p. 14). From such mythic beginnings as this, enhanced periodically by the rarely believed tales of early trappers, Yellowstone Lake emerged into clearer public consciousness in the 1870s, shedding most of its myths, but none of its wonder.

Yellowstone National Park was established in 1872. Even in the early days, when the park's primary reason for existing was its unparalleled collection of geological and geothermal features, the lake was a "must-see" stop for visitors. If they did not consider the lake the strangest in the world, they clearly thought of it as one of the most beautiful and satisfying (Schullery and Whittlesey 1992). Native Americans used the lake and its environment for thousands of years, but whatever effects they had on the area were evidently slight compared to what followed 1872. Although Native American influences on North American landscapes were often considerable, we see no evidence in the general or local archaeological record (as it is known so far) to suggest that these people heavily exploited or manipulated Yellowstone Lake's ecological components.

Yellowstone Lake is deep (maximum depth, 120 m; average depth, 40 m) and large (at 35,400 ha, it is the largest high-elevation lake in North America; Fig. 4.1). It has a capacity of 1,500 km^3 and takes more than 10 years to replace its water from inlet to outlet (Benson 1961). At this size, and under the peculiar political circumstances that surround the well-protected features of national parks, Yellowstone Lake has afforded both the public and the scientific community an extraordinary opportunity to learn. That learning has proceeded in many directions at any one time. In this paper we explore especially the advantages that resource managers and researchers have taken of that opportunity, especially in their attempts to preserve the Yellowstone Lake cutthroat trout population.

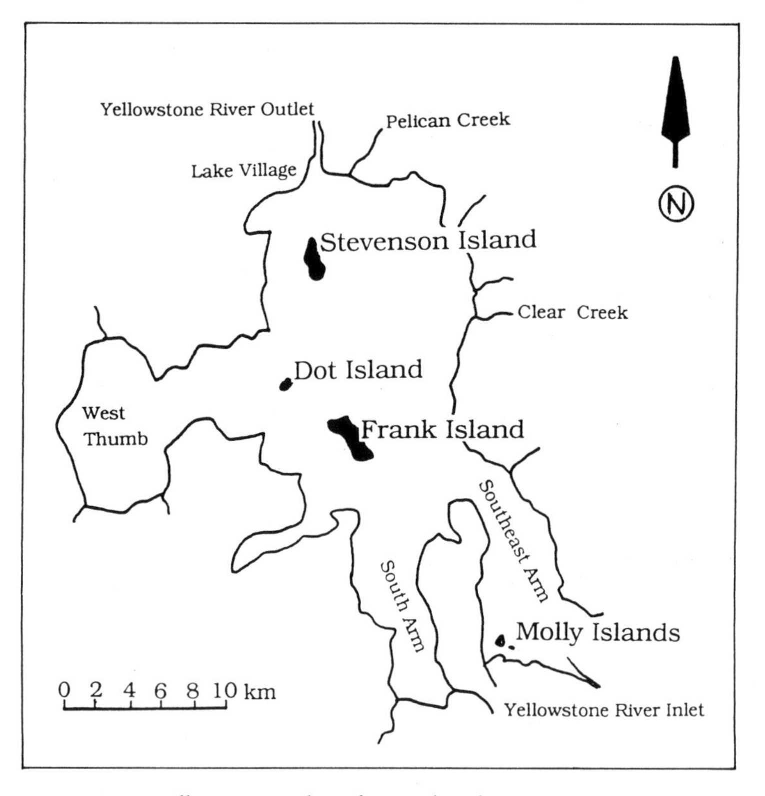

Figure 4.1 Sites on Yellowstone Lake referenced in the text.

We are not presenting here merely a chronicle of formal scientific response to management issues. The management and monitoring of an extremely large and complex resource, such as Yellowstone Lake, has not always lent itself to the straightforward constraints of traditional hypothesis-testing science. Frequently, too little was known or appreciated about the ecological processes of the lake and its surrounding country for managers and their staff to even pose a proper hypothesis; usually the system itself raised the important questions, which the human observers then had to be savvy enough to recognize and try to answer. We believe, however, that the very imperfection of this process offers priceless lessons in "real-world" research and monitoring of national park resources. The story we tell here is a complex blend of evolving management policies, evolving scientific understanding, and evolving public attitudes and needs. To piece together the big picture of Yellowstone Lake over the past century, we must resort not only to the most rigorously tested work of scientists, but also to the suppositions and best estimates of managers, as well as to our own interpretations of probable causes of certain events.

Because of its great attraction both to the public and to the scientific community, Yellowstone Lake is blessed with a rich history in research and monitoring. Scientists traveling with the first formal survey parties produced baseline information on the lake's ecology, establishing a tradition of study that continually enriches and refines understanding of not only the lake but also the piscine, avian, and mammalian inhabitants of its watershed.

Through this continued research and monitoring effort, even with its historical lapses and gaps, park managers are in many ways better able to fulfill the National Park Service (NPS) mandate to maintain natural systems in an unimpaired condition. The challenge to contemporary managers is to define the nature of "unimpaired condition" and to create the right device to measure the health of an ecosystem in a temporal sense. Indeed, the study of Yellowstone Lake, begun in an age when the husbandry of natural resources was directed more toward harvest than preservation of systems, now allows managers to take a holistic view not even imagined by the park's founders.

Studies have continued through distinct periods of park administration, beginning with a civilian administration from 1872 to 1886. From 1886 to 1916, the U.S. Cavalry worked in cooperation with the U.S. Department of the Interior. During an interim period of 2 years (1916–1918), NPS had been established, but the army still managed the park. NPS, however, has administered it since 1918. Research on the lake environment has been conducted by many parties, primarily the federal government through the bureaucratic lineage of the original U.S. Fish Commission, which grew into the U.S. Bureau of Sport Fisheries and Wildlife and the present U.S. Fish and Wildlife Service.

In the following narrative, we are going to speak generally about the findings of Yellowstone fisheries research and monitoring without necessarily specifying which agency or agencies conducted the research in question. When we, the authors, speak of "we" having observed a problem and then solved it, we are speaking in the broadest of terms of the agencies and the constituencies that have been concerned with these resources.

We are also going to speak generally of observed effects and results rather than of the often convoluted processes by which change is implemented. When we describe some event or research finding as leading to new management, we do not mean to portray this as a simple, linear process. The history of fisheries research and management in Yellowstone, like the history of just about everything else, has often been contentious and controversial. The 1961 closures of fishing and access on portions of the arms of Yellowstone Lake were, for example, enormously controversial at the time. In the interests of focusing on the actual research and monitoring and its benefits to the ecological setting, we are going to leave out the "war stories" that accompany any such resource issue in a place as popular and politically visible as Yellowstone.

The Yellowstone Cutthroat Trout as an Ecological Centerpiece

The Yellowstone cutthroat trout (*Oncorhynchus clarki bouvieri*) has turned out to be a useful indicator of the health of the greater aquatic and terrestrial system it inhabits. This is remarkable because it was chosen for study by early investigators with little regard for or awareness of the criteria by which such indicators might be selected (National Research Council 1986).

The history of Yellowstone Lake metaphorically reflects the dramatic saga of the greater history of American conservation. Because of environmental degradation, introduction of and subsequent competition with nonnative species, and human exploitation, the Yellowstone cutthroat trout, once the most abundant and widely dispersed subspecies of inland cutthroat trout, survives in only about 10% of its original range (Gresswell and Varley 1988; Varley and Gresswell 1988). The watershed of which Yellowstone Lake is a part is the last major stronghold of the species, but at various times over the past century, human activities have placed dire stresses even on that stronghold. Hatchery and spawn-taking operations, introduction of nonnative fishes, and both commercial and sport fishing have, singly or in combination, profoundly affected and even threatened this trout population.

The study of the lake's fish populations, limnology, and sport fishery began in the late 1880s under the leadership of David Starr Jordan and Barton Evermann (e.g., Jordan 1891), two of the foremost fisheries investigators of the day. Periodic creel surveys began in the 1930s and became annual in 1950; spawner escapement has been monitored annually since 1945. A substantial number of studies on the history (e.g., Varley 1979, 1981), fish ecology (e.g., Welsh 1952; Laakso and Cope 1956; Biesinger 1961; Raleigh and Chapman 1971), limnology (e.g., Benson 1961), and fishery (e.g., Moore et al. 1952; Ball and Cope 1961, Benson and Bulkley 1963; Gresswell and Varley 1988; Jones et al. 1991), many of which described the lake as environmentally unimpaired, provide a superb database for evaluating the health of the lake ecosystem and the influence of humans on it.

Fishery exploitation by European Americans began in earnest in the 1860s and grew with the popularity of the park. Except for the years during the World Wars and the Great Depression, angling pressure on Yellowstone Lake has increased steadily since 1872. Attitudes toward the harvest of fish were extremely liberal in the early days, and excessive harvest, with accompanying waste of killed fish, was common.

A decline in the quality of the fishing was noted shortly after the turn of the century. In 1908, a daily creel limit of 20 trout was imposed (Young 1908), and in 1919, commercial fishing, which supplied hotel restaurants, was eliminated (Albright 1920).

Hatchery operations began on Yellowstone Lake in 1899, eventually growing into a large system that included the permanent fish hatchery at Lake Village and 14 subsidiary spawn-collecting facilities at the largest of the lake's tributaries. The hatchery system became the world's foremost cutthroat-trout egg factory, peaking at 43.5 million eggs in 1940 and totaling more than 818 million eggs by the end of hatchery operations in 1957 (Gresswell and Varley 1988). These eggs were shipped to dozens of locations around the United States and abroad (Varley 1981). Although it was a well-intentioned program, it resulted in the mixing of scores of distinct in-lake races of cutthroat trout. This in all likelihood caused deterioration of their separate genetic integrities, and some evidence suggests that it also resulted in the demise of some small discrete spawning populations (archival spawning-run data, Yellowstone National Park). In short, the hatchery program was enormously disruptive of an ecosystem that has only existed since the last ice age, not only in gross effects on population size and extent, but also in more subtle effects on the population's evolving genetic diversity.

An additional perturbation of nearly equal significance was the introduction of nonnative fishes into what was an essentially one-fish system. Longnose suckers (*Catostomus catostomus*) were established by humans in the 1920s, and several small cyprinid species were introduced by the 1950s (Gresswell and Varley 1988). Other species, Atlantic salmon (*Salmo salar*) and rainbow trout (*Oncorhyncus mykiss*), were also introduced unsuccessfully (Varley 1981). If they had flourished, these species could have had disastrous effects on the native trout. Even today there is an enduring risk that a clandestine introduction of a nonnative fish could, if not discovered soon enough, dramatically affect the native fish population. Brook trout (*Salvelinus fontinalis*) were illegally introduced into a small tributary of Yellowstone Lake recently but were discovered and removed before they could spread along shoreline areas to other streams (Gresswell 1991).

By the late 1930s, concern over the decline of the lake fishery led to the hiring of a fishery biologist to assess the level of trout harvest and make management recommendations. World War II interrupted fisheries management, however, and with the postwar recreation boom, the Yellowstone Lake fishery continued to decline.

In the luxury of hindsight, we can say that the postwar (1948–1964) efforts in research and monitoring were both an immense success and a monumental failure. They were a success because by the 1950s and 1960s, few sport fisheries in the world could claim a knowledge base of the quality and depth that existed for Yellowstone Lake (more than 200 published works relevant to the cutthroat trout). Fisheries management concepts and modeling were innovative and at the forefront of scientific knowledge. The detailed and complex life-history, ecological, and exploitation factors were pieced together into a surprisingly realistic exploitation model. (Benson and Bulkley [1963] first predicted the annual maximum sustainable yield [MSY] for the lake population.)

On the other hand, these efforts were also failures because the knowledge was not being applied correctly and because harvesting any population at its MSY discounts inherent natural variations and is, therefore, a form of Russian roulette. For reasons too complex to consider here (see Gresswell and Varley 1988), those harvest limits were exceeded for many years. By the late 1950s and continuing into the 1960s, the Yellowstone Lake cutthroat-trout fishery was in a collapsed state. The trout standing crop and spawner escapements were meager; spawning populations were composed of few age classes poorly distributed in spawning habitat instead of widely distributed, multiple-age classes that existed historically. Though few people were aware of what all this meant for the ecosystem, these developments had already translated into another kind of failure, one of much more immediate concern: poor angler success (Figs. 4.2, 4.3, and 4.4).

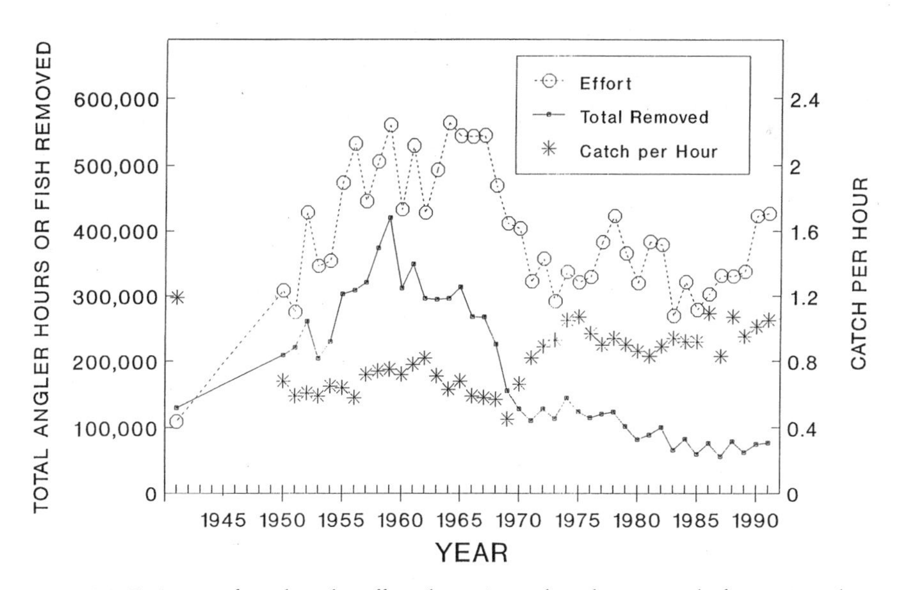

Figure 4.2 Estimate of total angler effort (hours), total angler removal of trout (number of trout killed), and catch per unit effort (trout landed per hour), Yellowstone Lake, 1941–1991.

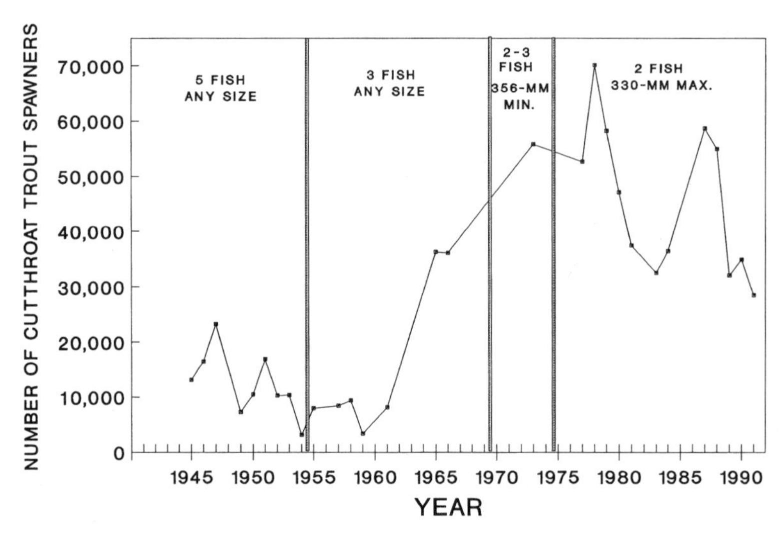

Figure 4.3 Number of cutthroat trout spawners entering Clear Creek, Yellowstone Lake, under various angling regulations, 1945–1991.

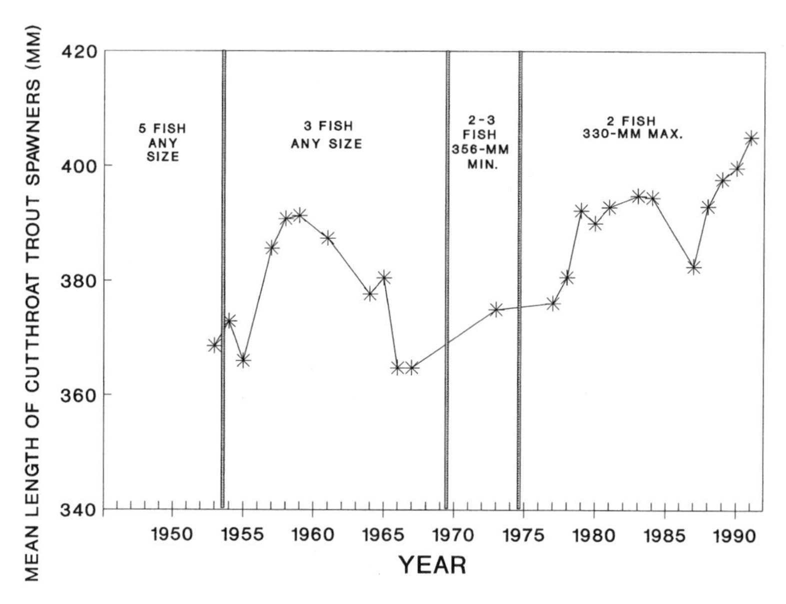

Figure 4.4 Mean length of cutthroat trout spawners entering Clear Creek, Yellowstone Lake, under various angling regulations, 1950–1991.

Table 4.1. Cutthroat-trout sanctuaries or refugia that contributed to the restoration of stocks in Yellowstone Lake by protecting populations from humans, besides general regulations governing the fishery.

Sanctuary or Refugia, Year Created	Stream or Lake	Type	Approximate Area (Hectares)
FISHING ELIMINATED			
Fishing Bridge, 1973	Stream	Spawning areas	58.9
Hayden Valley, 1965	Stream	General population	88.3
LeHardy's Rapids, 1965	Stream	Migration	2.1
Molly Islands, 1960	Lake	General population	97.1
Tributary streams, 1983	Stream	Spawning areas	197.9
Subtotal			444.3
RESTRICTED/CURTAILED FISHING			
Buffalo Ford, 1991	Stream	Spawning areas	1.0
Flat Mt. Arm, 1961	Lake	General population	47.3
South Arm, 1961	Lake	General population	457.1
Southeast Arm, 1961	Lake	General population	1,465.6
Subtotal			1,971.0
TOTAL			2,415.3

Fishery managers, challenged by necessity and public disappointment to restore the trout population, attacked the problem on several fronts:

1. They created fish sanctuaries or refugia, in which trout were protected either totally or at certain vulnerable times (Table 4.1).

2. They manipulated daily creel limits to reduce anglers' take while still allowing for some harvest for "campfire meals" (Figs. 4.4, 4.5, and 4.6).

3. They imposed maximum size limits to redirect exploitation from crucial spawning-age adults (current reproducers) to juvenile fish (potential reproducers; Figs. 4.2, 4.4, and 4.5).

4. They eliminated certain baits and lures to reduce the lethal effects and ensure a higher survival rate in "catch-and-release" fishing.

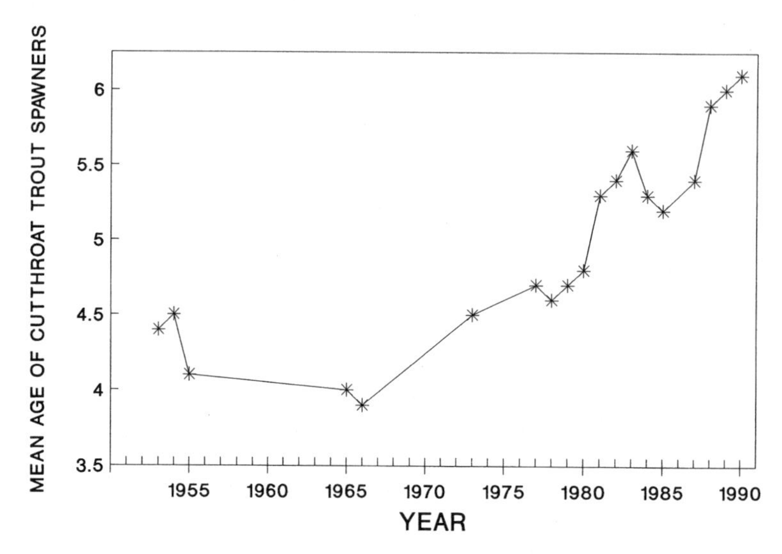

Figure 4.5 Mean age of cutthroat trout spawners entering Clear Creek, Yellowstone Lake, 1953–1990.

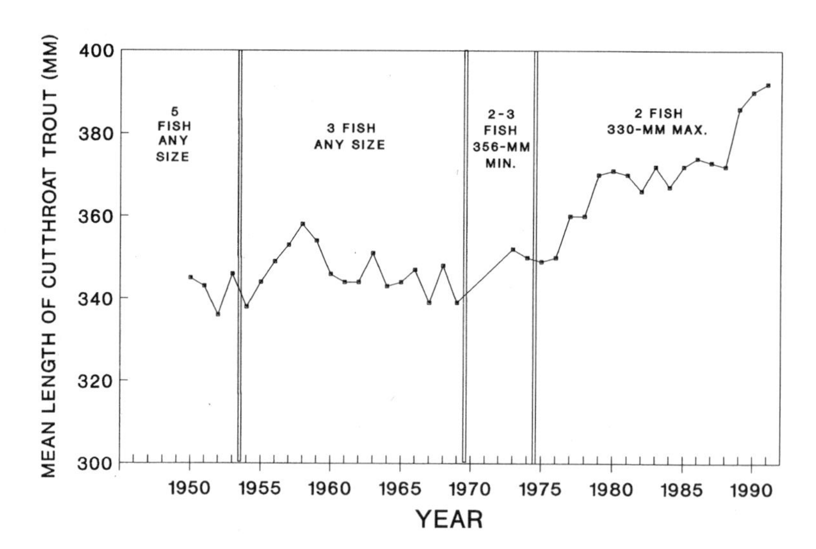

Figure 4.6 Mean length of cutthroat trout landed in Yellowstone Lake fishery under various angling regulations, 1950–1991.

5. They launched an ambitious educational program aimed at justifying and explaining the changes, which for the era, were radical and foreign to most park visitors.

Through trial and error, each of the fishery management actions was modified and fine-tuned until the desired effect—that of reversing the downward population trend and rebuilding the population size—was achieved (Figs. 4.2–4.6). Each fishery management action was intensively monitored so that the efficacy of every step could be tested. The long-term information base that was created, coupled with the ever-increasing awareness of the role and significance of these fish in park ecosystem processes, has led to growing enthusiasm for the program, both public and scientific.

The reversal of fortunes for Yellowstone Lake trout allowed the trout population to regain some semblance of its prehistoric abundance. The fishery management program had other ecological consequences that were perhaps even more significant in the world of long-term monitoring. These reveal not only the ecological "ripple effect" of restoring a trout population, but also the value of monitoring related life forms and other environmental factors in the same ecosystem.

The Yellowstone White Pelicans

The North American white pelican (*Pelecanus erythrorhynchos*) serves as an unusually sensitive environmental barometer of wilderness conditions because it requires, especially during its breeding season, nearly complete isolation from humans (Diem and Condon 1967). The Molly Islands in Yellowstone Lake (Fig. 4.1) are the site of a pelican rookery that is unique in several ways. It is the only continuously surviving colony of white pelicans known in a U.S. national park. It was for several decades the only surviving colony in Wyoming (by the 1930s there were only seven surviving colonies in North America). At 2,357 m, it is the highest known breeding colony for the species (Diem and Condon 1967). The autumn southerly migration splits the population, which winters on both coasts of North and Central America (Diem and Condon 1967). The colony has been studied and monitored periodically, and then more consistently, for more than 100 years, during which time it has been subjected to various management approaches that we can now evaluate for their effectiveness.

The colony has a rich research history, beginning with George Bird Grinnell (1876) in 1875 (see Linton 1891; Skinner 1917, 1925; Adams 1925; Miller 1932; Thompson 1932; Murphy 1960; Schaller 1964; Diem and Condon 1967).

The Molly Islands breeding colony has been the subject of controversy almost since it was discovered. The pelicans were early and correctly implicated as a primary host in the Yellowstone cutthroat trout's infamous tapeworm infestation (Linton 1891, and many others). They were further resented for their consumption of large numbers of trout that sport fishermen were thus denied. Promoters of the park's sport fisheries so disliked the parasite and so resented the trout consumption that they officially and successfully launched a pelican population-control program in 1924. There is also evidence of "unofficial" control even earlier (Wright 1934). That such a measure could even be proposed suggests the extent to which management philosophy of national parks has evolved, however far it still may have to go.

The control program, which was essentially an egg-destruction campaign, as well as the associated human disturbance on the nesting islands (to which public access was unlimited), caused the pelican population to decline (Fig. 4.7). At the end of the 1931 season, at least partly because of public objection, the control program was halted, and people were no longer allowed to land on the islands during the breeding season (Wright 1934). The pelican population rebounded.

After World War II, increased visitation to Yellowstone National Park was accompanied by an increase in the use of recreational powerboats. Though not permitted to land on the islands, the constant parade of boats nearby caused another population decline (Fig. 4.7). In 1960 NPS closed much of that portion of the lake to powerboats and created a 0.4-km "no human entry" zone around the islands themselves. Since 1966 the pelican population has undergone an impressive, if erratic, population increase (Fig. 4.7).

Over many decades, research and monitoring of the Molly Islands pelicans has revealed some noteworthy trends. First, it is clear that the pelicans were in serious decline at least twice and that management actions (sometimes correcting harmful earlier management actions) to reduce human disturbance reversed those declines (Table 4.1 and Fig. 4.7).

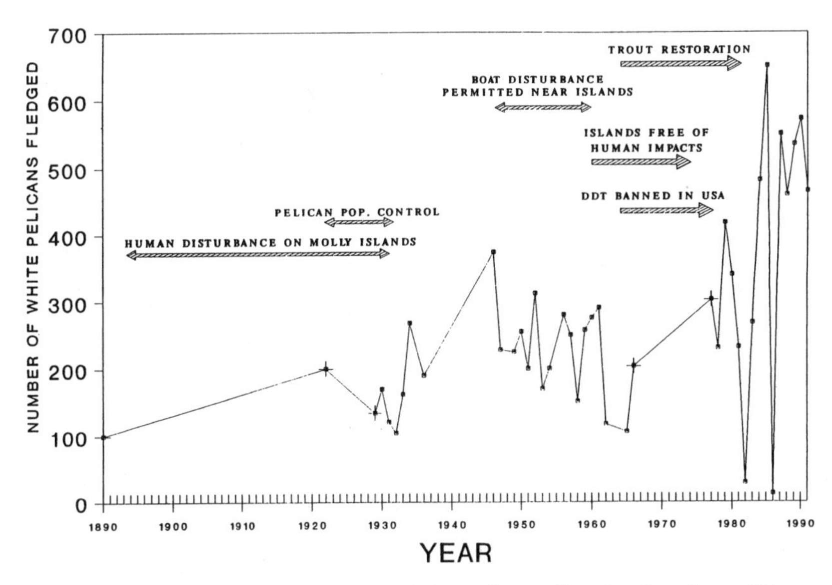

Figure 4.7 Number of white pelicans fledged from the Molly Islands colony, Yellowstone Lake, in response to various management scenarios, 1890–1991.

Second, although the insecticide DDT and its derivatives were never conclusively demonstrated to have a negative effect on pelicans in Yellowstone, the ban on it in the Greater Yellowstone Ecosystem in 1964 (Mussehl and Findley 1967) likely had a positive effect on pelican survival.

Third, inferential evidence suggests that the restoration of the pelican's primary food source, Yellowstone cutthroat trout (Figs. 4.2–4.6), may have facilitated the restoration of recruitment to the pelican population (Fig. 4.7).

Fourth, the importance of interrelated monitoring efforts is nowhere more clearly demonstrated than in the relationship between pelican numbers and geological/hydrological research. The erratic swings in the production of fledglings observed in the 1980s, if accepted with no knowledge of seemingly unrelated geological/hydrological research and monitoring on the lake, might cause a pelican biologist considerable confusion and alarm. However, geological monitoring of inflation and deflation of the Earth's crust at the lake's hydraulic outlet, caused by a deep fissure in the Earth called the "Yellowstone Hot Spot," (Pelton and Smith 1982; Smith et al. 1989) coupled with annual hydrological monitoring of water levels (U.S. Geological Survey 1923–1991), reveal that the pelican rookery nests were unavoidably flooded in some years (crustal uplift + high snowpack = few pelican fledglings). "Doming" of the landscape, caused by the movement of super-hot fluids or molten rock, altered the lake depth and flooded the rookery. These two factors are engaging examples of just how many and how subtle are the variables that control the fortunes of an ecosystem. Researchers, however, have been unable to determine to what extent these geological events played a role in the pelican population's condition at various times earlier in the century. Even when management apparently corrected past errors and helped restore a population of animals, it may be impossible to know to what extent other factors were involved.

The preservation of the Yellowstone pelican population demonstrates the remarkable marriage of data gained from many disciplines required to manage ecosystems. For these pelicans it ranged from historical documents to snowpack records to crustal uplift measurements to satellite technology to ongoing studies of the birds themselves. Although pelicans and trout are valued by humans, it was the joining of such apparently disjunct and seemingly irrelevant data that yielded important results, not only in the conservation of the species, but also in the understanding of how the ecosystem works.

The Complete Trout Angler: The Yellowstone Lake Osprey

Ospreys (*Pandion haliaetus*) have often been regarded as indicators of the ecological well-being of an area and have been observed to be in decline in many parts of the United States (Henny and Ogden 1970). The species has been present and apparently abundant on Yellowstone Lake from the time of the first explorers (Bonney and Bonney 1970). Populations have been researched on Yellowstone Lake (Skinner 1917; Turner 1968; Swenson 1975) and have been monitored periodically since 1924 and regularly since 1972.

Ospreys are migratory, obligate fish eaters. They home to the area where they were born and are thus susceptible to (and reflections of) changes in environmental conditions anywhere on their breeding, migratory, and wintering ranges. In the past century, Yellowstone osprey populations have declined in numbers and in extent of distribution (Swenson 1979). Osprey nests on Yellowstone Lake have declined approximately 46% since 1924 (Swenson 1975). The cause of this decline was apparently human disturbance of their breeding areas, although pesticides accumulated by birds while outside the park may have been another factor (Swenson 1979).

Management responses to the decline were accompanied by national responses, especially the nationwide ban on DDT in 1972. The most direct and probably most important management action in Yellowstone Park was the closing of all human campsites within 1 km of all known nesting sites (Fig. 4.8). Indirect actions included, besides the regional and then national bans on DDT, the restoration of the cutthroat trout population beginning in the early 1970s (Figs. 4.2–4.6).

By 1991 the osprey population had responded with an increased number of breeding pairs along the lakeshore and on the lake's crucial nesting islands (Fig. 4.8). The reduction of human disturbance of nesting sites was experimentally shown to be the most immediate and powerful factor in enhancing osprey reproduction in the park (Swenson 1979). The DDT ban and the tenfold increase in available trout were undoubtedly contributing factors as

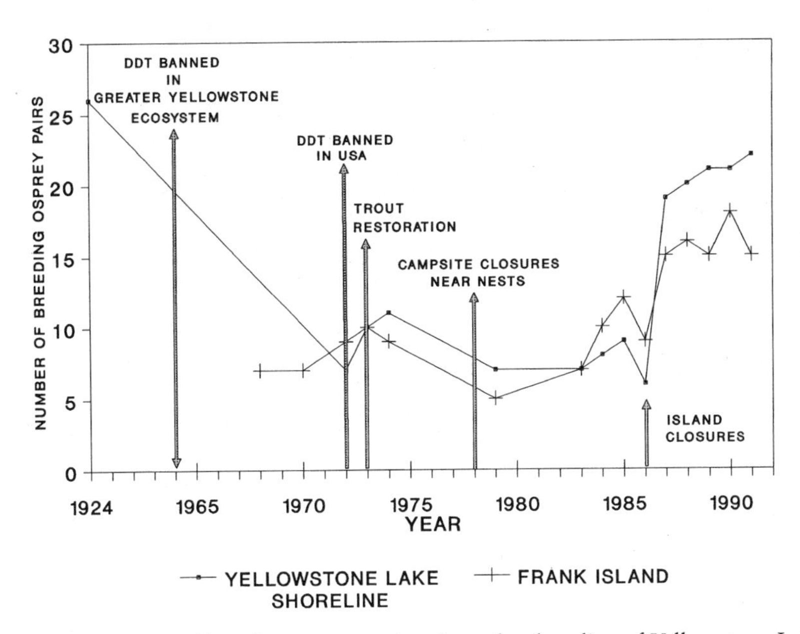

Figure 4.8 Number of breeding osprey pairs along the shoreline of Yellowstone Lake and on Frank Island, the largest island in the lake, in response to various management scenarios, 1924–1991.

well. An important lesson for long-term monitoring programs is that it is not always possible to measure the relative merits or effects of various management measures and environmental factors. An equally important lesson, however, is that only by measuring several factors can a researcher even become aware of the complexity of the system. Circumstances rarely provide a researcher with a background-noise-free, field-experiment situation.

Grizzly Bears and Trout: The Ripples Spread

Beginning in the late 1880s, park managers condoned and then encouraged the feeding of grizzly bears (*Ursus arctos*) and black bears (*Ursus americanus*) at garbage pits and dumps, a practice that was quickly institutionalized into a major visitor attraction between 1900 and 1940 (Schullery 1992). Continued feeding of garbage to bears was customary, out of sight of the public, until 1970 (Craighead et al. 1974; Cole 1976). After 1970, when this food source was eliminated, grizzly bears were weaned to natural foods, especially to vegetation and large mammals (Servheen et al. 1986; Schullery 1992). This process was accompanied by enormous public, scientific, and political controversy (Craighead 1979; McNamee 1984).

Each year, May through July, cutthroat trout ascend about 59 tributaries of Yellowstone Lake to spawn (Reinhart and Mattson 1990). Because of their relatively large size (averaging 0.8 kg), the quality of their flesh in protein and fats, their appearance in large numbers, and their vulnerability in narrow, shallow streams, these fish form a significant seasonal food source for both grizzly and black bears (Reinhart and Mattson 1990).

The historical record of bears eating spawning fish in the Yellowstone area is scanty, although the literature contains occasional anecdotal tidbits (Skinner 1927; Whittlesey 1988; Schullery 1991). Reasons for this are not well understood but probably include the earliness of the spawning runs in the tourist season; the blockage early in this century, by road culverts and other activities, of some of the most visible spawning streams; the general wariness of bears toward humans; and the light backcountry travel near many spawning streams. The available evidence in Yellowstone's ample literature and files (J. Craighead, pers. comm., 1984; Schullery 1991; R. B. Smith, pers. comm., 1992; U.S. Fish and Wildlife Service, unpubl. data) indicates that consump-

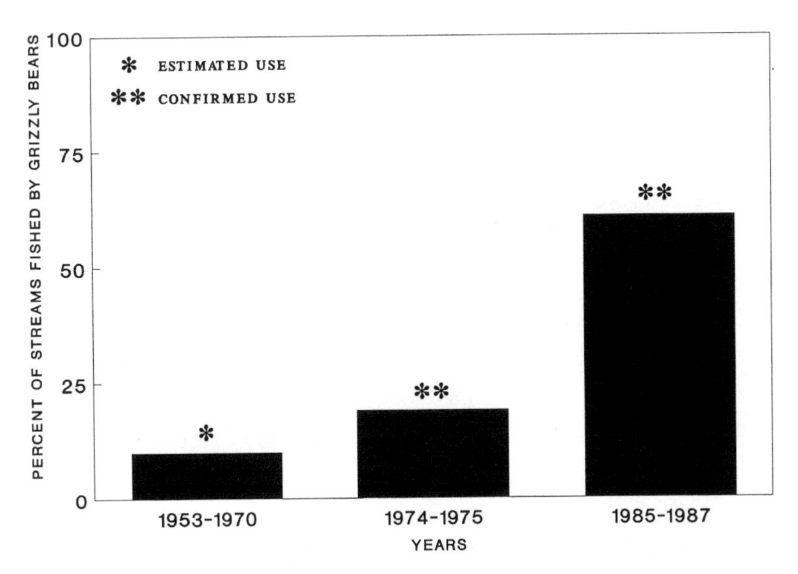

Figure 4.9 Estimated and confirmed use of cutthroat trout spawners by grizzly bears on streams tributary to Yellowstone Lake, 1953–1970 (J. Craighead, pers. comm., 1984; R. B. Smith, pers. comm., 1992; U.S. Fish and Wildlife Service, unpubl. data), 1974–1975 (Hoskins 1975), and 1985–1987 (Reinhart and Mattson 1990).

tion of fish by grizzly and black bears between 1953 and 1970 was slight (Fig. 4.9).

Surveys conducted in 1974 and 1975 found evidence of bear fishing on 11 of the 59 (19%) spawning streams (Hoskins 1975). This was only 4 years after the closure of the dumps and indicated at least a moderate and possibly increasing knowledge among bears of the fish as a food source (Fig. 4.9).

Fifteen years after the dumps were closed, bear use of spawning streams had increased to 61% (Reinhart and Mattson 1990). Reinhart and Mattson concluded that grizzly bear consumption of spawning trout was largely a positive function of spawner density. As a result of the change in trout harvest policies on the lake, the number of available spawning fish increased (Fig. 4.3), as did their average size (Fig. 4.4). Researchers have observed an individual adult female bear maintain an average harvest of 100 fish per day for 10 days (S. French, Yellowstone Grizzly Foundation, pers. comm., 1989). By any standards, this is a massive protein and fat intake and a major new element in the seasonal food habits of the Yellowstone grizzly bear.

Discussion

This is a continuing story. Ongoing monitoring and the ever-growing information base provided by science should help researchers to anticipate other problems—to foresee trends and their consequences. Such foresight may enable park personnel to function more often in a proactive, rather than reactive, role in resource management. The changes in fisheries management that were initiated in Yellowstone during the past 20 years have resulted in the park's program being heralded as a world model of progressive fisheries management and as an ecologically sensitive approach to a traditionally consumptive resource use (Wulff 1978; Varley 1984).

The success of this program depended upon many factors, including the park's unusual jurisdictional situation, which gave managers the necessary freedom to act decisively, and a growing awareness in the public and sportfishing communities of the need for better management. The program, however, would have been far more difficult, perhaps even impossible, to conceive and execute without an exceptional research and monitoring tradition. The institutional memory of Yellowstone Lake managers, enriched as it was by so much information and experience, allowed the program to set the kinds of goals and tests needed to succeed.

There is inevitable uncertainty in the management of natural resources. Wildlife populations pose special problems, not only because they are "fugitive resources" that may entirely abandon the manager's jurisdiction (and the researcher's study area) for periods of time, but also because they are part of very complex natural systems. Given recent revelations about the changing condition of the global environment, the level of uncertainty that managers and investigators must face will apparently increase.

Reversing the trend of declining populations offers especially interesting challenges. Traditionally, applied science was conducted to determine the cause of the decline, and then management actions, usually focused tightly on the key cause, were taken to correct the perceived dysfunction. Now, however, when confronted with a declining population, managers may act first rather than dwell on basic research and the inevitable delay that research causes.

> [M]anagement should be viewed as an adaptive process: we learn about the potentials of natural populations to sustain harvesting mainly through experience with

> management itself, rather than through basic research or the development of general ecological theory. . . . We keep running into questions that only hard experience can answer, and a basic issue becomes whether to use management policies that will deliberately enhance that experience. (Walters 1986, p. vii)

In the Yellowstone Lake experience with research and monitoring, Walters' (1986) management vision may be more of a future goal than a past reality; certainly, a foundation has been prepared for such an experiment. McNab (1983) made a useful suggestion in this regard, proposing that, in an appropriately experimental atmosphere, "the present distinction between research and management would blur, research and management being forced into a tighter working association (p. 401)."

The species associated with Yellowstone Lake treated in this chapter teach lessons about both the traditional and the experimental management (as in McNab 1983) techniques. First and most important, traditional research and monitoring on Yellowstone Lake demonstrated their worth. Declines in cutthroat trout, white pelican, and osprey populations were identified, and management actions were successful in reversing those declines.

Second, management actions were rigorously evaluated to determine their effectiveness. For example, fishing regulations on Yellowstone Lake evolved over more than a century, a responsive process based on monitoring, and researchers monitored pelican populations on the Molly Islands after the 1931 ban on human disturbance.

Third, in some cases a set of management actions produced satisfactory results, but it was impossible to determine the relative worth of each action in the set. For example, increased control over human disturbance, a ban on DDT, and restoration of the trout population all apparently contributed to the increased breeding success of the pelican population in the late 1970s and early 1980s.

Fourth, an interdisciplinary marriage of apparently incompatible sciences proved to be essential in understanding the complex interplay of the ecosystem's components. Communication among (to say nothing of the research being conducted by) hydrologists, geophysicists, and ecologists was necessary to explain the wild fluctuations in pelican fledgling rates during the past 10 years.

Fifth, Pasteur was right: chance really does favor the prepared mind, or, in this case, the prepared research and monitoring program. Research and monitoring are at their best when they can accommodate unanticipated consequences. Had we not been listening to several of the ecological stories that Yellowstone Lake was telling us, we would not have been in a position to even hear the story of the grizzly bear. Because of the spectacular effects that trout restoration had not only on grizzly bear activities but also on the general welfare of this threatened bear population, this is not a story we would have wanted to miss. A consequence of managing and monitoring as large, complex, and slowly changing a system as this is that the realistic manager must recognize that not all the issues can be resolved through straightforward hypothesis testing. It is not possible, with current knowledge of the ecosystem or current technology, to design a hypothesis or set of hypotheses that will embrace such an elaborate combination of ecological, geological, and cultural factors. Because of the current rate at which we are still learning about this system, we will for a long time manage it best and monitor it most successfully if we are prepared for the sort of seat-of-the-pants analyses that have so often characterized past management. This idea may bring considerable discomfort to scientists and managers who are devoted to solving all problems through hypothesis-testing techniques, but it is true. Many elements of the Yellowstone Lake setting can be treated by formal hypothesis testing, but some will require other approaches, including less formal but no less scientifically meaningful considerations and evaluations of various sets of seemingly unrelated data.

Sixth, we experience considerable uneasiness when we wonder how it is ever possible to manage resources intelligently and responsibly without long-term information. In the early 1970s, when managers and scientists were mapping out the restoration of the cutthroat trout, none predicted or even imagined that the population would still be "sorting things out" 20 years after the actions were taken (Figs. 4.2–4.6). For students of the "quick fix" school of management, the meaning and extent of the phrase "ecological ripples" take on a new and more powerful dimension here. The ecological ripples on Yellowstone Lake and its environs do not quickly dissipate and may continue to crisscross the water long after the "splash" of the initial event has disappeared.

The magnitude and durability of those ripples, both natural and human-induced, give us ample reason to avoid overly congratulating ourselves. We must not forget that the park's institutional history of 120 years, though long in terms of a human lifetime, is short in terms of many natural processes. Recent work in paleolimnology (Whitlock et al. 1991), paleoecology (Hadly 1990), fire ecology (Despain and Romme 1989; Knight and Wallace 1989), and paleoclimatology (Balling et al. 1992) suggests that most of Yellowstone's fundamental ecosystem processes operate on longer scales—centuries and millennia. "Long-term" monitoring of the fire-return interval in Yellowstone's lodgepole pine forests, as just one example, would require a scientific attention span of thousands of years. Thus, we enter our second century of research and monitoring on Yellowstone Lake well aware that it has much more to teach us.

Authors' Note

During the summer of 1994, NPS managers learned that nonnative lake trout (*Salvelinus namaycush*) had become established in Yellowstone Lake, almost certainly through an intentional and illegal introduction. Then-Superintendent Robert Barbee called this thoughtless introduction "an appalling act of environmental vandalism." Additional investigation and gillnetting in the fall of 1994 and the spring of 1995 revealed at least four and possibly as many as six age classes of lake trout in Yellowstone Lake. Park personnel suspect that a few lake trout were illegally introduced into the lake as early as 1984. These fish successfully established themselves and are now producing offspring.

The ecological, scientific, and economic consequences of this growing lake-trout population are profound. Based on fisheries management experience in other large western lakes, researchers believe that within a few decades, the lake trout could greatly reduce or destroy the cutthroat trout population. Many other species, including grizzly bears, bald eagles, osprey, and pelicans, depend heavily upon a robust population of cutthroat trout. A significant part of the regional recreational industry also depends upon this one species.

This situation appears to present both the trout and the humans who manage them with the greatest challenge they have faced since the creation of Yellowstone National Park. It will be interesting and instructive to observe how past monitoring and research on Yellowstone Lake will affect this process.

Literature Cited

Adams, C. 1925. The relation of wildlife to the public in national and state parks. Roosevelt Wildlife Bulletin 2(4):371–401.

Albright, H. M. 1920. Superintendent's annual report for Yellowstone National Park—1919. National Park Service, Yellowstone National Park.

Ball, O. P., and O. B. Cope. 1961. Mortality studies on cutthroat trout in Yellowstone Lake. U.S. Fish and Wildlife Service Research Report 55. 62 p.

Balling, R. C., G. A. Meyer, and S. G. Wells. 1992. Climate change in Yellowstone National Park: is the drought-related risk of wildfires increasing? Climate Change 22:35–44.

Benson, N. G. 1961. Limnology of Yellowstone Lake in relation to the cutthroat trout. U.S. Fish and Wildlife Service Research Report 56. 33 p.

Benson, N. G., and R. V. Bulkley. 1963. Equilibrium yield of cutthroat trout in Yellowstone Lake. U.S. Fish and Wildlife Service Research Report 62. 44 p.

Biesinger, K. E. 1961. Studies on the relationship of the redside shiner (*Richardsonius balteatus*) and the longnose sucker (*Catostomus catostomus*) to the cutthroat trout (*Salmo clarki*) population in Yellowstone Lake. Unpubl. M.S. thesis, Utah State University, Logan.

Bonney, O., and L. Bonney. 1970. Battle drums and geysers. Swallow Press, Chicago. 622 p.

Cole, G. 1976. Management involving grizzly bears in Yellowstone National Park, 1970–1975. National Park Service Natural Resources Report 9. 26 p.

Craighead, F. 1979. Track of the grizzly. Sierra Club Books, San Francisco. 261 p.

Craighead, J., J. Varley, and F. Craighead. 1974. A population analysis of the Yellowstone grizzly bears. Montana Forest and Conservation and Experiment Station, University of Montana, Bulletin 40. 20 p.

Despain, D. G., and W. Romme. 1989. Historical perspective on the Yellowstone fires of 1988. BioScience 39(10):695–699.

Diem, K., and D. D. Condon. 1967. Banding studies of water birds on the Molly Islands, Yellowstone Lake, Wyoming. Yellowstone Library and Museum Association, Yellowstone National Park. 39 p.

Gresswell, R. E. 1991. Use of antimycin for removal of brook trout from a tributary of Yellowstone Lake. North American Journal of Fisheries Management 11:83–90.

Gresswell, R. E., and Varley, J. D. 1988. Effects of a century of human influence on the cutthroat trout of Yellowstone Lake. American Fisheries Society Symposium 4:45–52.

Grinnell, G. B. 1876. Zoological report. P. 63–71 in Report of Reconnaissance from

Carroll, Montana Territory, on Upper Missouri, to Yellowstone National Park and Return, Made in the Summer of 1875. U.S. War Department.

Hadly, E. A. 1990. Late holocene mammalian fauna of Lamar Cave and its implications for ecosystem dynamics in Yellowstone National Park, Wyoming. Unpubl. M.S. thesis, Northern Arizona University, Flagstaff. 128 p.

Henny, C. J., and J. C. Ogden. 1970. Estimated status of osprey populations in the United States. Journal of Wildlife Management 34:214–217.

Hoskins, W. 1975. Yellowstone Lake tributary study. Interagency Grizzly Bear Study Team, Bozeman, Montana. 10 p.

Jones, R. L., R. Andrascik, D. G. Carty, R. Ewing, L. R. Kaeding, B. M. Kelly, D. L. Mahony, and T. Olliff. 1991. Fishery and aquatic management program in Yellowstone National Park. U.S. Fish and Wildlife Service, Yellowstone National Park, Technical Report for 1991. 200 p.

Jordan, D. S. 1891. A reconnaissance of streams and lakes in Yellowstone National Park, Wyoming, in the interest of the U.S. Fish Commission. Bulletin of the U.S. Fish Commission 9(1889):41–63.

Knight, R., and L. Wallace. 1989. The Yellowstone fires: issues in landscape ecology. BioScience 39(10):700–705.

Laakso, M., and O. B. Cope. 1956. Age determination in Yellowstone cutthroat trout by the scale method. Journal of Wildlife Management 20:139–153.

Linton, E. 1891. A contribution to the life history of *Dibothrium cordiceps* Leidy, a parasite infesting the trout of Yellowstone Lake. Bulletin of the U.S. Fish Commission 9(1889):337–358.

McNab, J. 1983. Wildlife management as scientific experimentation. Wildlife Society Bulletin 11(4):397–401.

McNamee, T. 1984. The grizzly bear. Alfred Knopf, New York. 308 p.

Miller, G. 1932. American white pelican, *Pelecanus erythrorhynchos*. Unpubl. report in Yellowstone National Park Library. 32 p.

Moore, H. L., O. B. Cope, and R. E. Beckwith. 1952. Yellowstone Lake creel censuses, 1950–51. U.S. Fish and Wildlife Service Special Science Report 81.

Murphy, J. R. 1960. The Molly Islands nesting colonies of Yellowstone Lake. Unpubl. report in Yellowstone National Park Library. 9 p.

Mussehl, T. W., and R. B. Findley, Jr. 1967. Residues of DDT in forest grouse following spruce budworm spraying. Journal of Wildlife Management 31:270–287.

National Research Council. 1986. Ecological knowledge and environmental problem solving—concepts and case studies. National Academy Press, Washington, D.C. 388 p.

Pelton, J. R., and R. B. Smith. 1982. Contemporary vertical surface displacements in

Yellowstone National Park. Journal of Geophysical Resources 87:2745–2761.

Raleigh, R. F., and D. W. Chapman. 1971. Genetic control of lakeward migrations of cutthroat trout fry. Transactions of the American Fisheries Society 100(1):33–40.

Reinhart, D., and D. Mattson. 1990. Bear use of cutthroat trout spawning streams in Yellowstone National Park. International Conference on Bear Research and Management 8:343–350.

Schaller, G. B. 1964. Breeding behavior of the white pelican at Yellowstone Lake, Wyoming. Condor 66(1):3–23.

Schullery, P. 1991. Yellowstone bear tales. Roberts Rinehart, Niwot, Colorado. 212 p.

———. 1992. The bears of Yellowstone. High Plains Publishing Co., Lander, Wyoming. 318 p.

Schullery, P., and L. Whittlesey. 1992. The documentary record of wolves and related wildlife species in the Yellowstone National Park area prior to 1882. P. 1-3–1-73 in W. Brewster and J. Varley, eds. Wolves for Yellowstone? IV. National Park Service, Washington, D.C.

Servheen, C., R. Knight, D. Mattson, S. Mealey, D. Strickland, J. Varley, and J. Weaver. 1986. Report to the Interagency Grizzly Bear Committee on the availability of foods for grizzly bears in the Yellowstone ecosystem. Unpubl. report in Yellowstone National Park Library. 21 p.

Skinner, M. P. 1917. The birds of Molly Islands, Yellowstone National Park. Condor 19(6):177–182.

———. 1925. The birds of the Yellowstone National Park. Roosevelt Wildlife Bulletin 3(1):1–192.

———. 1927. The predatory and fur-bearing animals of Yellowstone Park. Roosevelt Wildlife Bulletin 4(2):163–282.

Smith, R. B., R. E. Reilinger, C. M. Meertens, J. R. Hollis, S. R. Holdahl, D. Dzurisin, W. K. Gross, and E. E. Klingele. 1989. What's moving at Yellowstone! The 1987 crustal deformation survey from GPS, leveling, precision gravity and trilateration. American Geophysical Union, EOS 70:113–125.

Swenson, J. E. 1975. Ecology of the bald eagle and osprey in Yellowstone National Park. Unpubl. M.S. thesis, Montana State University, Bozeman. 146 p.

———. 1979. Factors affecting status and reproduction of ospreys in Yellowstone National Park. Journal of Wildlife Management 43(3):595–601.

Thompson, B. H. 1932. History and present status of the breeding colonies of the white pelican (*Pelecanus erythrorhynchos*) in the United States. National Park Service, Contributions to Wildlife Division, Occasional Paper 1. 82 p.

Turner, J. F. 1968. Preliminary report on the osprey, *Pandion haliaetus,* in northwestern Wyoming—1968. Unpubl. report, University of Michigan, Ann Arbor.

U.S. Geological Survey. 1923–1991. Water resources data for Montana, Wyoming, and Idaho. U.S. Geological Survey, Helena, Cheyenne, and Boise.

Varley, J. D. 1979. Record of egg shipments from Yellowstone fishes, 1914–1955. Yellowstone National Park, Information Paper 36.

———. 1981. A history of fish stocking activities in Yellowstone National Park between 1881 and 1980. Yellowstone National Park, Information Paper 35. 93 p.

———. 1984. The use of restrictive angling regulations in managing wild salmonids in Yellowstone National Park, with particular reference to cutthroat trout. P. 145–156 in J. M. Walton and D. B. Houston, eds. Proceedings of the Olympic Wild Fish Conference, March 23–25, 1983, Port Angeles, Washington.

Varley, J. D., and R. E. Gresswell. 1988. Ecology, status, and management of the Yellowstone cutthroat trout. American Fisheries Society Symposium 4:13–24.

Varley, J. D., and P. Schullery. 1983. Freshwater wilderness, Yellowstone fishes and their worlds. Yellowstone Library and Museum Association, Yellowstone National Park. 132 p.

Walters, C. 1986. Adaptive management of renewable resources. Macmillan Publishing Co., New York. 374 p.

Welsh, J. P. 1952. A population study of Yellowstone blackspotted trout (*Salmo clarki*). Unpubl. Ph.D. dissert., Stanford University, Stanford. 180 p.

Whitlock, C., S. C. Fritz, and D. R. Engstrom. 1991. A prehistoric perspective on the northern range. P. 289–305 in R. B. Keiter and M. S. Boyce, eds. The Greater Yellowstone Ecosystem. Yale University Press, New Haven.

Whittlesey, L. 1988. Yellowstone place names. Montana Historical Society Press, Helena.

Wright, G. 1934. The primitive persists in bird life of Yellowstone Park. The Condor 36:145–157.

Wulff, L. 1978. The bright future of trout fishing. Sports Afield 179(3):35–39.

Young, S. B. M. 1908. Report of the superintendent of the Yellowstone National Park to the Secretary of the Interior. U.S. Government Printing Office, Washington, D.C.

5

Wolf and Moose Populations in Isle Royale National Park

R. Gerald Wright

Biological studies of natural communities are often difficult owing to the inherent complexity of the environment. The task of understanding these communities is further complicated because important events occurring in such environments can sometimes be witnessed only over long intervals. These events may happen under conditions that represent a continuum of variations and may involve causes that are unknown or even unsuspected. The best approach to evaluate such factors is to continuously observe and record what is seen. As data accumulate, they may reveal phenomena that are repeated or conform to some pattern. This pattern, in turn, can lead to theories about what is going on and hence to experiments designed to test specific hypotheses.

Unfortunately, long-term observations of biological phenomena, particularly those involving animal life in natural communities, are rare. In an era when most ecological research involves short-term studies, the 34-year study of wolf and moose population dynamics on Isle Royale is unique. Few studies on animal populations have continued for so long, have produced so much useful information, or have achieved such notoriety.

There is much that scientists and organizations involved in or contemplating long-term ecological research can learn from the Isle Royale study. This paper examines the insights that can be gained from an evaluation of this study. The emphasis is on how the study was organized and conducted and how the study influenced park management. An evaluation of the actual data collected and the inferences drawn from them is largely excluded from this analysis because they have been adequately dealt with in numerous scientific publications.

The Setting

Isle Royale National Park is an archipelago consisting of one large island (Isle Royale) 72 km long and 14 km at its widest point (644 km^2) and a complex of many smaller islands lying in northern Lake Superior 24–29 km from the Ontario shore.[1] The highest point on the large island is 238 m above Lake Superior. The island's basaltic and conglomeritic bedrock is covered by shallow, largely organic soil and an almost continuous forest cover. Glacial activity some 9,000 to 11,000 years ago resulted in a longitudinal ridge and valley or "washboard" topography (Peterson 1977).

The island contains an interspersion of two major vegetation types: the boreal spruce-fir forest and the northern-lake-states hardwoods (Linn 1957). Typical boreal tree species grow near the lakeshore, where atmospheric moisture is greater and temperatures during the growing season are lower and less variable than in the interior of the island. Species include white spruce (*Picea glauca*), balsam fir (*Abies balsamea*), white birch (*Betula papyrifera*), and aspen (*Populus tremuloides*). At higher elevations in the interior of the island, species typical of the northern hardwood forest grow, such as sugar maple (*Acer saccharum*) and yellow birch (*Betula allegheniensis*).

The fauna is distinctly limited as compared with the adjacent mainland (Mech 1966). The most influential mammalian herbivores are the moose (*Alces alces*), beaver (*Castor canadensis*), snowshoe hare (*Lepus americanus*), and red squirrel (*Tamiasciurus hudsonicus*). The principal carnivores are the wolf (*Canis lupus*) and red fox (*Vulpes fulva*). Mammals found on the mainland but notably absent from the island are the white-tailed deer (*Odocoileus virginianus*), porcupine (*Erethizon dorsatum*), raccoon (*Procyon lotor*), and black bear (*Ursus americanus*).

The limited diversity of vertebrate life and the relatively small size of the island suggest that it would be an excellent example of biotic instability. The record is not entirely clear in this respect, and a longer period of observation may be necessary for reliable interpretation. Major fluctuations have occurred. However, at least some of the changes in the island's fauna are probably directly related to historical events and land-use changes in areas north of Lake Superior and to human activities on the island, rather than to the inherent characteristics of Isle Royale (Allen 1976). The island is not totally isolated; some species have colonized it over time by swimming or crossing on a temporary winter ice bridge from the Canadian shore.

Early in the twentieth century, logging and burning converted much of the region north of Lake Superior from a woodland caribou (*Rangifer tarandus*) habitat of forests and muskegs into brush and early tree successions favorable to moose. A small number of moose apparently colonized Isle Royale (probably by swimming) before 1910, and the caribou that had been residents or migrants disappeared in the 1920s (Mech 1966). At that time, no effective big-game predator lived on the island, and human harvest was limited.

The island was made a wildlife sanctuary by the state of Michigan in 1925. Moose rapidly multiplied after their initial colonization, eliminated much of the available browse, and altered the browse species composition. The herd may have been as large as 3,000 moose by the late 1920s and was the subject of national interest. The herd was drastically reduced, however, in the early 1930s, primarily because of malnutrition and disease (Murie 1934; Hickie 1936). By 1935 as few as 200 moose were left on the island.

A fire burned more than a fifth of the island in 1936. This burn, in turn, supported a regrowth of browse, which encouraged another increase in the moose population in the 1940s (Aldous and Krefting 1946). The population again declined in the winters of 1948 and 1949, primarily because of malnutrition attributed to a lack of nutritious forage (Krefting 1951). Krefting speculated that without human intervention, further cycles of browse depletion and moose die-off were inevitable. He also believed that moose were destroying the wilderness character of the island and argued forcibly for moose control either through wolf introduction or hunting by Native Americans from the Grand Portage Reservation.

Wolves probably occupied Isle Royale at various times since the last glacial recession. When Europeans began to use and occupy the island in the mid-1800s, however, they found no wolves (Adams 1909). A breeding pack of wolves reached Isle Royale in the winter of 1948–1949, evidently crossing on the ice from Ontario (Linn 1956). Feeding conditions on moose and beaver were good, and wolf numbers increased. The first winter surveys by the National Park Service (NPS) in 1956 estimated that about 25 wolves and 300 moose lived on the island (Cole 1956a).

Establishment of the Park

The idea for a national park on Isle Royale first arose in the early 1920s. In 1921 the Michigan State Conservation Commission passed a unanimous resolution endorsing the concept of a national park on Isle Royale. Soon thereafter a campaign promoting the idea began, led by Albert Stroll and the *Detroit News*. At the time, two companies, the Island Copper Company and the Minnesota Forest Products Company, owned about 70% of the main island. About 7% of the lands were still in the public domain, and the remainder was divided among the state of Michigan and small landowners. Isle Royale was already popular as a summer retreat; numerous small cottages were scattered along the coastline, and four lodges were in operation.

During the summer of 1924, an official party, led by NPS Director Stephen Mather, inspected the island with the aim of recommending that President Coolidge proclaim it a national monument. After this visit, Mather concluded that Isle Royale fully measured up to national-park standards and "would make the finest water and trail park I can think of" (Hakala 1955, p. 41). In early 1931, bills were introduced in the House and Senate to establish Isle Royale National Park. These bills quickly passed and were signed by President Hoover. Efforts to acquire the funds to buy the private lands on the island, however, took many years—a process complicated by the Depression. The park was formally established in April 1940 when Secretary of the Interior Ickes accepted the deeds to all lands on Isle Royale on behalf of the United States. Formal dedication of the park was delayed another 6 years (August 1946) because of World War II.

Genesis of the Wolf-Moose Study

The winter surveys of James Cole on Isle Royale in the mid-1950s alerted NPS to the importance of the wildlife situation in the park and its potential for becoming a dangerous public-relations problem. Cole felt that "if the experiment of allowing moose and wolves to coinhabit a relatively small, isolated island succeeds without intervention, NPS will justly deserve and receive commendations from many conservation groups. If it should fail with the ultimate loss of both moose and wolves, NPS would certainly be criticized" (Cole 1956a, p. 4). He concluded that it was unlikely that the reduced moose population would survive through the 1957 winter if predation increased and that, once the moose were gone, wolves would also die or leave the island (Cole 1956b). He also concluded that some type of wolf control was therefore necessary. As a result, NPS began to seriously consider a thorough study of the problem (Wirth 1956).

NPS, however, had no clear idea of how to conduct a study. Because of the myriad wildlife problems confronting the agency at the time, the situation may have never been addressed. (Cole was unavailable for future work on the island.) However, as is the case in many important studies, luck played an important role. Durward Allen, having learned that wolves lived on Isle Royale, saw at the park the opportunity to preserve a few wolves in a natural state and to develop a program of cooperative studies for graduate students (Allen 1957a). Allen was a professor of wildlife management at Purdue University. In July 1957 he took this idea to Gordon Fredine, principal naturalist in the NPS Washington office. At the time, Allen was unaware of Cole's work or the agency's concerns about the island (D. L. Allen, pers. comm., 1991).

Fredine, not surprisingly, was receptive to the idea, indicating that NPS had been planning on giving "more biological attention to the wildlife problems at Isle Royale and that . . . mutually agreeable arrangements could be made for [Allen's] proposed studies" (Fredine 1957a). In subsequent correspondence with Allen, Fredine indicated that he anticipated NPS would be receiving additional funds for wildlife study and that "it is reasonable for you to depend on enough concrete support from the NPS in one form or another which would help" initiate the project (Fredine 1957b). Allen thereupon wrote a proposal for a study on wolves and other wildlife on Isle Royale that was designed to last 3 years and to cost $19,000 (Allen 1957b).[2]

Overview of the Research Program

The initial organization of the research program took place in the summer of 1958 at a meeting to review program objectives. The meeting included representatives from NPS, U.S. Fish and Wildlife Service, Minnesota Department of Conservation, Michigan Department of Conservation, Ontario Department of Lands and Forests, and Purdue University.

The principal component of the research program was, from its inception, the winter studies. These studies normally took place in a 7-week period from mid-January to mid-March—an interval when the ice was firm enough to permit plane landings in protected bays or inland lakes. The studies were based out of the Windigo Ranger Station on the western end of the island. The normal complement of personnel in the winter included two university researchers, one park representative, and a pilot. Field research has also been conducted in the summers since 1961. Program administration can be divided into two phases. The first phase was directed by Durward L. Allen of Purdue from 1958 through 1975. The second phase has been directed by Rolf O. Peterson, one of Allen's Ph.D. students. Since 1975 Peterson's work has been based out of Michigan Technological University in Houghton. Table 5.1 provides a list of research personnel involved in the study.

The principal winter activities were counting wolves and moose and observing wolf behavior and predation patterns from the air. Flights were made every day that weather permitted during the study period. Very accurate records of the total number of wolves and moose on the island have been maintained since 1958 and 1969, respectively (Fig. 5.1). Researchers examined all possible moose kills in the field and collected jaw mandibles and other

Table 5.1. Research personnel involved in the Isle Royale study, 1958–1990.

Date	Season	Personnel
1958	Summer	L. David Mech
1959	Winter	L. D. Mech; Robert M. Linn
1959	Summer	L. D. Mech
1960	Winter	L. D. Mech
1960	Summer	L. D. Mech; Philip C. Shelton
1961	Winter	L. D. Mech; Durward L. Allen
1961	Summer	L. D. Mech; P. C. Shelton
1962	Winter	P. C. Shelton; D. L. Allen; R. M. Linn

continued

Table 5.1. *Continued*

Date	Season	Personnel
1962	Summer	P. C. Shelton
1963	Winter	P. C. Shelton; D. L. Allen
1963	Summer	P. C. Shelton; Peter A. Jordan
1964	Winter	P. A. Jordan; D. L. Allen
1964	Summer	P. A. Jordan
1965	Winter	P. A. Jordan; D. L. Allen
1965	Summer	P. A. Jordan
1966	Winter	P. A. Jordan; D. L. Allen
1966	Summer	P. A. Jordan; Wendel J. Johnson
1967	Winter	W. J. Johnson; Michael L. Wolfe; D. L. Allen
1967	Summer	W. J. Johnson; M. L. Wolfe
1968	Winter	W. J. Johnson; M. L. Wolfe; D. L. Allen
1968	Summer	W. J. Johnson
1969	Winter	M. L. Wolfe; D. L. Allen
1969	Summer	M. L. Wolfe; P. C. Shelton
1970	Winter	M. L. Wolfe; John C. Keeler
1970	Summer	Rolf O. Peterson
1971	Winter	R. O. Peterson; James M. Dietz; D. L. Allen
1971	Summer	R. O. Peterson
1972	Winter	R. O. Peterson; D. L. Allen
1972	Summer	R. O. Peterson
1973	Winter	R. O. Peterson; D. L. Allen
1973	Summer	R. O. Peterson; P. C. Shelton
1974	Winter	R. O. Peterson; D. L. Allen
1974	Summer	R. O. Peterson; P. C. Shelton
1975	Winter	R. O. Peterson; D. L. Allen
1975	Summer	R. O. Peterson
1976	Winter	R. O. Peterson
1976	Summer	Joseph M. Scheidler; Phillip W. Stephens
1977	Winter	R. O. Peterson; J. M. Scheidler
1977	Summer	R. O. Peterson; J. M. Scheidler; P. W. Stephens

Table 5.1. *Continued*

Date	Season	Personnel
1978	Winter	R. O. Peterson; J. M. Scheidler
1978	Summer	R. O. Peterson; J. M. Scheidler; P. W. Stephens
1979	Winter	R. O. Peterson; J. M. Scheidler
1979	Summer	R. O. Peterson; P. W. Stephens
1980	Winter	R. O. Peterson
1980	Summer	R. O. Peterson
1981	Winter	R. O. Peterson
1981	Summer	R. O. Peterson; Richard E. Page
1982	Winter	R. O. Peterson; R. E. Page
1982	Summer	R. O. Peterson; Kenneth L. Risenhoover; T. G. Laske
1983	Winter	R. O. Peterson
1983	Summer	R. O. Peterson; K. L. Risenhoover
1984	Winter	R. O. Peterson; R. E. Page
1984	Summer	R. O. Peterson; K. L. Risenhoover
1985	Winter	R. O. Peterson; K. L. Risenhoover; R. E. Page; Timothy N. Ackerman
1985	Summer	R. O. Peterson
1986	Winter	R. O. Peterson
1986	Summer	R. O. Peterson; K. L. Risenhoover; R. E. Page; T. N. Ackerman; Thomas A. Brandner
1987	Winter	R. O. Peterson; K. L. Risenhoover; R. E. Page; T. N. Ackerman
1987	Summer	R. O. Peterson
1988	Winter	R. O. Peterson
1988	Summer	R. O. Peterson
1989	Winter	R. O. Peterson
1989	Summer	R. O. Peterson; Joanne M. Thurber
1990	Winter	R. O. Peterson; J. M. Thurber
1990	Summer	R. O. Peterson; J. M. Thurber
1991	Winter	R. O. Peterson
1991	Summer	R. O. Peterson

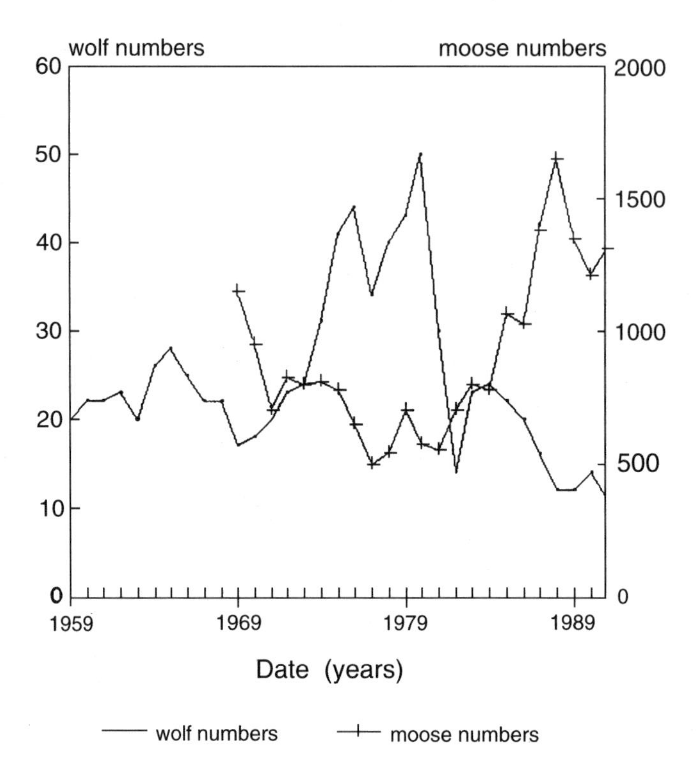

Figure 5.1 Wolf and moose numbers at Isle Royale National Park, 1959–1991.

bones to help determine the age and condition of the animals. Over the years, many hundred dead moose have been examined. Kill sites inaccessible during the winter were mapped, and material was collected where possible during the summer. Wolf scats were collected from the trails in early spring. In the early years of the program, several moose were killed each year and autopsied for condition, disease, and parasite load.

The research program has concentrated on three main topics over the years: wolf predation patterns, wolf behavior and ecology, and moose population dynamics. Studies of wolf predation patterns emphasized the age and sex of the moose killed, other prey species, hunting success, and the effect of snow depth on predation success and activities. Research on wolf behavior and ecology focused on social hierarchy in the packs, courtship and breeding, territoriality, communication, denning and rendezvous sites, reproduction, relationships with nonprey species, and movements. Field observations of moose population dynamics included population size, age and sex ratios and productivity, habitat relationships, food habits, and mortality factors.

One unique aspect of the research program was its long-standing policy that none of the living subjects were to be handled and all forms of human disturbance were to be kept to a minimum.[3] This policy was fully endorsed by the park as it created a situation in which activities of the animals could be observed in as natural a situation as possible. This factor alone has made the study unique in the annals of contemporary wildlife research. However, because no animals were captured and none were marked or radio-collared, some information that would have been valuable in interpreting events on the island was lost.[4]

One such event was a dramatic crash of the wolf population, which dropped from 50 to 14 between 1980 and 1982. Researchers later speculated that the decline was due to infection by canine parvo-virus disease (CPV). However, no wolves were actually tested for disease at that time. Red foxes were captured and blood-sampled beginning in 1985, but all have tested negative for CPV. Wolf numbers increased somewhat in 1983 and 1984 but have since undergone an almost steady decline. This factor, combined with the limited reproduction in the years following, led researchers and NPS resource managers to conclude that more intense investigation was needed. This idea brought about the first serious discussions of the capture and handling of wolves on the island (R. O. Peterson, pers. comm., 1991).

In 1988 a research proposal was developed that included the need to radio-collar wolves to evaluate hypotheses explaining the decline. Because of the controversial nature of the proposal,[5] a peer review was undertaken by the NPS Midwest Regional Office Science Division. This review strongly endorsed both the methods and the necessity for the study and therefore facilitated its initiation (R. J. Krumenaker, pers. comm., 1991). Between 1988 and 1990, eight wolves were live-captured, blood-sampled for disease and genetics studies, and released wearing radio-collars.

The new research was designed to evaluate three possible causes for the decline in wolf numbers:

1. food shortage caused by the relative lack of older moose that provide most of the food for wolves;

2. mortality from diseases such as CPV and Lyme disease;

3. loss of genetic variability, caused by a small founding population and genetic drift (Peterson 1990).

The study found no indication that the wolves suffered from a shortage of food. In the winter of 1989, moose mortality was high, providing abundant food for wolves. In fact, many moose carcasses were untouched by wolves (Peterson 1990; Peterson and Thurber 1990). The effect of disease was less clear. Data from the blood samples provided the first documentation that both CPV and Lyme disease are present on the island. However, only a small number of the tested animals were positive for either disease. Most experts reviewing the data remain unconvinced that CPV is currently an important factor in the wolf decline because of incomplete exposure and lack of mortality. The effect of Lyme disease on wolves remains unknown, although reproductive failure in affected animals is a possibility (Peterson 1990; Peterson and Thurber 1990).

Isle Royale wolves appear to have lost about half their genetic variability, and the examined wolves all have the same type of mitochondrial DNA. DNA fingerprinting indicated that all of the sampled Isle Royale wolves were as closely related as siblings or as parent and offspring. This is strong evidence that the entire population consists of descendants of a single female and that no new arrivals supplemented the original gene pool (Peterson 1990). The genetic losses of the magnitude seen in the wolves could explain their current low reproductive success; however, confirmation is still lacking, and the actual mechanism of reproductive failure is unknown.

The close monitoring made possible by the radio-collared wolves has provided new details of wolf predation patterns and social relationships and has enhanced and supplemented the knowledge gained from the aerial snow-tracking used in the study since its inception. In the winter of 1991, the radio-collars on 6 wolves allowed researchers to monitor 9 of the remaining 12 wolves; 2 of the 3 remaining wolves that were either uncollared or did not associate with a collared animal were observed only once during the 50-day study period, despite intense efforts to locate them through snow-tracking (Peterson 1991).

In the past, little has been known about wolf activities during the non-winter period because they have been so difficult to locate and observe. Radio-collars have substantially increased the researcher's ability to learn more about the wolves during this period.

Presently, the NPS does not plan to reverse the wolf decline through human intervention, even though the possibility of extinction of the wolves on Isle Royale is relatively high. Thus far, researchers and NPS feel that the scientific value of this natural experiment in population viability merits nonintervention. In the event of wolf extinction, restocking the island is an option NPS will consider; however, there is much debate on this issue, with little agreement (Peterson and Krumenaker 1989; Fink 1991; Peterson 1991).

The impact on the moose population caused by a demise of the wolves can only be speculated. Continued changes in the vegetal composition of the island with further declines in balsam fir can be expected. The moose population is slowly aging after high reproductive success in the early 1980s. In the 1990s the age structure will become very old, and the population may be vulnerable to high losses from malnutrition. Current trends in bone marrow fat of dead moose suggest continued deterioration in the forage base. Chronic poor nutrition, even if it reduces moose productivity, will probably not stabilize the population. The moose population will probably continue to grow until a new source of mortality appears. Mortality from starvation alone has not been an important regulator of Isle Royale moose since wolves arrived on the island, although it did have an effect before their appearance. Wolf predation at current levels has not stopped the growth of the moose population. In past years, however, when wolf numbers were higher, a classic predator-prey cycle did exist, indicating that wolves had a regulating influence on moose numbers (Peterson et al. 1984). Winter ticks can play an important role in moose mortality, as seen in the decline in moose numbers in 1989 and 1990. Tick populations, however, may be driven by weather patterns that are largely unpredictable.

Because park managers do not want to intervene to reverse the decline of wolves, what should they do if moose multiply so much they start dying of starvation (which appears to have happened in the winter of 1992 [R. O. Peterson, pers. comm., 1991])? NPS policies do not provide clear guidance, stating that "Natural processes will be relied on to control populations of native species to the greatest extent possible" (National Park Service 1988, chap. 4, p. 6). The many ambiguities facing NPS managers illustrate the importance of the insights from the research study and speak to the need for its continuation with as little human interference as possible.

What Has Been Learned from the Research Program?

The long-term research at Isle Royale has provided major scientific benefits, including a better understanding of the predator-prey process and of the movements and behavior of moose and wolves under a variety of environmental conditions. It has also provided important insights into the variability of ecosystems and has substantiated the value of long-term studies of natural ecosystems.

Treating Conclusions Based on Short-term Observations

One lesson the Isle Royale research teaches is that events in nature should always be interpreted cautiously and that any conclusion is subject to change. Throughout the 33 years of the project, there have been many instances in which interpretations about a phenomenon made during one period were later altered. For example, research conducted during the first decade of the program indicated that intensive wolf predation maintained the moose population below the level at which it would be restricted by food availability (Mech 1966). Subsequent research concluded that available forage played a major role in limiting the moose population. Likewise, in the first decade of the study, snow depth was not considered to be an important influence on moose activities or their vulnerability to predation. The researchers, however, came to realize that this was a period when snow depths were average or light. The next decade of above-average snow depths contradicted this finding (Allen 1979).

In the early 1980s, the changes observed in wolf and moose numbers led researchers to conclude that the pattern of fluctuations in these numbers that were noted in the late 1950s might again be repeated. They suspected that some broad 25-year cycle might be governing population changes on the island (Peterson et al. 1982). The continued decline of the wolf population in the late 1980s has since dramatically altered this interpretation.

Documenting the Influence of Wolf Predation

Predation by a variety of species has been a major force in shaping survival patterns, behavior, and physical characteristics of prey species throughout their evolutionary history. Isle Royale has offered a unique laboratory to follow changes in wolf winter predation rates and productivity through a series of increases and declines in moose population density and changes in age structure. The study has allowed researchers to evaluate the effect of these factors independent of habitat and human disturbance and under a variety of climatic conditions.

Over the course of the study, hundreds of moose kills have been documented and skulls and jaw bones collected, permitting the construction of a very accurate, long-term life-table. Likewise, the ability to accurately track the numbers of moose and wolves over long periods, as illustrated by the data displayed in Figure 5.1, is unprecedented in wildlife studies.

Learning about the Process of Extinction

The current wolf decline on Isle Royale offers scientists a unique opportunity to study the process of extinction. The number of extinctions of animal life on Earth is growing exponentially as the world's remaining natural habitat is being lost. The ecological status of Isle Royale is similar to that of many parks and natural areas, which are becoming islands of habitat surrounded by a hostile landscape. A common problem facing small populations of animals isolated in such areas may be genetic deterioration (Harmon 1990). Genetics and disease appear to be primary reasons for the current wolf decline on Isle Royale. If this population disappears, it will be one of the first places in which researchers could examine the causes and process of real-life extinction.

Managing Visitor Use of the Park

The research program has also had a profound influence on the way NPS manages Isle Royale and its resources. This is particularly true of the way visitor use is managed.

Park use in winter has always been very difficult, if not impossible, without special provisions because of access problems. With the advent of the research program and the importance of the winter study period, park managers actively discouraged individuals from using the island during the winter. Fortunately, few requests were made, thus few problems occurred. As the research program gained increasing public exposure, however, more and more requests were received to visit the island to view wolves. Winter is unique. During this period, the long-term lack of human disturbance on the island has allowed researchers to view wolf and moose behaviors rarely witnessed elsewhere (R. O. Peterson, pers. comm., 1991). To avoid disrupting the winter program, NPS legally closed the park between 1 November to 15 April. This is the only action of its kind in NPS.

The research program has also influenced where visitors can go in the summer. The park is zoned for visitor use; some 50% of the island is closed to camping because of potential conflicts with wolf denning and rendezvous areas. In the early 1970s, efforts to expand the trail system to new areas were halted because of potential conflicts with important wolf-use areas and because trailless areas are important to wolves; preserving them has remained a priority of park planning.

Molding the Park Perspective

Few national parks are as closely linked with a single natural resource (in this case an animal) as is Isle Royale with the wolf. Depending on one's viewpoint, this may or may not be a healthy situation, but at the present it is probably not alterable. The importance of the wolf on Isle Royale can be partially attributed to the publications of the research program and to other national publicity the program has received. The first question visitors ask park staff when they embark on the island usually concerns the status of the wolves. The wolf has essentially molded the visitors' perception of Isle Royale and is a major attraction.

Visitor services on the island are limited. Common activities are backpacking, canoeing, and viewing nature.[6] As a result, the average visitor appears to be very enlightened in knowledge and appreciation of park resources. This enlightenment extends to a concern for the wolves. For example, most visitors have shown a high degree of understanding over the need to close some areas to protect the wolf.

The close link between park management and the wolf leads one to consider what would happen if the wolf disappeared. Would support for the park (and the research program) suffer? Does the strong concern for the wolves come at the detriment of ignoring other park resources and their preservation needs? These are difficult questions to answer. A certain irony, however, pervades the current fascination with the wolf. Before the wolves arrived, moose symbolized the island and were the primary concern of management. Isle Royale National Park was often described as America's greatest moose refuge, and visitors often came just to photograph them (East 1941). To deal with this issue, the park has made a major effort to educate the public about the complexities of the current wolf decline and the pros and cons of intervention.

Why Has the Research Program Continued?

The study of the wolf and moose—the former a species that did not even live in the park when it was established—now dominates the park's natural-resource program, subsuming 40% of the $100,000 budget and monopolizing much of the time of the natural resource specialist, whose job description does not even include wildlife management. How—in an era when long-term research is often defined as a period of 5 years—has the program survived for so long? What lessons can it offer to those seeking to establish other long-term studies? Some potential reasons for the longevity of the program are offered below.

Continuity of Research and Support Personnel

Probably the single biggest reason the Isle Royale program has survived and been so successful has been the continuity maintained by personnel (Table 5.1). In 33 years, the program has had only two principal investigators, Durward L. Allen (1958–1975) and Rolf O. Peterson (1976–present). Peterson also served as a graduate student on the project between 1970 and 1975. These two individuals have provided the needed stability in direction and research protocols.

Another important factor is the continuity of pilots: only two have been employed during the winter throughout the project's history (Donald Murray, 1958–1979; Don Glaser, 1979–present). The winter studies focus on aerial observations of animals. Both pilots have been experts at such surveillance and have transferred this knowledge to several generations of graduate students who flew with them. Both Allen and Peterson have acknowledged the importance of the pilots to the program. Without the continuity they provided by using similar methods of tracking animals and making observations, the success of the program would be in doubt.

The program would also not have survived without the long-term support of park personnel, especially chief ranger Stuart Croll. Croll has been in his position for the past 16 years, unusual in an agency in which transfers every 3–4 years are common. For example, during the course of the study, the park has had 10 different superintendents. Croll has been a strong backer of the program and has provided the "glue" necessary to ensure NPS support of the program.

The Wolf Is a "Glamour Species"

The wolf is a species whose survival and activities concern a great many people. This factor has undoubtedly played an important role in public support and funding for the program. Even before there were wolves on Isle Royale, there was a strong interest in having them there. For example, in August 1952 two pairs of wolves from the Detroit Zoo were placed on the island. This transplant had the blessing of NPS director Conrad Wirth, Fish and Wildlife Service assistant director Clarence Cottam, and numerous conservation and scientific societies (Wirth 1952). The introduced wolves, however, were too tame and did not adapt to a natural environment. They became problems, and all but one were quickly killed or recaptured.

Today there is tremendous public interest in wolves and support for their reintroduction into parks from which they were extirpated. Surveys by McNaught (1985) in Yellowstone and Minn (1977) in Rocky Mountain National Park showed that 82% and 74% of the visitors questioned supported the reintroduction of wolves into the respective parks.

Strong Public and Administrative Support

Support from NPS administrators at park, regional, and national levels has been a key factor in assuring the continuity of the research program. The program was fortunate that Robert Linn was a naturalist at Isle Royale when the program started and was involved with the first winter studies. Linn subsequently became chief scientist of NPS and later directed the NPS Cooperative Park Studies Unit at Michigan Technological University. In both of these later positions, he was able to provide strong financial and moral support to the program.

Most NPS directors were very supportive of the program. It also had the support and personal interest of two Secretaries of the Interior, Stewart Udall and Rogers Morton, and Assistant Secretary Nathaniel Reed. Support at the park level was equally important. This was at no time more apparent than when Assistant Secretary of the Interior Ray Arnett terminated funding for the program in 1983—one of the most critical periods in the study, coming after the wolf population had undergone a 2-year, 72% decline. Funding was restored for that year primarily as a result of superintendent Don Brown's personal petition to the secretary and outside pressure from conservation organizations (National Parks and Conservation Association 1983).

Reflections on the Program

One legacy of the Isle Royale research program is that almost all of the projects, including the long-term monitoring of species such as moose, wolf, and beaver, have been contracted out to universities rather than being conducted by park personnel. Compared to many parks, this is a relatively unusual situation. Park managers have undertaken this course in recognition of their funding and staffing limitations, although they have recently sought to expand their in-house monitoring capabilities. Over the years, however, the involvement of park personnel in the moose-wolf research program, particularly during the important winter studies, has been limited.

Is this a problem? In this case, probably not. The relationship between NPS managers and scientists both in and out of the agency has never been smooth. When the relationship has been good, it has probably been more the result of personalities than professional affiliations. This seems to be true of Isle Royale. The relationship between the researchers and park personnel over the years has generally been good. Most of those involved in the study agree that Rolf Peterson has been chiefly responsible for this good will. He has been relatively easy to work with, is a very good research scientist, puts an incredible amount of personal effort into this and other park science and resource-management studies, and has been relatively successful in acquiring outside funding (50% of the total) for the work.

Outside scientists can also provide a scrutiny of park management practices that realistically cannot be provided by in-house scientists. However, if personnel or attitudes change and conflicts arise—for example, over differences of opinion on how the park should be managed—the interests of the park might be better served if its staff were more involved in the research program. Such conflicts are not unprecedented in the parks, particularly when the research has involved sensitive species.[7]

Conclusions

The Isle Royale study has produced invaluable scientific information and advanced the understanding of wolf winter predation rates and productivity. Likewise, it has provided unique information on the establishment, history, and disappearance of individual wolf packs. Whether it stands as a model for other studies is, however, difficult to determine. Certain factors have led to its success—some fortuitous, some planned—and some of these are worthy of emulation. In the final analysis, the true worth of the study may be in what it has yet to teach us. Durward Allen described the study as "one of those continuous searches into the unknown that has no foreseeable end" (Allen 1979, p. 371). Isle Royale National Park is a unique repository of primitive conditions. The study of wolves and moose on Isle Royale has, in turn, contributed a unique repository of information. Both, like precious antiques, will become more valuable in the future.

Acknowledgments

I am grateful for the helpful review comments and corrections offered by Rolf Peterson, Robert Krumenaker, and James Peek; for access to the records of Isle Royale National Park; and for the insights offered during personal visits with Durward Allen and Rolf Peterson.

Notes

1. Isle Royale National Park actually encompasses 212,215 ha, of which about 70% are the bordering waters of Lake Superior.

2. Although this budget was conservative, at the time the entire NPS research budget was only $25,000 (Wright 1992).

3. With very few exceptions, aerial observations in winter appeared to have no effect on wolf or moose behavior.

4. Rolf Peterson did seek permission to trap and radio-collar some wolves in 1974. This request was turned down by NPS personnel because they felt that the status of the wolf on Isle Royale as "free and untouched . . . will prove to be a most valuable asset to future wolf researchers" (Beal 1974).

5. Seventeen moose were radio-collared during the 1984–1985 field seasons as part of a study of foraging behavior. This proposal also stirred considerable controversy among park managers before it was approved (C. Axtell, pers. comm., 1991).

6. Total visitation to Isle Royale National Park is about 17,000 per year, one of the lowest totals for any national park outside of Alaska.

7. One well-known instance involved the conflict between the Craighead brothers (research scientists) and Yellowstone National Park over the best way to manage grizzly bears.

Literature Cited

Adams, C. C. 1909. An ecological survey of Isle Royale, Lake Superior. A report from the University of Michigan Museum, published by the State Biological Survey as part of the Report of the Board of the Geological Survey from 1908. 241 p.

Aldous, S. E., and L. W. Krefting. 1946. The present status of moose on Isle Royale. Transactions North American Wildlife Conference 11:296–308.

Allen, D. L. 1957a. Letter to Isle Royale superintendent John G. Lewis, 2 August 1957.

———. 1957b. A proposal for studies of the timber wolf and other wildlife in Isle Royale National Park. Unpubl. report, Purdue University, Lafayette, Indiana. 13 p.

———. 1976. The worth of wilderness: with interpretations from a study of wolves and moose on Isle Royale. P. 169–181 in National Park Service Symposium Series 1.

———. 1979. The wolves of Minong. Houghton Mifflin Co., Boston. 289 p.

Beal, M. D. 1974. Memorandum on proposal to trap and radio-collar on Isle Royale. Memorandum to superintendent, Isle Royale, from acting regional director, 24 December 1974.

Cole, J. E. 1956a. Winter wildlife study of Isle Royale National Park. Unpubl. report, National Park Service, Washington, D.C. 8 p.

———. 1956b. Memorandum to the superintendent, Isle Royale National Park, 28 March 1956.

East, B. 1941. Park to the north—Isle Royale. American Forests 47(6):245–250.

Fink, W. O. 1991. Wolves and us. Memorandum: Isle Royale National Park, 6 June 1991.

Fredine, C. G. 1957a. Letter to Durward L. Allen, 19 June 1957.

———. 1957b. Letter to Durward L. Allen, 28 October 1957.

Hakala, D. R. 1955. Isle Royale: primeval prince; a history. Unpubl. report, Isle Royale National Park.

Harmon, D. 1990. Last call for Isle Royale's wolves? Lake Superior Magazine (June/July):63–69.

Hickie, P. F. 1936. Isle Royale moose studies. Transactions of the North American Wildlife Conference 1:396–398.

Krefting, L. W. 1951. What is the future of the Isle Royale moose herd. Transactions of the North American Wildlife Conference 16:461–472.

Linn, R. M. 1956. Analysis of the current moose-wolf relationship on Isle Royale National Park. Memorandum to superintendent, 27 July 1956.

———. 1957. The spruce-fir, maple-birch transition on Isle Royale National Park, Lake Superior. Unpubl. Ph.D. dissert., Duke University, Durham, North Carolina.

McNaught, D. 1985. Park visitor attitudes toward wolf recovery in Yellowstone National Park. Unpubl. M.S. thesis, University of Montana, Missoula. 133 p.

Mech, L. D. 1966. The wolves of Isle Royale. National Park Service Fauna Series 7. 210 p.

Minn, B. P. 1977. Attitudes towards wolf reintroduction in Rocky Mountain National Park. Unpubl. M.S. thesis, Colorado State University, Fort Collins.

Murie, A. 1934. The moose of Isle Royale. University of Michigan Museum Zoological Miscellaneous Publication 25. 45 p.

National Parks and Conservation Association. 1983. Isle Royale wolf program now geared to monitoring. National Parks 57(7–8):37.

National Park Service. 1988. Management policies. U.S. Government Printing Office, Washington, D.C.

Peterson, R. O. 1977. Wolf ecology and prey relationships on Isle Royale. National Park Service Monograph Series 11. 210 p.

———. 1990. Ecological studies of wolves on Isle Royale. Annual Report 1989–90. Michigan Technological University, Houghton. 31 p.

———. 1991. Ecological studies of wolves on Isle Royale. Annual Report 1990–91. Michigan Technological University, Houghton. 33 p.

Peterson, R. O., and R. J. Krumenaker. 1989. Wolves approach extinction on Isle Royale: a biological and policy conundrum. George Wright Forum 6(4):10–16.

Peterson, R. O., R. E. Page, and K. M. Dodge. 1984. Wolves, moose, and the allometry of population cycles. Science 224:1350–1352.

Peterson, R. O., R. E. Page, and P. W. Stephens. 1982. Ecological studies of wolves on Isle Royale. Annual Report 1981–82. Michigan Technological University, Houghton. 28 p.

Peterson, R. O., and J. M. Thurber. 1990. Study of causes of wolf decline, Isle Royale National Park. Unpubl. progress report. Michigan Technological University, Houghton.

Wirth, C. L. 1952. Letter to Congressman John B. Bennett, 21 August 1952.

———. 1956. Moose and wolves relationship problems, Isle Royale National Park. Memorandum to the regional director, Region Five, 28 May 1956.

Wright, R. G. 1992. Wildlife management and research in the national parks. University of Illinois Press, Urbana. 224 p.

6

Saguaro Cactus Dynamics

Joseph R. McAuliffe

Of all the remarkable plants of the desert Southwest, the giant cactus, or saguaro (*Carnegiea gigantea*), stands out in the minds of many Americans as an icon representing the novelty and grandeur of the desert realm. In 1933 Saguaro National Monument[1] was established on the east side of Tucson, Arizona, to protect what was then one of the most awe-inspiring stands of saguaros to be found anywhere in the Sonoran Desert. However, within less than a decade, a precipitous decline of this particular saguaro population began. Today the giant, many-branched saguaros have all but disappeared from the original "Cactus Forest" of the 1930s (Fig. 6.1).

Since about 1940 and continuing to the present day, the considerable alarm generated by the specter of a cactus forest in Saguaro National Monument has spawned a variety of research investigations in an attempt to determine the cause of the decline. Unfortunately, the most sensational but seriously flawed research conclusions have had some of the strongest and long-lasting impacts on park management and information (or rather, misinformation) disseminated to the public by the press.

Several hypotheses, not all of which are necessarily mutually exclusive, have been proposed to explain the saguaro population decline in part of Saguaro National Monument. These hypotheses included the effects of pathogenic disease, mortality caused by catastrophic freezes, and lack of reproduction due to a variety of causes, including livestock impacts, climate variation, and demise of local pollinators. In addition, the saguaro decline has most recently been hypothesized to be at least partly due to air pollution and other recent atmospheric alterations by humans. Because of this bewildering array of potentially competing hypotheses, an understanding of the issues by some scientists, National Park Service (NPS) personnel, and the public has become increasingly muddled rather than clarified.

The purposes of this chapter are threefold. First, I trace the development of various research efforts that have addressed the saguaro population decline at Saguaro National Monument. Second, I evaluate both the rationale for these investigations and the conclusions derived from them. Third, I examine the ways in which the conclusions, whether warranted or not, have affected the management of Saguaro National Monument and the perceptions held by the public. To help the reader in understanding and evaluating these investigations, I will first briefly describe the environmental setting of Saguaro National Monument and the history of land use before the monument was created. As the reader will see, a common shortcoming of some investigations has been the researchers' failure to discern the relevance of this setting and history to the saguaro population's growth and decline.

Environmental and Historical Contexts

Saguaro National Monument includes two geographically separate districts: Rincon Mountain District (RMD) and Tucson Mountain District (TMD; Fig. 6.2). RMD on the east side of Tucson contains the area originally designated as the national monument in 1933 and the now vanished Cactus Forest (Fig. 6.1). This district, one of the highest areas in the Sonoran Desert, flanks the desert's eastern edge (Fig. 6.2). Elevation within the district ranges from approximately 820 to 2,640 m. The increasing severity of prolonged freezes with higher elevations limits the saguaro to elevations lower than about 1,280 m (Shreve 1911; Niering et al. 1963; Steenbergh and Lowe 1976, 1977, 1983). RMD can hardly be considered a "flagship" representation of typical Sonoran Desert vegetation, despite the presence of saguaros in lower elevations. Less than 20% of the district contains desertscrub vegetation and populations of saguaros. The once grand Cactus Forest that flourished dur-

Figure 6.1 Matched landscape photos from 1935, 1960, and 1985, Rincon Mountain District, Saguaro National Monument. H. Shantz took the 1935 photo; R. M. Turner took the 1960 and 1985 photos.

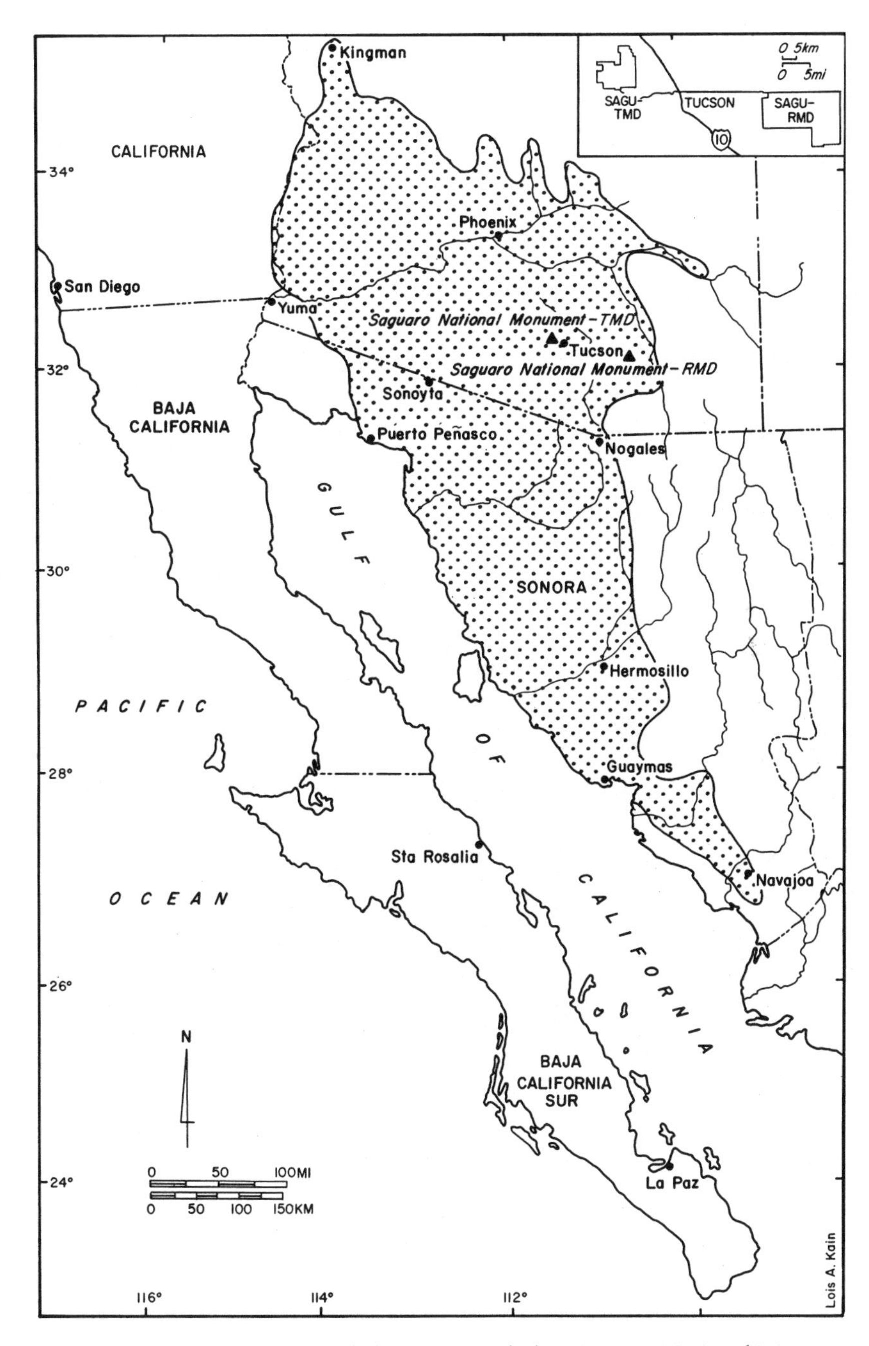

Figure 6.2 Sonoran Desert (stippled pattern), including Saguaro National Monument. Inset: detail of separately managed districts in the monument.

ing the early part of this century was restricted to an area less than 25 km^2 in the northwestern corner of the district. The marginal location of RMD in terms of the geographical and elevational distribution of the saguaro has important ecological consequences that have been ignored in some research efforts at the monument.

In contrast to RMD, TMD has a much lower and restricted elevational range (approximately 670–1,428 m). The vegetation in nearly 90% of TMD is classified as Sonoran desertscrub (Shelton 1985). Stands of saguaros in TMD have not catastrophically declined, as have those in RMD.

The two districts of Saguaro National Monument also differ considerably in the types and magnitudes of human impacts they sustained from the late 1800s through the middle of this century. Beginning in the early 1870s, the area at the foot of the Rincon Mountains was heavily settled because of the presence of perennial water in nearby Tanque Verde and Rincon Creeks. Settlement led to extensive use of lands for ranching and woodcutting in areas now contained in RMD. In 1880 an estimated 1,000 cattle from three ranches along Tanque Verde Creek grazed in and around the area that would become known as the Cactus Forest, located only 3 to 5 km to the south of the creek (Clemensen 1987). An 1893 map of Pima County showed 36 ranches and 2 schools distributed along Tanque Verde and Rincon Creeks near the present monument boundaries.

The use of the Cactus Forest by domestic stock did not end with the establishment of the national monument in 1933. Ownership of the area containing the most impressive stands of saguaros was not transferred to the federal government until 1958. Consequently, livestock continued to use the Cactus Forest until that date.

In contrast, TMD on the west side of the Tucson Mountains escaped these kinds of cultural impacts. This side of the Tucson Mountains is without perennial surface water; consequently, farming and ranching were naturally excluded. Not until the late 1920s did homesteaders settle in this area to any great extent. In response to local concerns for protecting the area, the U.S. government withdrew 11,731 ha in the late 1920s from further homesteading and mining claims. This withdrawn land was leased to Pima County in 1929 for development as a recreation area and preserve (Clemensen 1987).

Principal Lines of Research and Monitoring

Much of the research conducted on saguaros in the monument has focused on the saguaro population decline at RMD and can be divided into three chronological "chapters." The first, initiated in 1939 and continuing through the early 1960s, was primarily concerned with causes of mortality of old cacti and dealt with problems of plant pathology and disease control. The second chapter is more ecological and comprehensive in scope and involves basic research on many different factors that contribute to both mortality and reproduction. These investigations began primarily in the late 1950s and continue through the present. The third chapter has developed only within the last decade. It is related to federal air-quality legislation and has concentrated on hypothesized links between man-induced atmospheric alterations and saguaro vigor, mortality, and reproduction. Each chapter is discussed and evaluated below.

Saguaro Disease Investigations, 1939–1962

In 1939 and 1940, James G. Brown, a plant pathologist at the University of Arizona, reported a necrotic disease that was afflicting saguaros of southern Arizona. He reported that up to 20% of saguaros located in survey plots at the monument were infected, and diseased saguaros could be found over a wide area in southern Arizona. Reports of the malady in the scientific and popular press (e.g., Gill 1942) fueled a hysteria that saguaro populations of the Sonoran Desert, especially the Cactus Forest of Saguaro National Monument, could potentially succumb to the ravages of this disease.

A bacterium, *Erwinea carnegieana,* the supposed pathogen, was identified from tissue samples taken from decaying portions of afflicted saguaros (Lightle et al. 1942). From the earliest reports in the 1940s, the blackening necrosis of saguaros was referred to as a disease. The disease hypothesis was uncritically accepted, and personnel of the U.S. Department of Agriculture, Bureau of Plant Industry, immediately began a "disease suppression experiment" in Saguaro National Monument. The goal of the experiment was to determine if removal of infected plants could check the spread of the disease to healthy saguaros (Gill 1942; Gill and Lightle 1942).

An entire section of land (2.56 km^2) within the Cactus Forest was selected for the experiment. One-half of the area (129.5 ha) was a "treatment" area in which all diseased plants were removed, cut into sections, placed in large excavated trenches, drenched with dichlorobenzene dissolved in kerosene, and buried. The operation required 3 months in the winter of 1941–1942, during which 313 affected saguaros were removed, together with 12 healthy individuals that were accidentally injured in the operation. Steenbergh and Lowe (1983) documented the history of this project in detail in their inclusion of copies of original proposals and reports of the disease-control studies. Gill (1942) published a photo essay of the experiment. In 1941 the federal government provided $10,000 to implement the experiment (Bardsley 1957). Although small by today's research-grant standards, this amount was considered major funding 50 years ago, especially being allotted on the dawn of U.S. involvement in World War II.

Convinced of the seriousness of the disease situation by pathologists' reports, NPS management at Saguaro National Monument cooperated with these disease investigations. The entire Cactus Forest is contained within about 8–10 sections of land. The use of about 10% of the Cactus Forest for the experiment demonstrates the degree of concern that NPS personnel must have had to allow this scale of manipulation within the boundaries of the monument. In 1944, however, NPS biologist W. B. McDougal of Santa Fe visited the site and questioned the rationale behind the experiment. McDougal had discovered that the necrotic condition in saguaros was described in publications dating back to 1889. He also noted the "disease" seemed to affect only the older saguaros. He believed that, although a thinning of the saguaro stand might occur, the stand would eventually regenerate, and NPS should allow such natural phenomena to proceed without interference (Clemensen 1987). Despite this opinion, NPS performed no other direct scientific assessments of the project during the 1940s. The agency was not in a position to do so because its research branch was completely dismantled in the late 1930s and early 1940s (Chase 1986).

Monitoring of this experiment through 1950 showed no marked differences between the treated and untreated areas in the number of additional plants that succumbed to the "disease." The investigators reported this finding but also claimed that the experiments may have held the disease in check (Gill and Lightle 1946; Gill 1951). The repeated assertion of the benefit of these experiments, without corroborating data, demonstrated the uncritical acceptance of the disease hypothesis by plant pathologists.

Although disease-control measures on the scale of the 1940s' operation were never again attempted at Saguaro National Monument, research on various aspects of the bacterial necrosis maintained a high profile for at least 20 years (Boyle 1949; Bardsley 1957; Clemensen 1987). Alcorn and May (1962) extrapolated the losses from the disease and predicted that the saguaro forest would be extinct before the turn of the century.

Although research on the saguaro decline was for many years focused on plant pathology, scientists eventually rejected the hypothesis that a disease was the direct cause of the catastrophic mortality of saguaros at the monument. C. H. Lowe, an ecologist at the University of Arizona, and NPS scientist W. F. Steenbergh, working largely in the 1960s and 1970s, concluded that occasional, catastrophic freezes on the northern and eastern margins of the Sonoran Desert killed many saguaros. Furthermore, they noted that the delayed bacterial decomposition following freeze damage was mistakenly diagnosed as a "disease."

The mistaken conclusions regarding the "bacterial necrosis disease" had a variety of serious and lasting effects on park management and public perception. The manner in which newspapers and magazines reported disease investigations from the 1940s to the 1960s only tightened the hold of the disease perspective on management concerns.

Convinced that the saguaro population of the monument was doomed by the necrosis disease and frustrated by the federal government's inability to secure ownership of the lands containing the Cactus Forest, the regional NPS director recommended in 1945 that Saguaro be abandoned as a national monument (Clemensen 1987). Fortunately, the national NPS director did not accept that proposal and instructed that solutions to land-use problems stemming from the ownership issue be pursued more vigorously. The possibility that RMD would be abandoned continued, however, through the mid-1960s and had a detrimental effect on planning for the monument. Because of its uncertain future, Saguaro National Monument typically ranked low among western parks in receiving government funding and resources (W. Paleck, pers. comm., 1992).

Measures other than abandonment were proposed, and sometimes seriously considered, as management solutions to the problem of the saguaro "disease." In 1946 the principal pathologist of the Bureau of Plant Industry (the agency responsible for the saguaro disease investigations) offered to spray the saguaros of the monument with DDT by airplane at 6-week intervals to kill the moth suspected of spreading the bacterial disease (Clemensen 1987). The pathologist acknowledged that the spraying would eliminate all insects. He emphasized, however, that if the majestic saguaros were indeed valued above all else, then the spraying of DDT would allow them to survive for many more years. Because of the long-lasting ecological harm that this pesticide would have caused, it is fortunate the offer was declined.

Another misguided management strategy that was planned during this era was a "reforestation" of the declining Cactus Forest by transplanting juvenile saguaros throughout the area. Although this approach was strongly questioned by some NPS personnel (see Appendix 1 in Steenbergh and Lowe 1983), a lath house was constructed at RMD and saguaros were propagated for this purpose in the late 1950s. The "reforestation" program was never implemented, however, because a fire destroyed the lath house and its contents in the 1960s (R. M. Turner, pers. comm., 1992).

Concern about the demise of the saguaro population in RMD led NPS managers in the 1940s to consider acquiring land administered by Pima County in the Tucson Mountains to add to the monument. This interest was dropped, however. The reason for the eventual inclusion of TMD was not the saguaro decline in RMD, but rather, public outcry over the possibility of public lands in the Tucson Mountains being opened up to large-scale mining operations in the early 1960s (Clemensen 1987).

One of the most harmful and long-lasting impacts of the cactus disease investigations has been the fostering of incorrect notions in the public regarding the biology and ecology of the saguaro. The sensational reporting of the day did much to fan the flames of a mistaken hysteria that has yet to be extinguished (e.g., Gill 1942; Bardsley 1957; Ingle 1964; Robinson 1966). Occasional, inaccurate references to a "bacterial disease" responsible for deaths of saguaros continue to surface in present-day press reports (Yozwiak 1992) and have perpetuated the misunderstanding.

The failings of the cactus disease investigations that spanned more than two decades have been the source of much confusion among the public and have led to ill-conceived management solutions by NPS and other federal agencies. Steenbergh (1970) stated the situation rather bluntly: "The public has enjoyed the company of the National Park Service on their journey down this emotional primrose path as, at times, we proposed and supported desperate non-solutions to the non-problems derived from the non-valid interpretations of ecological non-sense."

Ecological Investigations, 1959–Present

Understanding the saguaro population decline in RMD required a far different research orientation than the narrow focus of plant pathology that had dominated earlier research. Two major lines of ecological investigations that largely followed the cactus disease investigations are discussed in this section: (1) catastrophic freezes as a cause of saguaro mortality, and (2) various factors that limit saguaro reproduction.

Effects of Catastrophic Freezes: An Alternate Explanation for "Bacterial Necrosis" In 1911 Forrest Shreve, the unrivaled and perceptive pioneer of Sonoran Desert plant ecology (Bowers 1988), published a short paper entitled "The influence of low temperatures on the distribution of the giant cactus." In that paper, Shreve suggested that the greatest number of consecutive hours of subfreezing temperatures was the most important factor controlling the northward and upper-elevation distributional limits of the saguaro. Shreve (1911) tested this hypothesis using simple freezers cooled with ice and salt. He demonstrated that saguaros subjected to -3°C to -11°C for periods longer than 29 hours were all killed, yet saguaros subjected to similar freezing regimes of only 6–15 hours did not die. Furthermore, he observed that within 2 weeks after the experiment had begun, the cacti subjected to the longer freezing period began to show a blackening of their tissues. After 4 weeks, one saguaro subjected to only -3°C for a 46-hour period had become black and soft throughout its lower half. None of the damaged individuals later exhibited any signs of recovery.

The conclusions of Shreve (1911), so important to understanding the ecology of the saguaro at the northern and upper elevational margins of its distribution (such as the case of RMD), went unnoticed for nearly half a century, until Steenbergh and Lowe (1969, 1977) explained the catastrophic loss of cacti in Saguaro National Monument and other marginal sites as a consequence of freezes. The studies of Steenbergh and Lowe (1969, 1977) and

Niering et al. (1963) indicated that prolonged, catastrophic freezes not only kill small saguaros, such as those studied by Shreve (1911), but also selectively kill a greater proportion of taller, older saguaros than those of intermediate height. The documentation of the consequences of the 11–13 January 1962 freeze on saguaro mortality on the slopes of the Santa Catalina Mountains immediately northwest of RMD convincingly demonstrated the impact of freezing on large saguaros and the increase in mortality with elevation (Niering et al. 1963). After the 1962 catastrophic freeze, some saguaros began to fail immediately, but other moribund individuals did not completely decay until several years later. This slow process of decomposition of freeze-killed cacti was identical to the condition that had previously been labeled the "bacterial necrosis disease." Steenbergh and Lowe (1976) recorded similar mortality in RMD after another severe freeze that occurred in 1971.

Steenbergh and Lowe (1976, 1977, 1983) argued that the "outbreak" of bacterial necrosis observed in 1939 through the early 1940s was the consequence of the 1937 freeze, during which temperatures dropped lower than they had for more than 20 years. They concluded that the "bacterial necrosis disease" of saguaros was an extremely mistaken interpretation and claimed that this condition was neither a disease nor a cause of death. Rather, it was the bacterial decay of tissues killed by extreme freeze damage.

Considerable mortality of saguaros following catastrophic freezes is a recurrent phenomenon in the Tucson area that has since been recorded by other biologists. For example, after an exceptionally hard freeze in December 1978, approximately 200 adult saguaros were lost from the grounds of the Arizona-Sonora Desert Museum, located near TMD. Most of the stricken saguaros remained standing but started to decay in the few years immediately after the freeze. In the first 5 years after that freeze, adult saguaros taller than 3 m succumbed to necrotic decay at the rate of 25–35 plants per year on the museum grounds. Losses rapidly declined thereafter (M. Dimmitt, pers. comm., 1992).

The contributions of studies by Lowe and his students and colleagues regarding the importance of freezing temperatures to the survival of seedling and juvenile saguaros, mortality of adults, and saguaro distributions provided NPS with the first comprehensive framework of saguaro ecology (Niering et al. 1963; Steenbergh and Lowe 1969, 1976, 1977, 1983; Soule and Lowe 1970). In addition to the research on the effects of freezes, these studies contributed a multitude of other ecological data on the age, growth, and demography of populations in both districts of Saguaro National Monument and in Organ Pipe Cactus National Monument (Steenbergh and Lowe 1977, 1983).

Although Steenbergh and Lowe (1977) declared a "requiem" for the "ecologically unsupportable myth" of the bacterial necrosis disease, this myth unfortunately has not died in the minds of many. To this day, many people still believe the saguaro is afflicted by some mysterious, lethal malady and consequently is declining steadily toward extinction. This confusion is not limited to the residents of Arizona. Reporters from national magazines and newspapers in phone interviews still express these misunderstandings. NPS must do much more to clear up the misconceptions that resulted from the disease-oriented research originally conducted at Saguaro National Monument. The works of Steenbergh and Lowe (1976, 1977, 1983) have provided a foundation on which a more realistic interpretation of saguaro ecology may be built.

Saguaro Reproduction The rarity of young saguaros in the Cactus Forest of RMD in the 1930s (Fig. 6.1) indicated a general failure of establishment of new saguaros for many decades before the monument was created. This lack of reproduction continued until the early 1960s (Turner 1992). Without reproduction, the eventual decline of the aged population was inevitable.

Investigations of factors limiting saguaro reproduction have focused on three general areas: (*a*) microhabitat requirements of seedlings; (*b*) impacts of climate variability, especially precipitation; and (*c*) pollination and seed dispersal. Research in each of these areas is discussed below.

MICROHABITAT REQUIREMENTS OF SEEDLINGS Protected environments beneath canopies of "nurse plants," especially small trees such as the foothills paloverde (*Cercidium microphyllum*), are necessary for establishment of saguaros in nonrocky habitats (McAuliffe 1984, 1988; Hutto et al. 1986). The ameliorated microclimate beneath nurse plants protects young saguaros on exposed soil surfaces from extremes of cold in winter (Steenbergh and Lowe

1976; Nobel 1980) and lethal high temperatures in summer (Turner et al. 1966; Despain 1974). The cutting of trees within the Cactus Forest for fuel and the use of the area by livestock until 1958 either directly eliminated nurse trees or trampled areas beneath remaining tree canopies.

Considerable woodcutting occurred within the Cactus Forest in the decades surrounding the turn of the century. Woodcutters supplied fuel for domestic use and for firing local lime kilns. Surveyors working in the area that would become RMD recorded numerous roads used by woodcutters in the 1890s. By 1905 wood became scarce near Tucson; reportedly every tree with a trunk diameter more than 17.8 cm had been cut within a radius of 16 km of town. Woodcutting to fuel lime kilns at the base of the Rincon Mountains had become so extensive that the cutting was ended by court order in 1920 in response to complaints by local ranchers (Clemensen 1987). Some woodcutting within the Cactus Forest continued after the monument was established in 1933 because of the problem of administering lands within the monument that were not actually owned by the federal government (Egermayer 1940a). Recognizing the implications of a lack of nurse-tree cover for saguaro reproduction, Turner et al. (1966, p. 102) concluded that in the Cactus Forest of RMD, "Saguaro stands that are chronically lacking in re-establishment will not recover until the nurse plant cover is restored."

The grazing of livestock within RMD from the late 1870s through 1958 also contributed to declining saguaro reproduction. For 25 years after the monument's creation, the Cactus Forest was owned by the University of Arizona and private individuals but was nominally included within the boundaries of the monument. NPS's lack of official jurisdiction over this area allowed livestock to use the Cactus Forest until the lands were transferred to NPS in 1958. Within the limited area of the Cactus Forest, 100 or more cattle were sometimes stocked in the early years of the monument (Clemensen 1987).

Concerns by NPS managers over possible deleterious impacts of cattle at the monument date to the first decade of the monument's existence (Egermayer 1940b). The 1947 master plan for Saguaro National Monument (cited in Clemensen 1987) pointed out an overgrazing problem but offered no solutions, primarily because of the lack of official jurisdiction over the area and the difficulties associated with the ownership issue.

Despite the repeated concerns expressed by superintendents of Saguaro

National Monument and other NPS personnel on the potential effects of livestock on the saguaro population, especially reproduction and survival of young cacti, this topic received very little early study at the monument other than range condition reports tallied by U.S. Forest Service (USFS) personnel (Clemensen 1987). The lack of any serious scientific investigation of this problem by NPS is a manifestation of the complete dismantling of the NPS Research and Education Division in the late 1930s and early 1940s. For nearly two decades after that time, NPS had no capacity to respond to many pressing research needs (Chase 1986, p. 239).

Limited studies of livestock impacts at Saguaro National Monument have been conducted by researchers from outside NPS. Niering et al. (1963) and Niering and Whittaker (1965) painted a hypothetical scenario of direct impacts of cattle (trampling and consumption of young saguaros) and a complex chain of cause-and-effect relationships: grazing-induced vegetation changes caused population expansions of rodents such as woodrats (*Neotoma albigula*). The researchers partly attributed the lack of saguaro reproduction to the consumption of young cacti by these rodents. Although logically deduced, this hypothetical scenario has never been adequately tested.

An opportunity to assess impacts of livestock grazing on saguaro reproduction in RMD presented itself as a fence-line contrast between an area that had been free of livestock since 1958 and an adjacent tract on which grazing had been permitted until 1978 (Abouhaidar 1989, 1992). This study revealed that the saguaro population on the more recently grazed side of the fence contained a significantly smaller proportion of young cacti than the side from which grazing had been excluded 20 years earlier.

Livestock grazing—especially at the stocking rates that existed on ranges at the base of Tanque Verde Ridge before the turn of the century and during the first 25 years of Saguaro National Monument—undoubtedly had impacts on the reproduction of saguaros. However, as one moves further from the time when grazing was allowed in the monument, it becomes increasingly difficult to decipher the amount of damage inflicted by livestock relative to that caused by other "natural" or human impacts. Because NPS conducted no scientific study on this issue, a large window of opportunity closed, preventing NPS personnel from learning much about the hypothesized direct and indirect impacts of livestock on saguaros after grazing was prohibited in the monument. This information would have proved extremely useful for the establishment of stocking policies and guidelines on other public lands in the Sonoran Desert.

IMPACTS OF CLIMATE VARIABILITY Hastings and Turner (1965) hypothesized that variation in the severity and duration of drought conditions in the Southwest has partially contributed to vegetation changes, including fluctuations in saguaro populations. Between 1959 and 1962, these investigators set up a regional network of 10 plots to study the effects of long-term climatic variation on saguaro populations in Arizona and adjacent Sonora, Mexico. Most of these plots are 2.5 ha in size; RMD and TMD each contain one plot. Turner and colleagues have monitored these plots at intervals of about 10 years to record data on plant mortality, growth, and reproduction. These studies have yielded several results that are important to the interpretation of the saguaro population in Saguaro National Monument.

One plot is located on the 1.6-km-wide floor of McDougal Crater in Sonora, Mexico, about 250 km west and 30 km south of Tucson. This site is in an area of the Sonoran Desert that is much hotter and drier than Saguaro National Monument. Nevertheless, the McDougal Crater site has provided data that are important to deciphering the relative impacts of climatic variability and other factors because the crater floor is inaccessible to livestock and woodcutters. Furthermore, extreme frosts are unlikely at this site. Using information on saguaro growth rates collected from the population over a period of three decades, Turner (1990) estimated the age and year of establishment of each saguaro. These data indicated a surge of reproduction in the late 1800s that waned after 1900. Turner (1990) suggested that the late-1800s' spate of saguaro establishment may have been due to summer rainy periods that were unusually moist before the turn of the century compared with those in subsequent decades. In the 30-year period from 1868 to 1898, July to August precipitation at Tucson exceeded 180 mm 11 times (180-mm precipitation is 1.75 times the long-term average). This arbitrary threshold was exceeded only seven times during the *90-year* period from 1898 to 1988. Turner furthermore suggested that these plentiful summer rains may have also contributed to a surge in saguaro reproduction in southern Arizona during the late 1800s.

Many of the saguaros that made up the original Cactus Forest in RMD were estimated to have become established in the latter half of the nineteenth century (Hastings 1961). In RMD, however, the many cultural impacts harmful to saguaro reproduction were synchronous with the period of lessened summer precipitation that Turner identified and that began at the turn of the century. Consequently, it is difficult, if not impossible, to decipher the relative contributions of each factor and the effects of several extreme freezes around the turn of the century (Steenbergh and Lowe 1976, 1977) to variation in past saguaro reproduction in RMD. As originally pointed out by Hastings and Turner (1965), it is probable that both cultural impacts and climatic variations have affected establishment of saguaros in RMD.

Another extremely important result of Turner's permanent plot studies is the finding of a marked increase in saguaro establishment since the early 1960s within the area of the original Cactus Forest. Most of these young saguaros are less that 0.5 m tall. The abundance of young cacti represents a "phenomenal population surge" and has the potential for becoming a new "Cactus Forest" in the next century (Turner 1992). This concrete documentation of abundant establishment in RMD counters statements such as that by Gladney et al. (1992) that the saguaro decline is continuing in RMD.

POLLINATION AND SEED DISPERSAL The possibility that inadequate pollination of saguaros and poor seed set has contributed to a decline in saguaro reproduction in RMD received considerable attention around 1960 and again during the last 5 years.

Alcorn et al. (1959, 1961) and McGregor et al. (1962) demonstrated that fruit set and production of viable seeds in saguaros required cross-pollination. The morphology of flowers and the large quantities of nectar they produced indicated that pollination was accomplished by nectivorous animals. Alcorn et al. (1961) and McGregor et al. (1962) conducted experiments in RMD to determine the relative importance of various pollinators to fruit set and seed production in saguaros. The researchers selectively covered receptive stigmas and styles of flowers with soda straws to prevent pollination during various periods. Although these studies showed that daytime pollinators were responsible for the majority of fruit set, the occurrence of substantial pollination during the night indicated that nectar-feeding bats may also be important pollinators.

Subsequent experiments conducted within a screened enclosure in RMD containing blooming saguaros examined the potential pollinating effectiveness of honeybees (*Apis mellifera*), white-winged doves (*Zenadia asiatica*), and lesser long-nosed bats (*Leptonycteris curasoae,* referred to as *L. nivalis* in Alcorn et al. [1961] and McGregor et al. [1962]). The three sets of animals were introduced into the enclosure during separate periods. The researchers tagged flowers that were open for potential pollination during each period and monitored them for fruit set and seed viability. Results of these trials showed that each of the potential pollinators visited flowers and were equally effective as cross-pollinators of saguaros. From these data, the researchers concluded that lack of pollination and seed production was not a limiting factor in saguaro reproduction in RMD.

NPS officials, however, became concerned about adequate saguaro pollination after the demise of a maternity colony of lesser long-nosed bats that originally occupied part of Colossal Cave, approximately 18 km south-southeast of the Cactus Forest. This colony may have at one time contained as many as 5,000 bats (Cockrum and Petryszyn 1991). Colossal Cave had been opened to commercial tours in the early 1920s, and the colony persisted until an exhaust-fan system was installed in the cave in 1961.

The elimination of this particular maternity colony, coupled with a reported general decrease in populations of this bat species in southern Arizona, led the U.S. Fish and Wildlife Service (USFWS) to list the bat as an endangered species in 1988 (Cockrum and Petryszyn 1991). Another factor that contributed to the endangered species listing was the bat's role as a pollinator of agaves, saguaros, and organ pipe cacti. The official justification for this listing stated that these plants may be affected by loss of the bat and "there is concern for the future of the entire southwestern desert ecosystem" (Shull 1988, p. 38458).

Earlier findings on pollination ecology (Alcorn et al. 1961; McGregor et al. 1962) were either ignored or overlooked. One USFWS spokesperson even stated that the bats were the major pollinators of saguaros (Hernandez and Erickson 1988). NPS personnel also expressed concern about a possible link between bat populations and the long-term viability of saguaros in Saguaro National Monument. The superintendent did recognize other pollinators of saguaros. However, because of the information generated at the time of the endangered species listing, he commented that if the bats were eliminated, the saguaro population would dramatically decline (Hernandez and

Erickson 1988). On the day the bat was listed as an endangered species, the newspaper article by Hernandez and Erickson (1988, p. A1) described the situation of the "tiny rare bat largely responsible for Tucson's unique saguaro landscape."

The rationale for listing the lesser long-nosed bat as an endangered species has recently been seriously challenged on the basis of more complete and accurate documentation of the status of bat populations (Cockrum and Petryszyn 1991). Regardless of the bat's actual population status, the following lines of evidence indicate that the long-term viability of saguaro populations throughout Arizona is not dependent on the presence of the bat, as expressed in the rationale for the endangered species listing.

1. The only data on the relative importance of bats to saguaro pollination collected before the maternity colony in Colossal Cave disappeared are those of Alcorn et al. (1961) and McGregor et al. (1962). These data collected in RMD, well within the foraging flight range of Colossal Cave, indicate that bats were minor pollinators of saguaros compared with daytime pollinators, which included a variety of bee and bird species.

2. Additional studies in the Tucson area, including Saguaro National Monument, have shown that honeybees from ubiquitous feral colonies are effective pollinators of saguaros (Buchmann 1992; Schmidt et al. 1992). These studies add to the previous findings of Alcorn et al. (1961) and McGregor et al. (1962) regarding the important ecological role played by honeybees.

3. Large parts of the distributional range of the saguaro lie well outside the area where nectar-feeding bats have been recorded (Cockrum and Petryszyn 1991). This observation demonstrates that the absence of bats as pollinators does not prevent saguaros from reproducing.

4. The considerable, recent surge of saguaro establishment documented in the Cactus Forest (Turner 1992) occurred after the demise of the maternity colony in Colossal Cave. Turner's data show an almost total lack of saguaro reproduction throughout the 1950s, when the breeding colony was active in the cave. These data indicate that conditions affecting survival of seedlings and young plants were the principal factors responsible for the lack of reproduction in the past and that establishment of new saguaros has not diminished with the elimination of the bat colony.

Some research on dispersal of saguaro seeds has also been conducted in RMD. Olin et al. (1989) demonstrated that white-winged doves are important seed dispersers. Transport of fruits and seeds to nests to feed the young results in the inadvertent deposition of seeds on the soil surfaces beneath trees. This pattern of seed dispersal was cited as one factor contributing to the common association of saguaros with nurse-tree canopies. No evidence suggests that reproduction of saguaros at Saguaro National Monument is limited by seed dispersal.

Air Quality Division Investigations, 1985–present

The most recent chapter of research and monitoring involving the saguaro at Saguaro National Monument has been prompted by federal clean-air regulations. Consequently, an understanding of those regulations is necessary to comprehend the rationale behind the research.

The Clean Air Act and Development of Monitoring Programs In the original Clean Air Act of the early 1970s, Congress concentrated on pollution in urban and industrial areas. Amendments to this act in 1977 included the protection of "refuges" of air or "airsheds" above large areas of federal wild lands. These Class I Airsheds include every national park exceeding 2,428 ha and all wilderness areas larger than 2,023.5 ha that existed when the amendments were passed in 1977 (Air Quality Division 1984). A Class I Airshed need not have "pristine" air; the air quality, however, cannot significantly deteriorate. For example, the air over Saguaro National Monument is designated a Class I Airshed, even though the city of Tucson, which borders the monument, is designated a "non-attainment" area for carbon monoxide levels. Tucson is also close to exceeding regulatory standards for ozone and particulates. Federal law requires that air quality and Air-Quality-Related Values (AQRVs) within Class I Airsheds be protected from further deterioration. Direct air-quality measures include levels of gaseous pollutants (e.g., ozone, carbon monoxide, sulfur dioxide, and nitrogen dioxide). AQRVs include the biota of an area and visibility within the designated airshed. For example, one AQRV that has received considerable attention throughout the United States is plant health as affected by ozone pollution (Karnosky et al. 1992).

The U.S. Environmental Protection Agency (EPA) is responsible for regulatory enforcement of the Clean Air Act and its amendments. EPA works with state and local agencies to develop programs to protect local air quality. Within the Department of the Interior, this responsibility lies with the assistant secretary for Fish, Wildlife, and Parks. In NPS, jurisdiction is passed from the assistant secretary through the NPS director to the Air Quality Division (AQD).

Local federal land managers (a superintendent in the case of a national park or monument) are required to take steps to prevent further deterioration of airshed under their jurisdiction. To accomplish this, a superintendent must know the baseline pollution level and its impact on AQRVs within the park. To understand air-quality issues, the superintendent must work with AQD, which offers additional help when any new source of pollution is proposed in the area (e.g., a coal-fired generating plant). Applications for permits to construct and operate such sources must be obtained from the appropriate state. These applications are sent to the NPS regional office or directly to AQD for review and not to the individual parks. Park personnel do not conduct the reviews because they generally lack the expertise.

Research Concerns at Saguaro National Monument The saguaro was included in AQD investigations at Saguaro National Monument in 1985. At that time the superintendent and resource management specialist discussed the long-occurring decline of saguaros in RMD with staff of the AQD Research Branch. Monument personnel were concerned that ozone might be affecting the saguaros, possibly contributing to their decline (R. L. Hall and K. Stolte, pers. comm., 1992). Suspected ozone-related damage to ponderosa pines was recorded in the higher elevations of RMD in 1985 (Duriscoe 1987; Hall 1991). However, AQD staff initially believed that cacti, in general, are extremely resistant to elevated ozone levels and that the relatively low levels of ozone pollution in the monument were probably not harmful to saguaros.

AQD research on saguaros began with reconnaissance and comparison of saguaro populations in the San Pedro Valley near San Manuel, Arizona, 40 km northeast of RMD. Although deleterious impacts were evident on saguaros near the San Manuel smelter, "green, fat saguaros" grew in areas farther from the smelter. The condition of these saguaros apparently contrasted greatly with the browning stems that were typical in RMD saguaros (K. Stolte and R. L. Hall, pers. comm., 1992). This reconnaissance led to

Figure 6.3 Epidermal browning on saguaros. A. Spine-covered epidermis of a saguaro showing no epidermal browning. B. Slight manifestation of epidermal browning characterized by discolored epidermis encircling areoles from which spines have been lost. Arrow indicates a single, black-colored areole. C. More extensive epidermal

the belief that the condition of the saguaros in RMD was similar to an "accelerated aging process" that ozone could produce and to the hypothesis that severe epidermal browning of saguaros and the decline of the Cactus Forest could be localized phenomena related to air pollution near Tucson. These concerns and hypotheses led to several areas of investigation involving saguaros. Funding for these projects started in 1987. The studies included (1) investigations of the epidermal browning phenomenon and its cause(s), (2) research on the potential effects of acid precipitation, (3) biomonitoring garden studies, and (4) a study on the impacts of nitrogen enrichment (McAuliffe 1993). The epidermal browning studies are discussed and evaluated below.

Epidermal browning is a brown discoloration of the normally green epidermis that originates at the areoles near the base of the saguaro stem (Air Quality Division 1991; Fig. 6.3). This condition is typically associated with the loss of spines and may completely cover lower parts of the stem on some saguaros. AQD staff formulated four principal lines of investigation to study the phenomenon:

1. establishment and collection of baseline data from a network of monitoring plots in both management districts of the monument;

2. studies of the morphological, histological, and histochemical characteristics of browned versus normal cacti and investigation of possible causes;

3. ecophysiological studies of affected versus normal cacti;

4. studies of elemental characteristics of soils and saguaro tissues.

Each of these areas of investigation involving epidermal browning is detailed below.

LONG-TERM SAGUARO TREND PLOTS From 1988 to 1989, 45 4-ha plots (200 m by 200 m each) were randomly located and established in Saguaro

---

browning characterized by discoloration of entire ridges. Area between arrows is a discolored zone on one ridge. Troughs of epidermal pleats are not discolored. D. Cactus showing recent epidermal browning together with dark, lignified bark (material with cracked appearance). Areas covered with bark probably indicate healing response to previous episodes of epidermal browning.

National Monument: 25 in RMD and 20 in TMD (Duriscoe and Graban 1992). Baseline data collected from those plots indicated two principal patterns of epidermal browning in saguaros: browning is a height- or age-related phenomenon, and south-facing surfaces of saguaro stems exhibited more browning than north-facing surfaces.

The researchers recorded degrees of epidermal browning ranging from complete absence to high levels in both RMD and TMD. Although two plots in RMD exhibited the highest "browning indexes," four plots in which saguaros exhibited little or no browning were also located in RMD. Three of these four plots were next to major washes; the significance of this association will be discussed later.

The extent of browning in individual saguaros was also correlated with the presence of associated perennial plants. Saguaros growing in areas with foothill paloverde (*Cercidium microphyllum*) had significantly greater amounts of browning than did saguaros growing in areas with mesquite. Mesquites in the monument are primarily limited to areas along major washes (runoff-collection areas), and paloverdes are the dominant tree on the considerably drier slopes and runoff-producing areas of mountainsides and bajadas. These data further indicate an association between epidermal browning and either water status of soils or plant growth rates.

Duriscoe and Graban (1992) also documented a strong negative correlation between spine retention and the severity of browning, but they were careful to note that this relationship provided no evidence that browning caused spine loss. Because spine retention and browning both vary with height, they suggested that decreased spine retention may be a normal part of the aging process.

Duriscoe and Graban concluded that epidermal browning is widespread in both districts of the monument, but some large cacti show no evidence of browning. Epidermal browning principally occurs on the southern exposures of saguaro stems. Because the condition varies greatly, the researchers suggested that a "plant-specific factor," such as age or growth rates, must be involved. This project did not provide the information necessary to determine cause-and-effect relationships regarding epidermal browning. However, the data obtained on the association between this condition and other factors are useful for evaluating the plausibility of hypothesized causes of browning suggested by other investigators (see following section).

MORPHOLOGICAL, HISTOLOGICAL, AND HISTOCHEMICAL STUDIES In studies conducted in RMD, the findings of Evans et al. (1992a) corroborated those of Duriscoe and Graban (1992): epidermal browning and spine loss in saguaros was greater on southern exposures than on northern exposures of stems. Additional work (Evans et al. 1992b) described morphological and histological changes associated with epidermal browning.

Although the initial work of Evans concentrated on histological descriptions, by 1990 he had begun work to establish a causal link between epidermal browning and solar ultraviolet radiation. Evans found epidermal browning in saguaros throughout the Sonoran Desert in both Arizona and Sonora. Because of its widespread occurrence, he concluded that the phenomenon could not be attributed to a local pollution problem, as AQD staff had originally suggested. Research that Evans conducted in the Southern Hemisphere (Argentina and Chile) indicated similar types of epidermal discoloration of other cactus species on the northern or equatorial exposures of stems. These observations led Evans to hypothesize that the browning may be a relatively recent phenomenon due to increases in ultraviolet radiation caused by stratospheric ozone depletion (L. S. Evans, pers. comm., 1992).

To test the hypothesis that epidermal browning of saguaros has recently increased, Evans examined archived photographs dating to 1903 that were taken near RMD. Based on examinations of 241 saguaros recorded in these photos, Evans et al. (1992a,c) concluded that severe epidermal browning is a recent phenomenon that has increased at a rate of 4.5% per decade from 1903 to 1987.

This conclusion, however, is questionable. According to the statistics presented in Evans et al. (1992a, c), the correlation between epidermal browning and time is extremely weak. Only 5–7% of the purported increase in epidermal browning can be explained by the passage of time between 1903 and 1987 (as estimated by r^2, the coefficient of determination). In other words, 93–95% of the variation in epidermal browning is related to factors other than the passage of time. The conclusions of Evans et al. (1992a,c) are further called into question because their claim of a linear increase in epidermal browning over time appears to be simply a consequence of their inappropriate use of linear regression to analyze data consisting of only binomial scores (0,1) that represent the presence or absence of browning characteristics. (See McAuliffe [1993] for further discussion.)

After concluding that epidermal browning has gradually increased over the last several decades, Evans attempted to establish a link with increases in ultraviolet radiation caused by stratospheric ozone depletion. In a letter justifying this research to the superintendent of Saguaro National Monument, Evans (1992b) stated "the UV-B hypothesis should not be rejected, and based on other data, is the only viable hypothesis relating to epidermal browning of saguaros and other columnar cacti in the Americas." In July 1991 Evans began a pilot study in RMD that involved the irradiation of two large saguaros with an electric ultraviolet light fixture (Evans 1991). This study was funded by the monument but was halted after 1 year because of design weaknesses and the researchers' failure to consider alternate hypotheses.

The research approach used in 1991 to establish a causal link between ultraviolet radiation and epidermal browning displayed serious flaws, not only in experimental design, methodology, and analysis, but also in the philosophical approach to scientific inquiry. Fifty years earlier, a similar failure, accompanied by a lack of relevant tests, led to the 20-year-plus path of misled investigations by plant pathologists studying the saguaro necrosis "disease."

Other hypotheses could explain the pattern of epidermal browning in saguaros, such as the impacts of infrared wavelengths. For example, cumulative heating of tissues above lethal temperatures must also be considered as a mechanism that may contribute to epidermal browning.

The localized pattern of discoloration associated with the initial stages of epidermal browning strongly suggests a connection to excessive (but localized) heat loading. The lightest forms of epidermal browning appear as discolored zones several millimeters wide that completely encircle areoles on the southern sides of stems, the lowest, oldest parts of which have lost their spines (Fig. 6.3B). These denuded, black areoles are 5–7 mm across. The lower albedo associated with this localized black color undoubtedly results in the absorbance of more solar radiation and in localized, higher, and possibly lethal tissue temperatures. A rise above the lethal limit could cause local tissue damage or death that leads to the browning condition. The healing of such injury would likely result in the formation of lignified "bark" at the site of the original damage (Fig. 6.3D; Steelink et al. 1967; Gibson and Nobel 1986).

More work must be done to test these hypotheses regarding high-temperature effects on saguaros. Nevertheless, because of the pattern initially exhibited by epidermal browning (Fig. 6.3B), the selective manifestation of browning in poorly hydrated saguaros, and the lethal and sublethal temperature impacts on saguaros (Smith et al. 1984), one can make a strong argument for a connection between epidermal browning in saguaros and heat-loading effects.

ECOPHYSIOLOGICAL STUDIES Lajtha and Kolberg (1992) studied carbon dioxide uptake as a function of the condition of saguaros in RMD. They conducted the study during the summer rainy season (August) of 1990 and 1991.

In 1990, a year of record-breaking summer precipitation, tissues on the southern sides of healthy saguaros had higher maximum rates of CO_2 uptake and remained active longer than healthy tissues on the northern sides of the same cacti. The reverse was true for saguaros with epidermal browning on more than 20% of their southern sides: CO_2 uptake was lower on the southern sides than on the northern sides.

In the dry summer of 1991 (one of the driest on record), CO_2 uptake was significantly lower than in 1990. Many plants, both with and without epidermal browning, experienced daytime respiration losses that exceeded nighttime CO_2 accumulation. In this drier year, CO_2 uptake was only marginally related to the degree of browning and was primarily controlled by the drought conditions. Further studies of water relationships, gas exchange, and tissue temperatures are needed to test the hypothesized cause-and-effect relationships among epidermal browning, water status, and potential heat loading.

ELEMENTAL ANALYSES OF SOILS AND SAGUARO TISSUE The reportedly greater local incidence of epidermal browning in RMD led Ken Stolte of AQD to believe that local pollution—perhaps emissions from copper smelters or even the turn-of-the-century lime kilns at the base of Tanque Verde Ridge—may have contaminated local soil, leading to reduced vigor of saguaros. This hypothesis led to studies of elemental content of soils and saguaro tissue.

The work of Gladney (1991) and Gladney et al. (1992) involved analytical assessments of elemental concentrations from 825 soil samples within RMD and sites near the San Manuel copper smelter approximately 40 km northeast of Saguaro National Monument. The concentration of 13 elements was also determined for tissues of four saguaros sampled in 1988 (Gladney et al. 1992).

The soil samples were analyzed for up to 46 elements, both common contaminants and rare earth elements. Researchers included the latter, unusual elements to identify enrichment (due to pollution) beyond the average elemental abundances in the Earth's crust.

Analyses indicated that concentrations of all the rare earth elements were "within normal ranges." No conclusive pattern of elemental enrichment of soils in the monument due to pollution could be detected other than slightly elevated levels of tin (Gladney et al. 1992). Although these detailed analyses generated tremendous amounts of information, they yielded few meaningful results beyond demonstrating that soils near the San Manuel smelter were enriched with a few metallic elements commonly associated with ore smelting. No connection can be made, however, between these results and the saguaro conditions in the monument.

Of the 13 elements analyzed in tissue samples, only manganese showed a trend (Gladney et al. 1992). One of the four saguaros exhibited considerable epidermal browning, and tissues from this cactus had slightly higher manganese concentrations than tissues from the three healthy saguaros. The finding of slightly elevated manganese levels in epidermally browned tissues is not surprising if one considers the normal location and function of this element in plant cells. Chloroplasts contain one or more proteins with bound manganese (Salisbury and Ross 1985). The destruction or damage of chloroplasts during epidermal browning could easily raise the manganese: tissue weight or volume ratio because saguaro stems typically shrink during browning.

Additional studies of elemental concentrations of saguaro tissues were limited to lead and cadmium, heavy metals known to inhibit plant physiological function, root growth, nutrient uptake, or soil nutrient cycling. The concentrations of these two metals were extremely low within the examined saguaros and were well within the range of concentrations seen in plants that are free of trace metal contamination (Lajtha and Kolberg 1992).

Formulation and Review of Research Programs Between 1986 and 1992, AQD provided more than $120,000 to fund various projects involving the saguaro at Saguaro National Monument. The preceding descriptions and reviews of these projects show that many of the research efforts failed to provide useful information on either air pollution effects or saguaro ecology. Some of these failures are due to the investigators' lack of rigor in scientific approach. However, the success of any research and monitoring program begins with the procedures by which (1) research questions and priorities are identified, and (2) proposals that address those issues are requested, reviewed, and awarded. These two areas are the responsibility of the Denver AQD and are considered in the following paragraphs.

DETERMINATION OF PRIORITIES The goal of AQD research at national parks and monuments is to provide the park managers with information that will help them with resource management problems, specifically questions regarding the responsibilities of managers under the Clean Air Act. Ideally, these problems would be identified by either park managers or by park and AQD personnel. This ideal was not achieved, however, with the plans for research programs at Saguaro National Monument.

Research priorities and plans concerning biological resources at the monument were drawn up almost unilaterally by a single individual within AQD between the mid-1980s and August 1991. Although the managers of Saguaro National Monument were interested and cooperated in developing and implementing these programs (see Hall 1991), this individual determined the research questions, priorities, and methodological approaches, even though he had many other responsibilities and lacked the time to be effective in any of them. One monument staff member commented that during the later studies, AQD failed to notify monument personnel about who would conduct the research and how it would proceed. Because AQD assumed too much responsibility for planning and implementing projects and the monument staff were denied meaningful involvement, much of the research was disconnected from real management concerns.

PROCEDURES FOR REQUEST, REVIEW, AND FUNDING OF PROPOSALS Inadequate procedures for the request and review of research proposals also contributed to the poor quality of some projects that AQD supported in Saguaro

National Monument. The NPS procurement system for research proposals affected the way AQD obtained funding. AQD sought big, nonspecific umbrella contracts from 1982 to 1987 because its contracts required final approval from the main NPS office in Washington, D.C. The NPS office did not consider requests of less than $100,000. AQD was therefore forced to push big funding packages through NPS administrative channels to get even small projects funded.

This funding procedure received a poor evaluation by an outside review panel. Consequently, in 1988 AQD turned to different procedures that involved other branches of NPS, academic institutions, and other federal agencies, such as USFS and EPA. To solicit saguaro-related research proposals for Saguaro National Monument, AQD staff sent a request for quotes to six potential researchers. The quotes from K. Lajtha and L. Evans were the only ones received in response to separate requests for ecophysiological and histological studies.

AQD had only one criterion for awarding a contract to a potential investigator: the lowest bid. A contract was sealed when the investigator received and signed a 3-page form listing the amount requested and containing administrative information and an agreement to conduct the project as described. Initial quotes upon which funding decisions were made did not include abstracts or statements describing the research or methods. Work plans were required from investigators only after the awards were made and usually contained no more information than that listed on the original request for quotes. This process was clearly inadequate. Research funded this way has been treated as little more than a business "product": questions and approaches are stipulated, and investigators lack any creativity in developing the "product."

The poor quality of some investigations conducted at Saguaro National Monument was the direct consequence of a lack of rigorous review of the qualifications of researchers and their proposals, as well as inadequate AQD staff time to communicate properly with researchers and monument staff. AQD did little follow-up to ensure that reports were on schedule, reviewed, and shared within NPS. In August 1991, personnel changes in the AQD Research Branch led to a second review of procedures and an improvement in the formulation and review of the air-quality research program.

Conclusions

Despite the impacts inflicted on Saguaro National Monument before and even decades after its establishment, this area unquestionably deserves the protection afforded by monument status. The diversity of biotic communities in RMD is unique. The Tucson Basin is one of the few places where Sonoran desertscrub communities with stands of saguaros thrive near forests of pine and fir in adjacent mountains. RMD preserves not only a saguaro population, but also the spectacular transition from hot, dry desert to cool, moist coniferous forest.

The close proximity of a city with more than half a million inhabitants and housing developments on the monument boundary will probably cause more perceived and actual environmental threats to Saguaro National Monument. For protection of this special environment, programs for ecological monitoring must be well planned and carefully implemented. Despite past impacts, irreparable ecological damage has generally not occurred. A striking demonstration of this recovery is the recent surge of young saguaros in an area where researchers once feared saguaros would disappear forever. The abundant, young saguaros, most of which are no more than waist-high, hold the realistic promise of forming another grand Cactus Forest. Monitoring efforts from this time forward can provide the tremendous opportunity to study the processes of environmental recovery as well as the potential threats to that recovery. First, however, NPS must considerably improve its administration of these scientific endeavors.

The shortage of research staff within or closely associated with parks has required that park managers rely on outside investigators or other NPS divisions. The research priorities and agenda of these other entities may differ greatly from those of the park managers. Divisions such as AQD with narrowly focused research and monitoring missions cannot be expected to provide park managers with information on all resource management and environmental issues.

As of 1989, only about 75 scientists were working for the 354 NPS units, and research comprised only about 2% of the entire NPS operating budget, compared to approximately 10% in other federal land-management agencies with similar responsibilities (Simon 1989). The problem of inadequate research staffing and expertise is especially great for small parks such as Saguaro National Monument. It is unreasonable to expect managers to find personnel who can fill a variety of specialized research needs. The need for scientists with various types of expertise can be partially filled by research staff within university-based Cooperative Park Studies Units (CPSUs). This solution is not without its problems, however. The CPSU at the University of Arizona is responsible for a geographic area that includes 11 parks and monuments, is chronically understaffed (three scientists), and has suffered from a lack of administrative direction. Despite its location in Tucson, this CPSU has had little involvement with research conducted on saguaros at Saguaro National Monument during the last 5 years. The CPSUs must be strengthened and their responsibilities more clearly identified before they can contribute in more meaningful ways to pressing research and monitoring concerns.

Even with a strengthened CPSU working in concert with local administrators and resource managers of Saguaro National Monument, many research problems will arise that require expertise from outside NPS. Establishment of fruitful liaisons with local, regional, and national experts is a responsibility that must be shared by local land managers and scientists of the CPSU. These liaisons are an important element in the formulation, implementation, and review of research programs. More involvement of the scientific community in evaluating research concerns, publicizing requests for research proposals, and reviewing proposals would contribute greatly to the quality and credibility of NPS research efforts.

Note

1. On 14 October 1994, the designation was changed to Saguaro National Park.

Literature Cited

Abouhaidar, F. 1989. Influence of livestock grazing on saguaro seedling establishment. Unpubl. M.S. thesis, Arizona State University, Tempe.

———. 1992. Influence of livestock grazing on saguaro seedling establishment. P. 57–59 in C. P. Stone and E. S. Bellantoni, eds. Proceedings of the Symposium on Research in Saguaro National Monument, 23–24 January 1991.

National Park Service, Rincon Institute, and Southwest Parks and Monuments Association.

Air Quality Division. 1984. National Park Service air resource management manual. National Park Service, Air Quality Division, Denver.

———. 1991. Request for quotes for AQD histology studies in Saguaro National Monument, 9 July 1991.

Alcorn, S. M., and C. May. 1962. Attrition of a saguaro forest. Plant Disease Reporter 46:156–158.

Alcorn, S. M., S. E. McGregor, G. D. Butler, Jr., and E. B. Kurtz, Jr. 1959. Pollination requirements of the saguaro (*Carnegiea gigantea*). Cactus and Succulent Journal 31:39–41.

Alcorn, S. M., S. E. McGregor, and G. Olin. 1961. Pollination of saguaro cactus by doves, nectar-feeding bats, and honey bees. Science 133:1594–1595.

Bardsley, W. A. 1957. Will science save the saguaros? Pacific Discovery (May):24–29.

Bowers, J. E. 1988. A sense of place: the life and work of Forrest Shreve. University of Arizona Press, Tucson. 195 p.

Boyle, A. M. 1949. Further studies of the bacterial necrosis of the giant cactus. Phytopathology 39:1029–1052.

Buchmann, S. 1992. Bees, the forgotten pollinators for saguaros and other columnar cacti. Paper presented at the Special Symposium of the Species Survival Commission, 6 April 1992, Arizona Historical Society Museum, Tempe, p. 4–5 (abstract).

Chase, A. 1986. Playing God in Yellowstone: the destruction of America's first national park. Atlantic Monthly Press, Boston. 446 p.

Clemensen, A. B. 1987. Cattle, copper, and cactus: the history of Saguaro National Monument. National Park Service, Denver.

Cockrum, E. L., and Y. Petryszyn. 1991. The long-nosed bat, *Leptonycteris:* an endangered species in the Southwest? Occasional Papers of the Texas Tech University Museum 142:1–32.

Despain, D. G. 1974. The survival of saguaro (*Carnegiea gigantea*) seedlings on soils of differing albedo and cover. Journal of the Arizona Academy of Science 9: 102–107.

Duriscoe, D. M. 1987. Evaluation of ozone injury to selected tree species in the Rincon Mountains of Arizona. Final report to NPS/AQD, September 1987. Contract C-001-4-0058. 130 p.

Duriscoe, D. M., and S. L. Graban. 1992. Epidermal browning and population dynamics of giant saguaros in long-term monitoring plots. P. 237–258 in C. P.

Stone and E. S. Bellantoni, eds. Proceedings of the Symposium on Research in Saguaro National Monument, 23–24 January 1991. National Park Service, Rincon Institute, and Southwest Parks and Monuments Association.

Egermayer, D. W. 1940a. Monthly report—February 1940. Saguaro National Monument files.

———. 1940b. Monthly report—April 1940. Saguaro National Monument files.

Evans, L. S. 1991. Effects of UV-V on epidermal browning of stem surfaces of saguaro cacti. Research proposal submitted to Saguaro National Monument. Saguaro National Monument files.

Evans, L. S., V. A. Cantarella, and K. W. Stolte. 1992b. Phenological changes associated with epidermal browning of saguaro cacti at Saguaro National Monument. P. 35–46 in C. P. Stone and E. S. Bellantoni, eds. Proceedings of the Symposium on Research in Saguaro National Monument, 23–24 January 1991. National Park Service, Rincon Institute, and Southwest Parks and Monuments Association.

Evans, L. S., K. A. Howard, and E. J. Stolze. 1992c. Epidermal browning of saguaro cacti (*Carnegiea gigantea*): Is it new or related to direction? Environmental and Experimental Botany 32:357–363.

Evans, L. S., K. A. Howard, and K. H. Thompson. 1992a. Epidermal browning of saguaro cacti: is it new or related to direction? P. 23–34 in C. P. Stone and E. S. Bellantoni, eds. Proceedings of the Symposium on Research in Saguaro National Monument, 23–24 January 1991. National Park Service, Rincon Institute, and Southwest Parks and Monuments Association.

Gibson, A. D., and P. S. Nobel. 1986. The cactus primer. Harvard University Press, Cambridge, Massachusetts. 286 p.

Gill, L. S. 1942. Death in the desert. Natural History (June):22–26.

———. 1951. Mortality in the giant cactus at Saguaro National Monument, 1941–1950. Official report of the U.S. Department of Agriculture, Bureau of Plant Industry. Saguaro National Monument files (also reproduced in Steenbergh and Lowe [1983]).

Gill, L. S., and P. C. Lightle. 1942. Cactus disease investigations: an outline of objectives, plans, and accomplishments on project J-2-8. U.S. Department of Agriculture, Bureau of Plant Industry, Forest Pathologist, Albuquerque. Saguaro National Monument files (also reproduced in Steenbergh and Lowe [1983]).

———. 1946. Analysis of mortality in saguaro cactus. Official report from the U.S. Department of Agriculture, Bureau of Plant Industry. Saguaro National Monument files (also reproduced in Steenbergh and Lowe [1983]).

Gladney, E. S. 1991. Origins and effects of dry-deposited materials in desert ecosystems: some atmospheric chemistry considerations. P. 52–67 in D. Mangis, J. Baron, and K. Stolte, eds. Acid Rain and Air Pollution in Desert Park Areas. National Park Service Technical Report NPS/NRAQD/NRTR-91/02.

Gladney, E. S., R. W. Ferenbaugh, and K. W. Stolte. 1992. An investigation of the impact of inorganic air pollutants on Saguaro National Monument. P. 223–234 in C. P. Stone and E. S. Bellantoni, eds. Proceedings of the Symposium on Research in Saguaro National Monument, 23–24 January 1991. National Park Service, Rincon Institute, and Southwest Parks and Monuments Association.

Hall, R. L. 1991. A history of the air quality program at Saguaro National Monument. P. 120–122 in D. Mangis, J. Baron, and K. Stolte, eds. Acid Rain and Air Pollution in Desert Park Areas. National Park Service Technical Report NPS/NRAQD/NRTR-91/02.

Hastings, J. R. 1961. Precipitation and saguaro growth. Arid Lands Colloquium, University of Arizona 1959–60/1960–61:30–38.

Hastings, J. R., and R. M. Turner. 1965. The changing mile. University of Arizona Press, Tucson. 317 p.

Hernandez, M., and J. Erickson. 1988. Bat noted for desert role is placed on U.S. endangered list. Arizona Daily Star (31 October):A1.

Hutto, R. L., J. R. McAuliffe, and L. Hogan. 1986. Distributional associates of the saguaro (*Carnegiea gigantea*). Southwestern Naturalist 31:469–476.

Ingle, G. 1964. Will King Cactus lose his throne? San Diego Union (9 August):H1.

Karnosky, D. F., P. C. Berrang, and J. P. Bennett. 1992. Regional air pollution threatens biodiversity in national parks. P. 213–216 in C. P. Stone and E. S. Bellantoni, eds. Proceedings of the Symposium on Research in Saguaro National Monument, 23–24 January 1991. National Park Service, Rincon Institute, and Southwest Parks and Monuments Association.

Lajtha, K., and K. A. Kolberg. 1992. Ecophysiological and toxic element studies of the saguaro cactus in Saguaro National Monument. P. 47–55 in C. P. Stone and E. S. Bellantoni, eds. Proceedings of the Symposium on Research in Saguaro National Monument, 23–24 January 1991. National Park Service, Rincon Institute, and Southwest Parks and Monuments Association.

Lightle, P. C., E. T. Standring, and J. G. Brown. 1942. A bacterial necrosis of the giant cactus. Phytopathology 32:303–313.

McAuliffe, J. R. 1984. Sahuaro-nurse tree associations in the Sonoran Desert: competitive effects of sahuaros. Oecologia 64:319–321.

———. 1988. Markovian dynamics of simple and complex desert plant communities. American Naturalist 131:459–490.

———. 1993. Case study of research, monitoring, and management programs associated with the saguaro cactus (*Carnegiea gigantea*) at Saguaro National Monument, Arizona. National Park Service Technical Report NPS/WRUA/NRTR-93/01. 50 p.

McGregor, S. E., S. M. Alcorn, and G. Olin. 1962. Pollination and pollinating agents of the saguaro. Ecology 43:259–267.

Niering, W. A., and R. H. Whittaker. 1965. The saguaro problem and grazing in southwestern national parks. National Parks Magazine 39:4–9.

Niering, W. A., R. H. Whittaker, and C. H. Lowe. 1963. The saguaro: a population in relation to environment. Science 142:15–23.

Nobel, P. S. 1980. Morphology, nurse plants, and minimum apical temperatures for young *Carnegiea gigantea*. Botanical Gazette 141:188–191.

Olin, G., S. M. Alcorn, and J. M. Alcorn. 1989. Dispersal of viable saguaro seeds by white-winged doves (*Zenadia asiatica*). Southwestern Naturalist 34:282–284.

Robinson, L. W. 1966. Decline of the saguaro. American Forests (May):46, 69.

Salisbury, F. B., and C. W. Ross. 1985. Plant physiology, 3rd ed. Wadsworth Publishing Co., Belmont, California. 540 p.

Schmidt, J. O., S. L. Buchmann, and S. C. Thoenes. 1992. Saguaro and cardon cacti: nectar sources *par excellence* for native and honey bees. P. 73–78 in C. P. Stone and E. S. Bellantoni, eds. Proceedings of the Symposium on Research in Saguaro National Monument, 23–24 January 1991. National Park Service, Rincon Institute, and Southwest Parks and Monuments Association.

Shelton, N. 1985. Saguaro. Official national park handbook. U.S. Government Printing Office, Washington, D.C. 98 p.

Shreve, F. 1911. The influence of low temperatures on the distribution of the giant cactus. Plant World 14:136–146.

Shull, A. M. 1988. Endangered and threatened wildlife and plants; determination of the endangered status for two long-nosed bats. Federal Register 53(190): 38456–38460.

Simon, D. J. 1989. Prestigious commission urges new NPS vision. Park Science: A Resource Management Bulletin 9:10.

Smith, S. D., B. Didden-Zopfy, and P. S. Nobel. 1984. High-temperature responses of North American cacti. Ecology 65:643–651.

Soule, O. H., and C. H. Lowe. 1970. Osmotic characteristics of tissue fluids in the sahuaro giant cactus (*Cereus giganteus*). Annals of the Missouri Botanical Garden 57:265–351.

Steelink, C., J. Yeung, and R. L. Caldwell. 1967. Phenolic constituents of healthy and wound tissues in the giant cactus (*Carnegiea gigantea*). Phytochemistry 6: 1435–1440.

Steenbergh, W. F. 1970. The saguaro problem: information, misinformation, and desperate enterprise. Proceedings of the meeting of research scientists and resource managers of the National Park Service, sponsored by Office of Natural Science Studies, Horace M. Albright Training Center, Grand Canyon National Park, Arizona, 18–20 April 1970.

Steenbergh, W. F., and C. H. Lowe. 1969. Critical factors during the first years of life of the saguaro (*Cereus giganteus*) at Saguaro National Monument. Ecology 50:825–834.

———. 1976. Ecology of the saguaro: I. The role of freezing weather in a warm-desert plant population. P. 49–92 in Research in the Parks. Transactions of the National Park Centennial Symposium. National Park Service Symposium Series 1.

———. 1977. Ecology of the saguaro: II. Reproduction, germination, establishment, growth, and survival of the young plant. National Park Service Monograph Series 8. 242 p.

———. 1983. Ecology of the saguaro: III. Growth and demography. National Park Service Monograph Series 17. 228 p.

Turner, R. M. 1990. Long-term vegetation change at a fully protected Sonoran Desert site. Ecology 71:464–477.

———. 1992. Long-term saguaro population studies at Saguaro National Monument. P. 3–11 in C. P. Stone and E. S. Bellantoni, eds. Proceedings of the Symposium on Research in Saguaro National Monument, 23–24 January 1991. National Park Service, Rincon Institute, and Southwest Parks and Monuments Association.

Turner, R. M., S. M. Alcorn, G. Olin, and J. A. Booth. 1966. The influence of shade, soil, and water on saguaro seedling establishment. Botanical Gazette 127: 95–102.

Yozwiak, S. 1992. Grandad of all saguaros is dying. Arizona Republic (21 November):A1, A11.

7

Alien Species in Hawaiian National Parks

Charles P. Stone and Lloyd L. Loope

The ecosystems of Hawaii are extremely vulnerable to biological invasions of continental organisms, in part because they evolved in isolation. Forces that helped shape most continental systems—foraging and trampling by ungulates; predation by ants, wasps, and small mammals; virulent plant and animal diseases; frequent and intense fires; and extremes of temperature and humidity—were reduced or absent on these remote islands. High endemicity and localized and small populations are characteristics of oceanic islands, making losses to perturbations more likely than on continents. The "evil quartet" of threats to biological diversity (habitat loss and fragmentation, overkill, alien species, and extinction chains) is especially prevalent on the Hawaiian Islands (Diamond 1989).

Early Introductions

By the time Captain James Cook arrived in Hawaii in 1778, much of the lowland vegetation had been altered by more than 1,000 years of occupation by Polynesian settlers (Kirch 1982; Cuddihy and Stone 1990). These early human colonists brought with them 32 species of plants from their homelands (Nagata 1985), as well as domestic pigs, chickens, and dogs. The Polynesian rat (*Rattus exulans*), four species of geckos, three skinks, two snails, and unknown numbers of species of other invertebrates also arrived by canoe

(McKeown 1978; W. C. Gagné and Christensen 1985). Although chickens may have ranged up to 2,100-m elevation (Schwartz and Schwartz 1949), and Polynesian rats undoubtedly took a toll on native invertebrates, plants, and land birds (Ramsay 1978; W. C. Gagné and Christensen 1985; Kirch 1985), the effects of all the introductions on native flora and fauna were probably minor compared with land clearing, burning, planting, and irrigating as practiced by early human settlers (Cuddihy and Stone 1990). Only above 760-m elevation was the vegetation left essentially undisturbed by Polynesian peoples (Kirch 1982).

Post-European Contact

Biological Invasions

When continental humans arrived in Hawaii in the eighteenth century, biological impoverishment accelerated with the introduction of large browsing and grazing ungulates, such as goats, pigs, cattle, horses, and sheep. Hawaiians were forbidden to kill many of these animals, and feral ungulate populations soon began to open, degrade, and destroy upslope forests. Plantations and large ranches developed over vast upland areas in the last part of the nineteenth century, biological introductions for agriculture and human amusement increased, and commerce accelerated rates of accidental introductions (Table 7.1). In addition to the large herbivorous mammals mentioned above, the following organisms significantly affected native ecosystems:

1. a host of fire-adapted grasses that alter fire regimes in many lowland areas;

2. many invasive plants with fleshy fruit (e.g., banana poka [*Passiflora mollissima*], faya tree [*Myrica faya*], strawberry guava [*Psidium cattleianum*], and various species of *Rubus*) that have been widely spread by birds and pigs;

3. alien game and song birds that may serve as reservoirs for bird diseases and parasites and compete with native species;

4. arboreal black rats (*R. rattus*) that destroy rare native birds and plants;

5. Indian mongooses (*Herpestes auropunctatus*) that prey on ground-nesting birds;

Table 7.1. Number of native, endemic, and established alien species in major groups in the Hawaiian Islands.[1]

Taxonomic Group	No. of Extant Native Species	No. (%) of Endemic Species	No. of Invasive Alien Species
Flowering plants	970	883 (91%)	800
Ferns and allies	143	105 (73%)	21
Hepaticae (liverworts)	168	ca. 112 (67%)	2
Musci (mosses)	233	112 (48%)	3
Lichens	678	268 (38%)	0
Resident land birds	57	44 (77%)[2]	38
Mammals (on land)	1	0	18
Reptiles (on land)	0	0	13
Amphibians	0	0	4
Freshwater fish	6	6 (100%)	19
Arthropods	6,000–10,000	(98%)	ca. 2,000
Mollusks	ca. 1,060	(99%)	9

[1] From Loope and Mueller-Dombois 1989. Native species include both indigenous (found elsewhere) and endemic (found only in Hawaii) species. Alien species were brought to Hawaii by humans.

[2] James and Olson (1991) consider 40–55 historically known species and 35 fossil species of passerines endemic (i.e., at least 75–90 endemic species). An additional 39 nonpasserine endemic species are known (Olson and James 1991). At least 8 additional fossil forms of passerines are undescribed, bringing the total to 140 species.

6. the avian malaria protozoan (*Plasmodium relictum*) and the mosquito (*Culex quinquefasciatus*), which serves as the vector for the protozoan and the avian pox virus;

7. parasitoids of native and alien insects that have doubtlessly impoverished the invertebrate fauna, removed pollinators of native plants, and diminished important native bird prey.

On the average, some 25 new alien species become established each year in Hawaii (The Nature Conservancy of Hawaii and the Natural Resources Defense Council 1992). A number of alien birds, insects, and plant diseases have taken hold in recent years, despite growing awareness of the problem.

A few recent examples are the western yellowjacket *Vespula pensylvanica*, previously present in low numbers in the state, but now spreading widely after a new strain was introduced in 1978 (Nakahara 1980); brown tree snake (*Boiga irregularis*) from Guam and the Solomon Islands, which has been recorded six times; European rabbit (*Oryctolagus cuniculus*), which has multiplied in subalpine shrubland of Haleakala National Park; and *Miconia calvescens*, an invasive ornamental tree from South and Central America, which has become established on Hawaii and has been found in at least seven locations on Maui. Alien reptiles, amphibians, and fresh- and brackish-water fishes continue to escape or are deliberately released by their owners. Various species of plants and animals established in one area have been spread to disjunct areas, sometimes on different islands. Many aggressive invaders are still expanding their geographic ranges and increasing in density within occupied ranges. Future threats include the introduction of more virulent bird diseases and more cold-tolerant species of mosquitoes as vectors for these diseases; additional alien fruit-eating birds such as the parrots (*Psittidae*); snakes, land snails, and various small mammals such as ferrets and hamsters; and a multitude of invasive plants and invertebrates.

Early Biological Studies

Historical information is essential in evaluating change and providing goals for ecosystem restoration. European botanists and naturalists who visited Hawaii in the first few decades after Captain Cook's arrival collected and studied the plants introduced by Polynesians. Most of the plants brought by the first settlers, such as the coconut palm, still grow in Hawaii's lowlands. Only a few have found their way into the native forests (Fosberg 1972; Cuddihy and Stone 1990).

A few weedy plants once thought to be introduced by Europeans were recently found in the collections of notes of David Nelson, a naturalist who accompanied Captain Cook. Except for indigo (*Indigofera suffruticosa*), none are invasive in natural areas. British and French expeditions in the early nineteenth century also made important contributions to Hawaiian botanical understandings, but after the mid-nineteenth century, local residents provided most of the specimens (Sohmer and Gustafson 1987).

Two insects were described by naturalists on Captain Cook's initial voyage, but entomological collections did not really begin in Hawaii until the late 1800s (Howarth and Mull 1992). By that time, human activity and alien introductions had destroyed most of the native biological communities in the lowlands. Rev. Thomas Blackburn, collecting in the late 1800s, took many species that were never collected again. From 1892 to the early 1900s, R. C. L. Perkins made the remarkable collections that provide the baseline knowledge for Hawaiian arthropods today. The B. P. Bishop Museum, founded in 1889, encouraged other surveys. Agriculturists also supported considerable early work on invertebrates, including numerous introductions for biological control (Howarth and Mull 1992).

Eleven taxa of birds were described from Cook's visits to Hawaii (Medway 1981), and many new species were collected by the time of S. B. Dole's listings of Hawaiian birds in 1869 and 1879. At the end of the nineteenth century, Hawaiian ornithological interest increased, just when many species were apparently in rapid decline (Scott et al. 1986). R. C. L. Perkins made critical bird collections at this time to accompany his insect and mollusk emphases. Considerable worldwide interest in the Hawaiian avifauna did not really begin until the 1950s and has accelerated since then. The research efforts of James and Olson (1991) and Olson and James (1991) on fossil birds throughout the islands have been especially valuable in illuminating the past diversity of the Hawaiian avifauna and the extent of the losses during the Polynesian and European eras.

Extinction and Endangerment

Hawaii is a major center of extinction. Although the state encompasses only 0.2% of the land area of the United States, 75% of the nation's historically documented plant and bird extinctions have occurred in the islands. An estimated 10% of the Hawaiian flora has become extinct, and another 40–50% is threatened with extinction (Wagner et al. 1985; Vitousek et al. 1987). Hawaii's 52 federally listed endangered plants comprise 25% of the U.S. total; more than 150 additional Hawaiian plants will be added to the list in the next 2 years. At least 70 of 140 species of native birds known historically or from the fossil record are already extinct (Hawaii State Department of Land and Natural Resources, U.S. Fish and Wildlife Service, and The Nature Conservancy of Hawaii 1991). Thirty extant species of birds in Hawaii are federally endangered—41% of all U.S. birds listed.

More than 50% of the endemic land snails, including two dozen species of Hawaiian tree snails, and a large but unknown percentage of arthropods are already extinct. Hawaii currently has an estimated 10,000 total species of native terrestrial arthropods, of which 5,548 have been described (Howarth et al. 1988).

W. C. Gagné (1988) estimated that less than 10% of the land on Kauai, Maui, Oahu, and Hawaii continues to support undisturbed or intact forest. Nearly two-thirds of Hawaii's forest cover has been lost, including an estimated 90% of dry forests, 61% of moist forests, and 42% of wet forests. Fewer than five examples (most of them small in size) remain for 85 (57%) of 150 identified natural communities (Hawaii State Department of Land and Natural Resources, U.S. Fish and Wildlife Service, and The Nature Conservancy of Hawaii 1991).

Hawaii's National Parks

Hawaii National Park, which contained the major portions of present-day Haleakala and Hawaii Volcanoes National Parks, was established 1 August 1916 to preserve the spectacular geological and biological resources of Haleakala and its southeastern flank on Maui and the more geologically active Kilauea Volcano on Hawaii. In 1960 separate status was given to the two areas. Although biological concerns were definitely secondary to geological considerations in establishing the parks, both parks contain endangered, threatened, candidate, and rare species and a remarkable diversity of some taxa. Each park protects a continuous sea-level–to–mountaintop landscape. Both parks were designated International Biosphere Reserves in 1980, and Hawaii Volcanoes National Park received World Heritage status in 1988.

Despite serious biological invasions, relatively intact ecosystems still survive in Hawaii, especially at high elevations, on new volcanic substrates, in caves, and in bogs. Hawaii's 2 national parks, 19 state Natural Area Reserves, 3 U.S. Fish and Wildlife Service (USFWS) National Wildlife Refuges, and 12 preserves owned or managed by The Nature Conservancy of Hawaii, together with other protected areas, comprise about 24% of the land area of the state. The national parks have provided leadership in conservation biology in Hawaii, not only because funding for management and research on alien species has been relatively great compared with that for other agencies and organizations, but also because the parks comprise a proportionately large percentage of the total protected natural area in the state.

Early Inventories at Haleakala National Park

A Nature-Conservancy-sponsored expedition into the Kipahulu Valley (Warner 1968) stimulated considerable biological inventory work in the vicinity, including exploration of the Northwest Rift Zone and the Manawainui area of the park. Valuable baseline information was obtained in these efforts. Collections included a previously unknown species of Hawaiian honeycreeper (Drepanidinae), the poouli (*Melamprosops phaeosoma;* Casey and Jacobi 1974).

A Resources Base Inventory project was initiated in Haleakala in 1975. Species inventories were produced for insects (Beardsley 1980), flowering plants (Stemmermann et al. 1981), ferns and fern allies (Herat et al. 1981), birds (Dunmire 1961; Conant and Stemmermann 1980; Smith 1980; Conant 1981; Scott et al. 1986), and vegetation (Whiteaker 1983). These inventories, although not comprehensive, helped bring the biological resources of Haleakala to the attention of an international community of scientists and conservationists.

Early Inventories at Hawaii Volcanoes National Park

An ecological atlas for Hawaii Volcanoes National Park, prepared by Doty and Mueller-Dombois (1966), provided an excellent resource for further research. The park was used during the 1970s as a research study site for the International Biological Program (Mueller-Dombois et al. 1981). Researchers collected data in plots and along transects for vascular plants, terrestrial algae, soil and leaf fungi, birds, rodents, and invertebrates. Community structure, temporal organization in ecosystems, relationships of native and alien components, and genetics were emphasized. The authors reached the important conclusion that island communities can usually be considered biologically stable until humans enter the picture. The effort also helped draw the attention of a wider audience to the importance of evolutionary ecology and conservation biology in the Hawaiian Islands.

Baldwin (1941, 1944, 1953) did the first systematic censuses of forest birds in the park. Dunmire (1961, 1962) determined the status of birds in the park between 1959 and 1961. Conant (1980) surveyed birds in the Kalapana Extension of the park from 1976 to 1979, and Banko and Banko (1980) summarized historical trends.

Management of Natural Resources

Feral animals in both parks have been controlled by various means from the time of park establishment in 1916. The removal of plant cover and acceleration of soil erosion by goats were early topics of concern, especially because goats also lived on state and private lands adjacent to both parks. Yocum (1967) and others recommended fencing park boundaries and eliminating all goats from the parks. A study exclosure erected in the lowlands of Hawaii Volcanoes National Park in 1968 resulted in the appearance of an endemic leguminous vine that was new to science (*Canavalia kauensis*). It came from the seed bank in the area protected from goats (St. John 1972; Mueller-Dombois and Spatz 1975). This dramatic event, together with an active National Park Service (NPS) program to gain public support, resulted in funding and laborers to erect fences and control goats in the park (Baker and Reeser 1972).

Once a systematic control program was in place, long-term monitoring of goat numbers and vegetation was started. Monitoring vegetation recovery and surveying goat trespass continue to this day (Baker and Reeser 1972; Katahira and Stone 1982). Only a few goats now remain in unfenced higher elevations (2,130 m) of the park. Additional long-term management, monitoring, and research during the past 20 years at Hawaii Volcanoes National Park have emphasized the ecology of Hawaiian goose or nene (*Nesochen sandvicensis*); control of feral pigs and small mammals, including mongooses (*Herpestes auropunctatus*) and black rats (*R. rattus*); ecology and control of invasive alien plants; and studies of fire effects.

Long-term Research on and Monitoring of Alien Species

As the 1980s began, the stage was set for a highly productive era of research and management in Hawaii's national parks. NPS scientists selected for expertise necessary for park management were working in the parks and at the NPS Cooperative Park Studies Unit at the University of Hawaii on Oahu. Natural resource managers at the park were receiving regional support that was proportionately greater than that given to most mainland parks to tackle the severe problem of biological invasions. State and federal agencies and an increasing number of researchers from Hawaii and elsewhere were building on past data and working together more frequently. Results of the statewide Hawaii Forest Bird Survey organized by USFWS were becoming available, and The Nature Conservancy was working closely with USFWS to identify land-acquisition priorities for endangered bird habitat in the state. The Forest Bird Survey resulted in a statewide database on vegetation, native and alien plants, bats, mosquitoes, and ungulates.

General Approaches

Alien species problems in Hawaii are ranked according to anticipated or actual severity of biological threats, ecological importance of the area at risk, status of individual native species, legislative mandates, or a combination of these. Funding for alien species emphases has usually been tied to particular threats and supported short-term (1- to 3-year) efforts. The money has been inadequate, however, even for short periods. Thus, long-term, systematic, and complete monitoring and control of alien species have not been possible.

Alien species can destroy or degrade entire native communities. Priorities for control are based on degree of urgency, ecological values, realistic management possibilities, and available human resources. The concept of Special Ecological Areas was developed to focus on a few areas of high ecological value and a few invasive alien species at a time (Tunison and Stone 1992).

The most important alien-species problems call for long-term monitoring for four purposes:

1. to determine the species' importance in natural communities. Distribution of the species in an area, density or intensity of damage, and population structure can all provide information about the severity of the problem.

2. to determine success of control. The number of individuals remaining after control; their condition, age, and sex structure; and the potential for repopulation are important factors.

3. to determine the recovery of managed areas after removal of aliens. The same aliens or others may invade an area after control; native species diversity and abundance may remain lower than before the alien invasion.

4. to prevent establishment of new alien species in an area.

Managers and researchers must evaluate control results on a long-term basis to recognize where efforts with limited human resources will result in the greatest ecological returns. We will discuss several long-term, research and monitoring efforts for alien species management.

Feral Pig Control

Early Emphasis The earliest study of feral pigs (*Sus scrofa*) in Hawaii began in 1961 and was designed to provide data for a game-management program (Giffin 1978). The study addressed distribution and density, reproductive cycles, food habits, and diseases. Giffin noted that (1) large, mid-elevation wet forests provide the ideal pig habitat, and hunters generally only penetrate the fringes of these areas; (2) hunting had no apparent effect on pig distribution; and (3) the best time to harvest animals was June through September. Giffin (1978) stated that control was necessary in some forested areas to hold pig numbers in check. He believed that eradication in rain forest habitat would not be economically feasible, but that reduction of populations to a point where damage was tolerable seemed possible. He set a maximum density in wet forest habitat at 8–10 animals/km^2 to allow only minor damage to vegetation and soil. Giffin conducted no vegetation studies and did not quantitatively compare biological communities with and without pigs.

In 1979 Diong (1983) began a study on feral pig biology and management in the Kipahulu Valley wet forest in Haleakala National Park. Diong found that feral pigs consumed mostly native plants and especially relied on starch and sugar in ferns. Alien earthworms were the most important source of animal protein, and alien strawberry guava fruits were eaten and dispersed seasonally. Breeding of pigs was observed throughout the year, and high juvenile mortality occurred in this unhunted population. Although he conducted no direct vegetation analyses, Diong recommended an eradication strategy to control six factors: widespread strawberry guava dispersal; reduction of native trees and herbaceous plants; disruption of forest subcanopy; establishment of weeds; soil erosion; and an increase in the number of sites with standing water, which encourages mosquitoes.

During the 1980s, knowledge about the effects of feral pigs on plant communities in Hawaii increased. Park managers understood that if they followed the NPS mission to protect the native natural resources of the parks, they had to control the pigs (Spatz and Mueller-Dombois 1975; Katahira 1980; Diong 1983).

Control in the Parks In Hawaii Volcanoes National Park, the citizen hunter program started in 1972. Beginning in the 1980s, the state required a hunter safety course; the park also required this course, as well as registration before hunts and information on party-hours hunted, number of hunters, and success of the hunt. NPS personnel also trapped, snared, and hunted pigs in the park during this period. More than 15,000 animals were removed from the park between 1930 and 1982, but feral pigs continued to be as abundant as ever, and hunting success increased.

An analysis of data collected during the years of the citizen hunter program (Barrett and Stone 1983) showed that the program was effective only in easily accessible (peripherally roaded) areas of less than 500 ha. The major benefits of this hunting appeared to be that public participation reduced objections to park animal-removal programs and that more hunting opportunities were provided on the island. However, the NPS mandate to remove or control significant aliens on park lands (National Park Service 1988) was not being fulfilled.

Successful feral-goat control in the 1970s—including acquiring funds to build exclusionary and barrier fences and overcoming objections to eliminating a popular game species in a natural area—encouraged NPS to consider systematic control of feral pigs. Pig control, however, was more difficult than goat control for the following reasons: (1) pigs are most abundant in wet forests, which are not easily accessed by road, trail, or air; (2) pigs are rarely found in large groups in Hawaii, especially in wet forests; (3) pigs are mostly nocturnal in habit; (4) pigs' rooting abilities make fencing them out problematic; and (5) pigs are the most popular big-game species in the state; objections to their removal are considerable.

Fencing in Hawaii Volcanoes National Park began in 1966, with a pig control fence around Kipuka Puaulu (100 a.). A large, pig-free kipuka near Napau Crater was also fenced, and a pig fence was built between the branches of the 1968 lava flow in Makaopuhi Crater in the late 1970s.

Systematic feral-pig control, however, did not really begin until 1980, with the construction of fences around small areas (a few hundred acres) of rain forest. Researchers chose an accessible area of 525 ha with high pig density (> 10 animals/km^2) to test removal methods and monitoring of pigs and their habitat. They tested different types of traps, baits, and snares and evaluated systematic hunting with dogs (Kikuta and Stone 1986). Hunting with dogs and snaring with customized wire-cable snares were the most cost-effective and the least biased methods (they removed all sizes in proportion to their abundance; Stone and Loope 1987; Stone and Anderson 1988; Stone 1991). Both control methods also resulted in removal rates that were rapid enough to overcome reproduction rates and eliminate populations in a few months to a few years (Stone and Anderson 1988; Anderson and Stone 1993; Katahira et al. 1993). Researchers also determined the sizes of areas that could reasonably be managed in open and closed wet and mesic forests by fencing areas of different sizes. They used studies of pig movements to determine the boundaries of some management units for fencing and to develop strategies for more effective control by hunting and snaring (Stone 1991).

Monitoring of Control Effectiveness Researchers monitored pig activity by establishing belt transects across management units. At least 5% of each unit was sampled. Researchers recorded the frequency of occurrence of feces, digging, plant feeding, tracks, etc., and read transects at 4- to 6-month intervals. Changes in pig activity due to removal or movements of animals were useful in evaluating control success and planning continued control (Stone 1991). Monitoring pig activity was also useful in determining whether fences were breached by animals, falling trees, or earthquakes. Once pig density had been reduced to about one animal per km^2, scouting with dogs was found to be superior to systematic transects in detecting pig activity.

Monitoring of Community Recovery In rain forest and mesic forest habitat, invasive alien plants, such as strawberry guava, kahili ginger (*Hedychium gardnerianum*), faya tree, yellow Himalayan raspberry (*Rubus ellipticus*), and blackberry (*Rubus argutus*), can continue to be established in the absence of pigs (Katahira 1980; Stone et al. 1992). In grassland habitat, disturbance favors alien grass establishment and higher alien cover values (Spatz and Mueller-Dombois 1975; Stone et al. 1992). Because propagules of aggressive alien species are increasingly widespread in Hawaii, even natural disturbances, such as volcanic ash fall, storms, and some fires, can enhance establishment of introduced plants in natural areas. We believe that controlling alien plants after natural events occur and feral pigs are removed will be a continuing and necessary aspect of park management (Stone et al. 1992).

Alien Plant Control

Effective reduction of invasive alien plants in Hawaii's national parks is much more difficult than reduction of pigs and goats. Although alien plant management has often been delayed until ungulates have been removed from an area, concurrent management of plants and animals is becoming more important in natural areas and in some areas of the park that are not amenable to fencing. Integrated Pest Management for alien plants, which considers all control techniques as appropriate, is practiced in Hawaiian parks.

Weeds that cannot tolerate shade can sometimes be managed by simply removing pigs and goats, which encourages the closing of the forest canopy. Planting native species to control alien plant invasions deserves more attention. Park personnel, however, are careful to plant only those forms previously found in a given area. Availability of seeds or seedlings, nurturing of young, and high labor requirements are problems with this restoration strategy. Mechanically controlling plants with machinery and even manually pulling plants can open native forests to additional weed invasion, sometimes by more aggressive species. It can also be labor intensive and is impossible for removing some trees. Biological control in natural areas is a costly, long-term, and risky proposition: one never knows for sure what released insects and pathogens will do in natural areas (Gardner 1990; Markin et al. 1992). Prescribed burning to remove alien species often stimulates them and harms most native species that have not evolved with fire. Herbicide control is labor intensive, may harm native plant species through drift or translocation, and may damage other organisms, including invertebrates. All alien-plant-control methods are useful, but all have disadvantages.

Alien-Plant-Management Strategies With about 900 species of naturalized plants in the state, including at least 86 invaders of native plant communities

(Smith 1985), managers need to make choices. In Hawaii Volcanoes National Park, 29 of some 600 aliens are disruptive species, and 24 are too widespread to control parkwide by conventional methods (Cuddihy et al. 1988). Fire-adapted alien grasses dominate lowland habitats (610 m). Above 1,350 m the park supports primarily native vegetation, although several very disruptive species found elsewhere in the park could invade these areas. The area of most alien-plant-management activity is between 610 and 1,350 m (Cuddihy et al. 1988).

Until 1980 alien plant control at Hawaii Volcanoes National Park was done sporadically, was likely to depend upon changing concerns about particular species, and was not conducted within the framework of a parkwide plan. Current strategies (Fig. 7.1) include the following recommendations (Tunison 1992a): (1) feral ungulates should be controlled before or during alien plant control; (2) fire should be prevented; (3) the distribution of important alien plant species should be mapped; (4) localized alien plant species should be controlled throughout the park; (5) biological control should be developed for some widespread species; (6) all disruptive species should be controlled in areas of high ecological value in the park (Special Ecological Areas); (7) fountain grass (*Pennisetum setaceum*), a potentially devastating widespread species, should be confined to areas it now infests; (8) herbicidal control research should be conducted to develop safe and effective chemical control methods; (9) the ecology of important alien plant species should be studied; and (10) the public should be educated to the importance of alien plant control.

Biological Control Biological control is a possible solution for restraining some widespread, invasive alien plants that can continually reinvade natural areas. A cooperative, state-and-federal, interagency steering committee coordinates this effort in Hawaii. Eight species have been targeted to date. Park managers released insects to control four of the species (banana poka, gorse [*Ulex europaeus*], glorybush [*Tibouchina urvilleana*], and faya tree) and a fungus to control another (clidemia or Koster's curse [*Clidemia hirta*]). Research and exploratory activities in host countries of the target species have been headed by a research entomologist from the U.S. Department of Agriculture, Forest Service (USFS), an NPS plant pathologist, and state scientists.

Studies have shown that biological control in Hawaii has realized about a

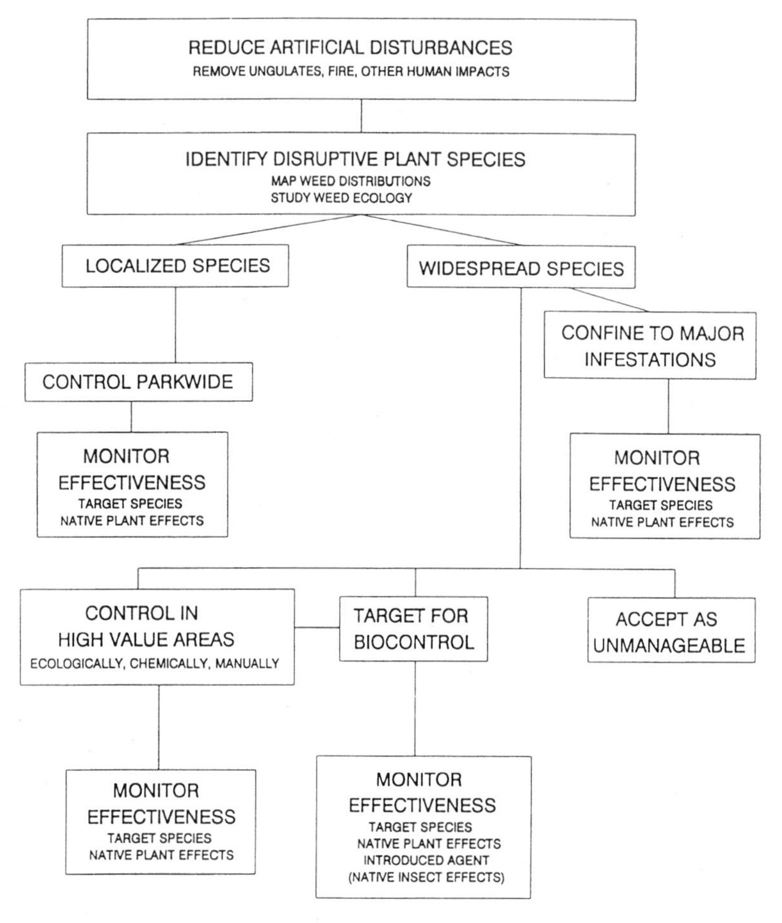

Figure 7.1 Strategies in the alien-plant-control program at Hawaii Volcanoes National Park.

50% success rate (including agricultural introductions) and that costs and time lags are high (Markin et al. 1992). Conflicts of interest with agriculture, ecological risks, complex regulations, the difficulties of foreign exploration and study, expensive testing and quarantine facilities, difficulties in predicting results, and long-term monitoring needs are other problems (Howarth 1983; Gardner 1990; Markin et al. 1992; Tunison 1992a). For some species, however, no reasonable alternative seems to exist. Researchers in Hawaii will be heavily involved in biological control for a long time.

Herbicidal Control Herbicides have been successfully used to control several woody and nonwoody species of plants in Hawaii's natural areas (Cuddihy et al. 1988; Tunison 1992a; Tunison and Zimmer 1992). Operational control with herbicides optimally requires mapping weed distribution; research to determine effective and ecologically safe chemicals, concentrations, and application methods; careful application in ecologically important areas; and monitoring to determine results and the need for retreatment. Herbicides cause less disturbance to an area than mechanical control methods and reduce invasion of alien plants after treatment. Herbicides can also be more cost-effective in operational programs. However, long-term studies of the effects of herbicides on target and nontarget plants must be made in small areas before operational use. Trial-and-error management over larger areas is not a good approach to herbicide use in natural areas.

Herbicide treatments for seven invasive plants were found in research conducted from 1984 to 1986 in Hawaii Volcanoes National Park. These species included olive (*Linociera ligustrina*), kahili flower (*Grevillea banksii*), silk oak (*G. robusta*), glorybush, yellow Himalayan raspberry, kahili ginger, and blackberry (Santos et al. 1992). From 1987 to 1990, researchers emphasized additional herbicides, concentrations, and techniques and spent more time determining effects of herbicides on native plant species. During this period, researchers considered many invasive target species: strawberry guava, banana poka, faya tree, yellow Himalayan raspberry, blackberry, kahili ginger, fountain grass, nasturtium (*Tropaeolum majus*), and several species of grasses. Many new and safer chemicals were available for this round of testing, as was an individual who was dedicated only to alien plant research. Park managers used most of the research results in operational programs as soon as these results were informally shared and before they were published. Researchers, however, were unable to monitor the results of this management use. NPS no longer funds herbicide research in Hawaii.

Fountain Grass Control Fountain grass, one of the most disruptive alien plants in Hawaii (Smith 1985), is capable of colonizing bare lava flows, thereby disrupting primary succession. It was introduced on the kona (western) side of Hawaii Island in the early twentieth century as an ornamental and now grows in dry to moist areas from sea level to 2,740-m elevation.

Because this large bunchgrass grows in monospecific stands, increases fuel loading, and is stimulated by fire, it is a threat to many native species. It is readily dispersed by vehicles, wind, water, and people. In Hawaii Volcanoes National Park, fountain grass could potentially occupy all areas outside closed-canopy rain forests, from coastal to subalpine zones. At present, about 8,500 ha in the park lowlands are infested with this species (Cuddihy et al. 1988; Tunison 1992a).

Park managers initiated a control program along park roadsides when the plant was first perceived as a threat in the 1960s. They soon expanded these efforts to a 400-ha infestation with the highest density of the species in the southwestern part of the park. From 1979 to 1983, they extended the program to roadsides outside the park and to some isolated and scattered populations inside the park. Helicopter reconnaissance indicated that fountain grass was widespread within the park. Approximately $125,000 was spent from 1976 to 1983 and $75,000 in 1984 and 1985 to collect data incidental to control, including distribution maps (Tunison 1992b).

Based on the information gathered in the studies and monitoring programs, managers decided to control fountain grass only in Special Ecological Areas, along roadsides, and in populations on the periphery of the main infestation. Control of the periphery is systematic and creates a buffer zone that confines fountain grass to its current range in the park lowlands. The buffer zone will be extended toward the center of the infested area when funding becomes available. Monitoring and early control continue to prevent encroachment of fountain grass elsewhere in the park (Tunison 1992b).

Preventing Establishment of Aliens

Preventing invasions of aggressive aliens depends on the availability of experts to identify the species, an ongoing monitoring program to detect invasions, and the capability and will to remove newly found individuals and populations. For many potentially damaging species of plants, insects, and disease-causing organisms, quarantine methods and laws are inadequate for preventing entry into Hawaii. If detected early enough, however, some invasions can be curtailed. Movements of introduced species from place to place on the same island and from island to island are as difficult to prevent as initial entry to the state, but again, expertise, awareness, and action may reduce the spread. Two examples of the importance of long-term monitoring and research in this effort follow.

Miconia calvescens *on Maui* An alien tree, *Miconia calvescens* has great invasive potential in Hawaii (Loope 1991a). An attractive plant with large, dark green leaves with maroon undersides, it grows from 340- to 1,800-m elevation in its native habitat from southern Mexico to Bolivia and Brazil. In the last two decades, it has extensively invaded forests on Tahiti. An estimated 60% of the forests are severely affected (Birnbaum 1991), and native forests and wildlife are being replaced at an alarming rate. Horticultural shipment in the late 1970s probably brought this invader to Oahu, Maui, and Hawaii (B. H. Gagné et al. 1992).

Miconia is "the one plant that could readily destroy the native Hawaiian forest" (R. Fosberg, Smithsonian Institution, pers. comm., 1991). Tiny, bird-dispersed seeds are produced after 4 to 5 years of vegetative growth, and a single tree can produce tens of thousands of seeds annually. Unlike most alien species, *M. calvescens* seedlings establish in both open sunlight and moderately dense shade. The tree is relatively well established on the island of Hawaii, and biological control is being considered there. However, the state has yet to include the species on the Noxious Weed List. Only after listing will it become illegal to bring the plant into the state, and only after listing will the state have authority to remove the weed with the cooperation of landowners.

In January 1991, Haleakala National Park personnel contacted the owner of a botanical garden near Hana, Maui, to explore the possibility of removing an *M. calvescens* tree, 30 cm in diameter and 10 m tall. This was the first record of the species on Maui of which they were aware. The owner of the garden gave full cooperation in destroying the tree and provided information about the locations of others. Park managers then initiated a program to publicize the problem and raise community consciousness and concern. "Wanted" posters were prepared and distributed in cooperation with various conservation groups and state agencies, and information on local distribution in East Maui was sought. An article was written for *Hawaii Landscape Industry News* (Loope 1991a). Other agencies and organizations (including plant nurseries) and private landowners were contacted; all were interested in eradication. Seven populations were found on the island of Maui. NPS staff and volunteers began eradication efforts in June and July 1991 near Hana, with the full cooperation of local landowners. Some 10,000 plants were removed, 97% of which were less than 5 cm in diameter.

As of January 1992, an estimated 90% of *Miconia* plants on Maui have been removed, but a vigorous seed bank necessitates a sustained removal effort during the immediate future. Further monitoring, educational efforts, and control of this species outside the park will require expertise, awareness, and action. Haleakala National Park researchers and managers have provided leadership in this effort and will continue to do so.

European Rabbits in Haleakala National Park Domestic European rabbits (*Oryctolagus cuniculus*), although established elsewhere in Hawaii, had not established themselves on one of the eight main Hawaiian Islands in more than a century. In the summer of 1990, however, a reproducing population of rabbits was discovered in high-elevation (2,070–2,130 m) native shrubland in Haleakala National Park. The invasion may have originated from as few as six animals that a pet owner released in October 1989, based on an interview with that individual.

Park researchers—knowing the destruction to native biota that introduced rabbits caused elsewhere in the world—were successful in obtaining highest priority in the park for rabbit eradication. A removal and monitoring program initiated in July 1990 showed that the animals had infested 25 ha. An educational campaign in the park increased visitor and employee awareness. Statewide publicity alerted others to the problem and made the community aware of the danger of releasing rabbits elsewhere on Maui. Other populations and releases were discovered in the process. From August 1990 to March 1991, 93 rabbits were removed from the area of initial infestation. Snaring, shooting, and trapping were the methods of control, with snaring proving most effective. Monitoring verified eradication of rabbits from the area of initial infestation but detected movement up to 2.5 km upslope (2,440-m elevation) from the main area of infestation. Four more animals were eventually taken from this disjunct group. As of May 1991, the European rabbit has been declared eradicated (Loope 1991b), but monitoring will continue.

Because public awareness has been raised, the chance of a repeat invasion in the park is no doubt lessened. This effect will be temporary, however, unless educational efforts continue beyond park boundaries. The important reasons for successful control of this invasion in a national park were (1) a mandate from the superintendent to emphasize immediate control; (2) availability of control expertise and people knowledgeable about the severity of the threat; (3) vulnerable target animals (diurnal, burrowing, and naive); (4) park personnel flexibility and attention to detail, from careful and rigorous monitoring to media relations and expeditious environmental-impact protocols; (5) availability of a variety of effective control techniques; and (6) a rate of increase and spread that could be overcome by the available control methods (Loope 1991b).

Conclusions

Long-term research has been especially valuable in the continuing process of managing alien species to protect native ecosystems in Hawaii. The availability of experienced NPS research personnel has facilitated the detection and monitoring of invasions. They have helped to establish the threat that aliens pose in natural areas by gathering and interpreting data to convince decision makers. Their efforts have made the development of control strategies more efficient. In some areas, NPS researchers have led in the prevention of alien establishment and have evaluated control effectiveness on a continuing basis. NPS researchers have also been important in the transfer of technology to other agencies and organizations and to the general public in Hawaii, where NPS research on alien species has held a leadership role. (The USFWS research effort is focused on endangered birds, USFS emphasizes biological control of alien plants with insects, universities are less management-oriented, and the state conservation agency has no research program.)

Alien species monitoring and research in Hawaii have generally been initiated in the midst of active management programs. They have been accomplished in 1- to 3-year increments on target species or areas and have been tied to short-term funding. Long-term monitoring and research must usually be bootlegged, supported from one short-term project to the next, or conducted at intervals that often preclude retaining personnel.

Monitoring and researching aliens while removing them from ecosystems have not severely affected the knowledge needed to evaluate control methods and short-term management measures. An incremental approach to target species and areas has similarly worked well in the control of important aliens and seems logical. For example, removing pigs and goats that disperse alien plant propagules before significantly investing in alien-plant-control programs has proven efficient. In most cases, a good strategy has been to concentrate first on the least invaded and most recoverable areas.

Reliance on fluctuating, highly problem-directed, and short-term-funded research has resulted in very little information about the responses of native species to alien species and their management. Researchers have assumed that removing highly invasive aliens will benefit native communities and that they must act without delay to prevent deterioration of ecosystems. Although these assumptions are valid, understanding long-term interactions among native and alien species would be a better strategy. If the intervals between significant research funding increase, management reversals may occur and long-term understandings will be further postponed. In Hawaii, where biological change is rapid (both declines in native species and increases in alien invasions), long-term monitoring and research are vital.

Literature Cited

Anderson, S. J., and C. P. Stone. 1993. Snaring to control feral pigs (*Sus scrofa*) in a remote Hawaiian rain forest. Biological Conservation 63(2):195–201.

Baker, J. K., and D. W. Reeser. 1972. Goat management problems in Hawaii Volcanoes National Park: a history, analysis and management plan. National Park Service Natural Resource Report 2. 22 p.

Baldwin, P. H. 1941. Checklist of the birds of the Hawaii National Park, Kilauea-Mauna Loa section, with remarks on their present status and a field key for their identification. Hawaii National Park Natural History Bulletin 7. 38 p.

———. 1944. Birds of Hawaii National Park. Audubon Magazine 46:147–154.

———. 1953. Annual cycle environment and evolution in the Hawaiian honeycreepers (Aves: Drepanididae). University of California Publications in Zoology 52:285–398.

Banko, P. C., and W. E. Banko. 1980. Historical trends of passerine populations in Hawaii Volcanoes National Park and vicinity. P. 108–125 in Proceedings of the 2nd Conference of Scientific Research in National Parks, vol. 8. National Park Service, Washington, D.C.

Barrett, R. H., and C. P. Stone. 1983. Hunting as a control method for wild pigs in

Hawaii Volcanoes National Park. Unpubl. report for resource management, Hawaii Volcanoes National Park. 37 p. + apps.

Beardsley, J. W. 1980. Haleakala National Park Crater District resources basic inventory. Insects. Technical Report 31. University of Hawaii Cooperative National Park Resources Studies Unit, Honolulu. 49 p.

Birnbaum, P. 1991. Exigences et tolerances de *Miconia calvescens* à Tahiti. Centre ORSTROM de Tahiti. Papeete, Polynesie Française. 66 p.

Casey, T. L. C., and J. D. Jacobi. 1974. A new genus and species of bird from the island of Maui, Hawaii (Passeriformes: Drepanididae). Occasional Papers B.P. Bishop Museum 24(12):215–226.

Conant, S. 1980. Birds of the Kalapana Extension. Technical Report 36. University of Hawaii Cooperative National Park Resources Studies Unit, Honolulu. 43 p.

———. 1981. Recent numbers of endangered birds in Hawaii's national parks. `Elepaio 41:55–61.

Conant, S., and M. A. Stemmermann. 1980. Birds in the Kipahulu District of Haleakala National Park. P. 67–75 in C. W. Smith, ed. Proceedings of Hawaii Volcanoes National Park Science Conference, vol. 3.

Cuddihy, L. W., and C. P. Stone. 1990. Alteration of native Hawaiian vegetation: effects of humans, their activities and introductions. University of Hawaii Cooperative National Park Resources Studies Unit. University of Hawaii Press, Honolulu. 138 p.

Cuddihy, L. W., C. P. Stone, and J. T. Tunison. 1988. Alien plants and their management in Hawaii Volcanoes National Park. Transactions of the Western Section of the Wildlife Society 24:42–46.

Diamond, J. 1989. Overview of recent extinctions. P. 37–41 in D. Western and M. C. Pearl, eds. Conservation for the Twenty-first Century. Oxford University Press, New York.

Diong, C. H. 1983. Population biology and management of the feral pig (*Sus scrofa* L.) in Kipahulu Valley, Maui. Unpubl. Ph.D. dissert., University of Hawaii, Honolulu. 408 p.

Doty, M. S., and D. Mueller-Dombois. 1966. Atlas for bioecology studies in Hawaii Volcanoes National Park. University of Hawaii Botanical Science Paper 2. 507 p.

Dunmire, W. W. 1961. Birds of the national parks in Hawaii. Hawaii Natural History Association, Honolulu. 36 p.

———. 1962. Bird populations in Hawaii Volcanoes National Park. `Elepaio 22:65–70.

Fosberg, F. R. 1972. Guide to excursion III. Tenth Pacific Science Congress. University of Hawaii, Botany Department, and Hawaiian Botanical Gardens Foundation, Inc., Honolulu. 249 p.

Gagné, B. H., L. L. Loope, A. C. Medeiros, and S. J. Anderson. 1992. *Miconia calvescens*: a threat to native forests of the Hawaiian Islands [abstract]. Pacific Science 46(3):390–391.

Gagné, W. C. 1988. Conservation priorities in Hawaiian natural systems. Bioscience 38(4):264–270.

Gagné, W. C., and C. C. Christensen. 1985. Conservation status of native terrestrial invertebrates in Hawai`i. P. 105–126 in C. P. Stone and J. M. Scott, eds. Hawai`i's Terrestrial Ecosystems: Preservation and Management. University of Hawaii Cooperative National Park Resources Studies Unit. University of Hawaii Press, Honolulu.

Gardner, D. E. 1990. Role of biological control as a management tool in national parks and other natural areas. National Park Service Technical Report NPS/NRUH/NRTR-90/01. 41 p.

Giffin, J. 1978. Ecology of the feral pig on the island of Hawaii. Final Report, Pittman-Robertson Project W-15-3, Study 11, 1968–1972. Hawaii Department of Land and Natural Resources, Honolulu. 122 p.

Hawaii State Department of Land and Natural Resources, U.S. Fish and Wildlife Service, and The Nature Conservancy of Hawaii. 1991. Hawaii's extinction crisis: a call to action. Honolulu. 16 p.

Herat, T., P. K. Higashino, and C. W. Smith. 1981. Haleakala National Park Crater District resources basic inventory. Ferns and fern allies. Technical Report 39. University of Hawaii Cooperative National Park Resources Studies Unit, Honolulu. 17 p.

Howarth, F. G. 1983. Classical biocontrol: panacea or Pandora's box? Proceedings of Hawaiian Entomological Society 24(2–3):239–244.

Howarth, F. G., and W. P. Mull. 1992. Hawaiian insects and their kin. University of Hawaii Press, Honolulu. 160 p.

Howarth, F. G., S. H. Sohmer, and W. D. Duckworth. 1988. Hawaiian natural history and conservation efforts. Bioscience 38(4):232–237.

James, H. F., and S. L. Olson. 1991. Descriptions of thirty-two new species of birds from the Hawaiian Islands: Part II. Passeriformes. Ornithological Monograph 46. American Ornithological Union, Washington, D.C. 88 p.

Katahira, L. 1980. The effects of feral pigs on a montane rain forest in Hawaii Volcanoes National Park. P. 173–178 in C. W. Smith, ed. Proceedings of the

3rd Conference of Natural Science, Hawaii Volcanoes National Park. University of Hawaii Cooperative National Park Resources Studies Unit, Honolulu.

Katahira, L., P. Finnegan, and C. P. Stone. 1993. Eradicating feral pigs in montane seasonal habitat at Hawaii Volcanoes National Park. Wildlife Society Bulletin 21:269–274.

Katahira, L., and C. P. Stone. 1982. Status of management of feral goats in Hawaii Volcanoes National Park. P. 102–108 in Proceedings of the 4th Conference of Natural Science, Hawaii Volcanoes National Park. University of Hawaii Cooperative National Park Resources Studies Unit, Honolulu.

Kikuta, A. H., and C. P. Stone. 1986. Food preferences of captive feral pigs: a preliminary report. P. 27–38 in Proceedings of the 6th Conference of Natural Science, Hawaii Volcanoes National Park. University of Hawaii Cooperative National Park Resources Studies Unit, Honolulu.

Kirch, P. V. 1982. The impact of prehistoric Polynesians on the Hawaiian ecosystem. Pacific Science 36(1):1–14.

———. 1985. Feathered gods and fishhooks: an introduction to Hawaiian archaeology and prehistory. University of Hawaii Press, Honolulu. 349 p.

Loope, L. L. 1991a. *Miconia calvescens*: an ornamental plant threatens native forests on Maui and Hawaii. Hawaii Landscape Industry News (Sept./Oct.)18–19.

———. 1991b. Haleakala rabbits declared eradicated (for now). Park Science 11(4): 17.

Loope, L. L., and D. Mueller-Dombois. 1989. Characteristics of invaded islands, with special reference to Hawaii. P. 257–280 in J. A. Drake, H. A. Mooney, F. di Castri, R. H. Groves, F. J. Kruger, M. Rejmànek, and M. Williamson, eds. Biological Invasions, a Global Perspective. SCOPE 37. John Wiley & Sons, New York.

Markin, G. P., P. Y. Lai, and G. Y. Funasaki. 1992. Status of biological control of weeds in Hawai`i and implications for managing native ecosystems. P. 466–482 in C. P. Stone, C. W. Smith, and J. T. Tunison, eds. Alien Plant Invasions in Native Ecosystems of Hawai`i: Management and Research. University of Hawaii Cooperative National Park Resources Studies Unit. University of Hawaii Press, Honolulu.

McKeown, S. 1978. Hawaiian reptiles and amphibians. Oriental Publishing Co., Honolulu. 80 p.

Medway, D. G. 1981. The contribution of Cook's third voyage to the ornithology of the Hawaiian Islands. Pacific Science 35:105–175.

Mueller-Dombois, D., K. W. Bridges, and H. L. Carson, eds. 1981. Island ecosystems: biological organization in selected Hawaiian communities. Hutchinson Ross Publishing Co., Stroudsburg, Pennsylvania. 583 p.

Mueller-Dombois, D., and G. Spatz. 1975. The influence of feral goats on the lowland vegetation in Hawaii Volcanoes National Park. Phytocoenologia 3(1): 1–29.

Nagata, K. M. 1985. Early plant introductions in Hawai'i. Hawaiian Journal of History 19:35–61.

Nakahara, L. M. 1980. Survey report on the yellowjackets, *Vespula pensylvanica* (Saussure) and *Vespula vulgaris* (L.) in Hawaii. Unpubl. report, Hawaii Department of Agriculture, Honolulu. 6 p.

National Park Service. 1988. Management policies. U.S. Department of the Interior, Washington, D.C.

The Nature Conservancy of Hawaii and the Natural Resources Defense Council. 1992. The alien pest species invasion in Hawaii: background study and recommendations for interagency planning. Honolulu. 123 p.

Olson, S. L., and H. F. James. 1991. Descriptions of thirty-two new species of birds from the Hawaiian Islands: Part I. Non-Passeriformes. Ornithological Monograph 45. American Ornithological Union, Washington, D.C. 88 p.

Ramsay, G. W. 1978. A review of the effect of rodents on the New Zealand invertebrate fauna. P. 89–97 in P. R. Dingwall, I. A. E. Atkinson, and C. Hay, eds. The Ecology and Control of Rodents in New Zealand Native Reserves. New Zealand Department of Lands Survey Information Serial 4.

Santos, G. L., D. Kageler, D. E. Gardner, L. W. Cuddihy, and C. P. Stone. 1992. Herbicidal control of selected alien plant species in Hawaii Volcanoes National Park. P. 341–375 in C. P. Stone, C. W. Smith, and J. T. Tunison, eds. Alien Plant Invasions in Native Ecosystems of Hawai'i: Management and Research. University of Hawaii Cooperative National Park Resources Studies Unit. University of Hawaii Press, Honolulu.

Schwartz, C. W., and E. R. Schwartz. 1949. The game birds in Hawaii. Board of Commissioners of Agriculture and Forestry, Honolulu. 168 p.

Scott, J. M., S. Mountainspring, F. L. Ramsey, and C. B. Kepler. 1986. Forest bird communities of the Hawaiian Islands: their dynamics, ecology and conservation. Studies in Avian Biology 9. Cooper Ornithological Society, Berkeley. 431 p.

Smith, C. W., ed. 1980. Resources base inventory of Kipahulu Valley below 2,000 feet. University of Hawaii Cooperative National Park Resources Studies Unit. The Nature Conservancy, Honolulu. 175 p.

———. 1985. Impact of alien plants on Hawai`i's native biota. P. 180–250 in C. P. Stone and J. M. Scott, eds. Hawai`i's Terrestrial Ecosystems: Preservation and Management. University of Hawaii Cooperative National Park Resources Studies Unit. University of Hawaii Press, Honolulu.

Sohmer, S. H., and R. Gustafson. 1987. Plants and flowers of Hawai`i. University of Hawaii Press, Honolulu. 160 p.

Spatz, G., and D. Mueller-Dombois. 1975. Succession patterns after pig digging in grassland communities on Mauna Loa, Hawaii. Phytocoenologia 3(2/3): 346–373.

Stemmermann, L., P. K. Higashino, and C. W. Smith. 1981. Haleakala National Park Crater District resources basic inventory. Conifers and flowering plants. Cooperative National Park Resources Studies Unit, University of Hawaii at Manoa, Department of Botany, Technical Report 38. 77 p.

St. John, H. 1972. *Canavalia kauensis* (Leguminosae), a new species from the island of Hawaii. Pacific Science 26:309–314.

Stone, C. P. 1991. Feral pig (*Sus scrofa*) research and management in Hawaii. P. 141–154 in R. H. Barrett and F. Spitz,, eds. Biology of Suidae. Institute Nationale de Recherche Agronomique, Tolosan, Cédex, France.

Stone, C. P., and S. J. Anderson. 1988. Introduced animals in Hawai`i's natural areas. Proceedings of Vertebrate Pest Conference 13:134–140.

Stone, C. P., L. W. Cuddihy, and J. T. Tunison. 1992. Responses of Hawaiian ecosystems to removal of feral pigs and goats. P. 666–704 in C. P. Stone, C. W. Smith, and J. T. Tunison, eds. Alien Plant Invasions in Native Ecosystems of Hawai`i: Management and Research. University of Hawaii Cooperative National Park Resources Studies Unit. University of Hawaii Press, Honolulu.

Stone, C. P., and L. L. Loope. 1987. Reducing negative effects of introduced animals on native biotas in Hawaii: what is being done, what needs doing, and the role of national parks. Environmental Conservation 14(3):245–258.

Tunison, J. T. 1992a. Alien plant control strategies in Hawaii Volcanoes National Park. P. 485–505 in C. P. Stone, C. W. Smith, and J. T. Tunison, eds. Alien Plant Invasions in Native Ecosystems of Hawai`i: Management and Research. University of Hawaii Cooperative National Park Resources Studies Unit. University of Hawaii Press, Honolulu.

———. 1992b. Fountain grass control in Hawaii Volcanoes National Park: management considerations and strategies. P. 376–393 in C. P. Stone, C. W. Smith, and J. T. Tunison, eds. Alien Plant Invasions in Native Ecosystems of Hawai`i: Management and Research. University of Hawaii Cooperative National Park Resources Studies Unit. University of Hawaii Press, Honolulu.

Tunison, J. T., and C. P. Stone. 1992. Special ecological areas: an approach to alien plant control in Hawaii Volcanoes National Park. P. 781–798 in C. P. Stone, C. W. Smith, and J. T. Tunison, eds. Alien Plant Invasions in Native Ecosystems of Hawai`i: Management and Research. University of Hawaii Cooperative National Park Resources Studies Unit. University of Hawaii Press, Honolulu.

Tunison, J. T., and N. G. Zimmer. 1992. Success in controlling localized alien plants in Hawaii Volcanoes National Park. P. 506–524 in C. P. Stone, C. W. Smith, and J. T. Tunison, eds. Alien Plant Invasions in Native Ecosystems of Hawai`i: Management and Research. University of Hawaii Cooperative National Park Resources Studies Unit. University of Hawaii Press, Honolulu.

Vitousek, P. M., L. L. Loope, and C. P. Stone. 1987. Introduced species in Hawaii: biological effects and opportunities for ecological research. Trends in Ecology and Evolution 2(7):224–227.

Wagner, W. L., D. R. Herbst, and R. S. N. Yee. 1985. Status of the native flowering plants of the Hawaiian Islands. P. 23–74 in C. P. Stone and J. M. Scott, eds. Hawai`i's Terrestrial Ecosystems: Preservation and Management. University of Hawaii Cooperative National Park Resources Studies Unit. University of Hawaii Press, Honolulu.

Warner, R. E., ed. 1968. Scientific report of the Kipahulu Valley expedition. Unpubl. report, The Nature Conservancy, Honolulu. 184 p.

Whiteaker, L. D. 1983. The vegetation and environment in the Crater District of Haleakala National Park. Pacific Science 37:1–24.

Yocum, C. F. 1967. Ecology of feral goats in Haleakala National Park, Maui. American Midland Naturalist 77(2):418–451.

3

No Park Is an Island

8

Water Rights and Devil's Hole Pupfish at Death Valley National Monument

Owen R. Williams, Jeffrey S. Albright, Paul K. Christensen,
William R. Hansen, Jeffrey C. Hughes, Alice E. Johns,
Daniel J. McGlothlin, Charles W. Pettee, and Stanley L. Ponce

In 1849 William Manly recorded the first observation of the Devil's Hole pupfish, *Cyprinodon diabolis*. He remembered the "night [he] camped near a hole of clear water [Devil's Hole] which was quite deep and had some little minus [*sic*] in it" (Johnson and Johnson 1987, p. 55).

Located in the Death Valley drainage basin of southern Nevada and southeastern California in the Great Basin physiographic province (Kilroy 1991), Devil's Hole is a planar fissure at the base of a carbonate-rock hill (Fig. 8.1). Faulting formed Devil's Hole in Ash Meadows (Hoffman 1988) on the edge of a 94.7-km^2 oasis that constitutes relic habitat for at least 23 endemic taxa of plants and animals, the largest such assemblage in so small an area in the United States (Cook and Williams 1982; Hershler and Sada 1987; Deacon and Williams 1991).

Biologists originally described specimens of *Cyprinodon* at Medbury Spring, Ash Meadows, Saratoga Springs, and Amargosa Creek in California on the Death Valley Expedition of the U.S. Biological Survey in 1891. Charles H. Gilbert (1893) described the pupfish as *C. macularius*. At that time, he thought the Devil's Hole fish were young individuals; therefore, he did not describe them as a separate species (Wales 1930). In 1930 Wales sampled fish from five apparently distinct populations of *Cyprinodon* in Ash Meadows and the central part of the Amargosa Desert. He determined that the Devil's Hole pupfish was a separate species and named it *C. diabolis*.

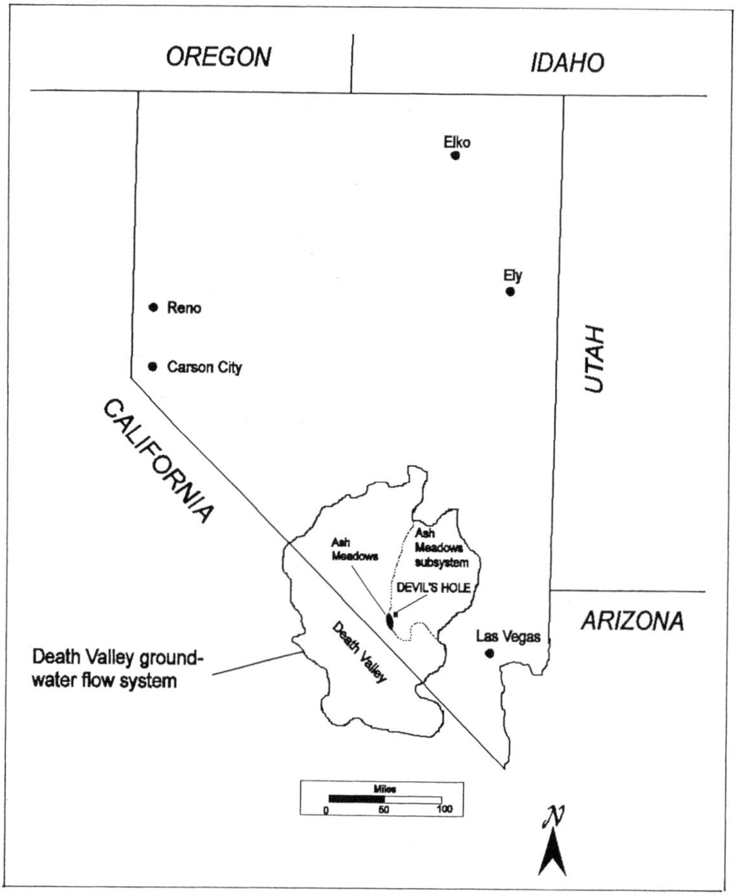

Figure 8.1 Death Valley National Monument.

Desert pupfish are remnants of larger populations that once inhabited the ancient lakes of Death Valley. As the lakes dried and deserts appeared, the pupfish adjusted to living in reduced habitats where unusual and extreme conditions exist (U.S. Department of the Interior 1970). Several species of pupfish in Inyo County, California, and adjacent Nye County, Nevada, are extinct or endangered (U.S. Department of the Interior 1970). The entire world's population of *C. diabolis* dwells in this unique pool in Ash Meadows, Nevada. The survival of *C. diabolis* requires enough water in the pool to inundate a rock ledge that the pupfish depend upon for spawning and feeding. Extinction of *C. diabolis* as a result of human impact would cause an incalculable loss to man's understanding of desert ecology.

Carl L. Hubbs, a professor of biology at the University of California, Scripps Institution of Oceanography, had the foresight to recognize the special nature of Devil's Hole. In 1950 and 1951, Hubbs, through his research and that of colleagues, convinced the National Park Service (NPS) that Devil's Hole and *C. diabolis* deserved the designation and protection of a national monument. He contended that without protection "the unique fish might well be exterminated." As a result of his efforts, Devil's Hole was added to Death Valley National Monument in January 1952 as a 16.2-ha detached unit.

Response to Groundwater Pumping

Before the mid-1960s, residents of the area in and around Ash Meadows used natural spring flow for irrigation. In the early 1960s, the U.S. Geological Survey (USGS) investigated the effects of groundwater pumping on springs in the Ash Meadows area. USGS researchers determined that, as of April 1963, no net change in water level at Devil's Hole had occurred. They concluded, however, that pumping nearby or in areas with good hydraulic connection to Devil's Hole could affect the water level there, with response times ranging from less than a year to more than several decades (Worts 1963). This prediction was borne out in 1969. Coincident with the pumping of groundwater through irrigation wells in the area of Devil's Hole (see Figs. 7 and 17 of Dudley and Larson 1976), the water level in Devil's Hole began to decline (Fig. 8.2). The pool reached its lowest level in 1972. After nearby groundwater withdrawals ended, the pool level gradually rose until 1988.

In 1988 the pool level began a gradual decline, which continued into the 1990s. The two most probable causes for the gradual decline were (1) a drought from about 1985 through 1990 in the Great Basin and (2) groundwater withdrawals located farther away than those that caused the rapid decline of the pool in 1969 and the early 1970s. The drought may have caused groundwater levels, and thus the pool level, to decline because less water infiltrated to the water table in recharge areas. The rate of groundwater withdrawals in distant areas has shown a substantial increase over the past 30 years with large fluctuations. Because of the distance of the groundwater

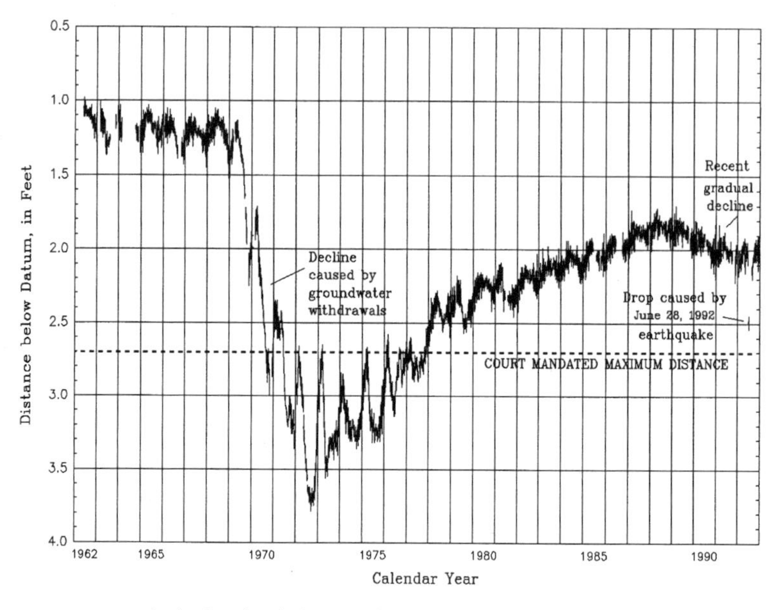

Figure 8.2 Mean daily level of the pool in Devil's Hole, 23 May 1962 through 30 September 1992 (provisional data).

withdrawals from Devil's Hole, tens or even hundreds of years might pass before the drawdown caused by the withdrawals reaches Devil's Hole.

Earthquakes may also affect the pool level, further complicating determination of the reason for the gradual decline. In late June 1992, a strong earthquake with its epicenter near Landers, California, caused a 1.8-m surge wave in Devil's Hole. The wave damaged the primary monitoring system and caused the pool level to fall about 0.16 m (see Fig. 8.2). Within 3 months of the earthquake, however, the pool level recovered to about its pre-earthquake level.

The Physical Setting

Complex, interbasin, regional groundwater-flow systems typify the eastern and southern parts of the Great Basin, including Devil's Hole and Ash Meadows. Groundwater moves from one basin to another along complex pathways through basin-fill aquifers, carbonate-rock aquifers, or both (Harrill et al. 1988).

A central corridor of the carbonate-rock aquifers in southern Nevada is contained within the carbonate-rock province. The corridor consists of a north-south-trending "block" of thick, laterally continuous carbonate rocks that probably contains the principal conduits (fractures, fault zones, and connected solution cavities) for regional groundwater flow from east-central to southern Nevada. Flow eventually discharges through springs at Ash Meadows (Winograd and Thordarson 1975; Dettinger 1989).

The Death Valley groundwater-flow system (Fig. 8.1) underlies an area of about 40,922 km^2 and contains about 30 basins, including Devil's Hole (Harrill et al. 1988). The Ash Meadows groundwater-flow subsystem or groundwater basin (Winograd and Thordarson 1975; Dudley and Larson 1976) is a subsystem of this interbasin, regional groundwater-flow system. This subsystem underlies about 11,655 km^2 and contains 10 basins. It discharges in the Ash Meadows area through springs, bare-soil evaporation, evapotranspiration, and underflow to an adjacent groundwater-flow subsystem (Winograd and Thordarson 1975; Kilroy 1991).

Faults near the eastern edge of Ash Meadows interrupt the continuity of the carbonate-rock aquifers and terminate the Ash Meadows subsystem. This causes groundwater to move upward and westward into a complex of shallow basin-fill aquifers to discharge along a spring line (Dudley and Larson 1976). The water level in Devil's Hole (pool level) is the local surface expression of the Ash Meadows subsystem.

The Legal Setting

NPS Mandates

President Herbert Hoover established Death Valley National Monument[1] in 1933 "for the preservation of the unusual features of scenic, scientific, and educational interest therein contained" (Proc. 2028, 47 Stat. 2554). President Franklin D. Roosevelt expanded the boundaries in 1937 to add areas of historic and scientific interest.

In January 1952, President Harry S. Truman added Devil's Hole to Death Valley National Monument because "the said pool . . . is unusual among caverns . . . and . . . in this pool . . . a peculiar race of desert fish . . . which is found nowhere else in the world, evolved only after the gradual drying up of the Death Valley Lake System isolated this fish population from the original ancestral stock that in Pleistocene times was common to the entire region" (Proc. 2961, 66 Stat. c. 18, 17 Fed. Reg. 691).

NPS manages Death Valley, including Devil's Hole, under the authority of the Organic Act (39 Stat. 535; 16 U.S.C. 1-3). The act created NPS to supervise, manage, and control specific areas "to conserve the scenery and the natural and historic objects and the wild life therein and to provide for the enjoyment of the same in such manner and by such means as will leave them unimpaired for the enjoyment of future generations."

Western Water Law

The Doctrine of Prior Appropriation is the basis of water law in most of the western states. Under this doctrine, the person who first diverts water for a state-defined beneficial use (i.e., appropriates the water) has a prior right to use, against all other appropriators: "first in time, first in right." An appropriative water right is a proprietary right. It can be bought and sold, and its place of use, purpose, and point of diversion can generally be changed without loss of priority. This right is also usufructuary; i.e., it applies only to the *use* of the water; the body of water belongs to the public. If the water is not put to beneficial use for several years, specified by statute, the water-right holder may relinquish (abandon) it or lose it through action by the state (forfeiture).

The federal government also holds reserved water rights. When the federal government reserves land for a particular purpose, it also reserves enough water from unappropriated sources to accomplish the purposes as defined by Congress or the president. The right to the water starts as of the date of the reservation, regardless of actual water use.

A general adjudication of water rights is the means by which the federal government files its claims to its reserved water rights and waves its immunity from suit following the Act of 10 June 1952 (66 Stat. 560; 43 U.S.C. 666 [McCarran Amendment]). Commonly, in a general adjudication, all water users on a stream and its tributaries must claim and defend their rights to use water. After considering the evidence, the court issues a decree that sets forth all the rights within the adjudicated area, including federal reserved water rights. Adjudications usually occur in state courts, but federal courts have concurrent jurisdiction.

Nevada Water Law

Nevada follows the Doctrine of Prior Appropriation in allocating both its surface and groundwater resources. To obtain a permit to appropriate water, one must file with the state engineer and advertise in a paper of general circulation, giving all parties notice and the opportunity to protest. If warranted, the engineer can hold a hearing on the application. At that time, the applicant, protestors, and other interested parties offer information (evidence) for the engineer's consideration.

The engineer will reject the application only if (1) no unappropriated water is available, (2) the appropriation would conflict with existing rights, or (3) the appropriation may be detrimental to the public interest. Otherwise, the engineer has a statutory duty to approve the application if the prescribed fees have been paid and the application is completed properly. The applicant or protestors can appeal the engineer's ruling. However, the court will not accept new factual evidence. The engineer's decision is considered correct, with the appealing party bearing the burden of proving otherwise.

Ecosystem Management at Devil's Hole

Research Basis for Action

The water level at Devil's Hole has been measured continuously since 1956. However, because of vandalism and quality-control problems, data collected before 1962 are of questionable value for most purposes. A copper washer and nail driven into the rock headwall above the pool established an arbitrary datum in 1962. The washer itself has since disappeared, but its point of attachment serves as the reference for all reported water-level measurements.

At the request of the U.S. Department of the Interior (USDI), the University of Nevada Desert Research Institute used existing data to conduct an initial study of the hydrogeology of Devil's Hole. Fiero and Maxey (1970) found a high correlation between maximum pumping by Spring Meadows, Inc. and water-level declines in Devil's Hole. They recommended (1) using surface storage of natural spring discharge as an alternative to pumping, (2) conducting additional studies to determine the level of pumping that would not materially interfere with the water level at Devil's Hole, and (3) setting up a detailed monitoring system, including pumping tests, to develop a water-management system for the Ash Meadows area.

NPS referenced Fiero and Maxey's report in its protests to water-right applications by Spring Meadows, Inc. as the basis for indicating that "further lowering of the water level in Devil's Hole, Ash Meadows, Nevada would result from the pumping of underground water in this area."

Upon review of this report and that of Worts (1963), a USDI Desert Pupfish Task Force directed USGS to examine in detail the causes of the water-level decline in Devil's Hole and in springs in Ash Meadows. The task force also asked the University of Nevada's Center for Water Resources Research to prepare a study of groundwater management to minimize environmental impact (see Dudley and Larson 1976).

During the early 1970s, research on changes in the pupfish population as a function of water-level changes intensified. Based on his field observations, James E. Deacon, University of Nevada–Las Vegas, indicated that a small, natural rock shelf served as the limited feeding area for the fish and that the amount of water covering the shelf directly affected the size of the pupfish population (Memorandum of Minutes, task force meeting of 3/21/72). Research by Deacon and his coworkers described the relationship among water levels, nearby pumping, and reductions in pupfish numbers and was the foundation upon which the federal government based its request for a preliminary injunction against further pumping of groundwater (Devil's Hole Pupfish Recovery Team 1980; Findings of Fact accompanying 24 March 1978, Order Modifying Final Decree, *United States of America v. Francis Leo Cappaert et al.*, Civil No. LV-1687).

Administrative Actions

USDI formed the Desert Pupfish Task Force in 1970 because of the lowered Devil's Hole pool level in 1969. The task force, composed of representatives from the NPS, U.S. Bureau of Land Management, U.S. Bureau of Reclamation, U.S. Bureau of Sport Fisheries and Wildlife, and USGS, recommended actions to protect the Devil's Hole pupfish as well as other desert pupfish in the area.

The task force outlined steps to save the pupfish. These included studying the hydrogeology of the Ash Meadows area, identifying the requirements of the pupfish, and acquiring lands to protect pupfish habitat. USDI also planned to seek, through the state of Nevada, restrictions on pumping that affected pools containing pupfish (U.S. Department of the Interior 1970).

In May 1970, USDI requested that Nevada take the necessary actions to prohibit groundwater pumping by Spring Meadows, Inc. The federal government maintained that groundwater pumping by this company was lowering the water level of Devil's Hole, threatening the extinction of the endangered pupfish.

In June 1970, Spring Meadows, Inc. filed applications for additional well development. In the fall of 1970, the Desert Pupfish Task Force urged the engineer to delay approving applications in the Ash Meadows area near Devil's Hole for 2 or 3 years to allow a full evaluation of the hydrological regime and asked for 1 year to complete specific studies. The engineer responded that actions to prevent the unauthorized use of underground water had been taken, appropriations exercised under Nevada water law would be allowed, and hearings on the applications for appropriations in the Ash Meadows area would be scheduled.

In December 1970, the engineer held a public hearing on applications by Spring Meadows, Inc. As a result of that hearing, he entered a decision that concluded (1) water was available for appropriation; (2) there were no water rights for Devil's Hole; (3) the water table would not be unreasonably lowered (thus the applications would not adversely affect existing rights); and (4) it was in the public interest to grant the water rights, which in turn would provide revenue to the county and state.

The Legal Challenges

In 1970 and 1971, the water level in Devil's Hole dropped below the rock shelf upon which the pupfish breeds (Dudley and Larson 1976). In August 1971, the United States filed suit in U.S. District Court for the District of Nevada (*United States of America v. Spring Meadows, Inc., et al.,* Civil No. LV-1687), seeking to forbid Spring Meadows, Inc. from pumping specific wells. The suit also asked (p. 4) "[t]hat the court determine and decree that the plaintiff United States has the rights to the use of so much of the waters from the springs and acquifers [*sic*] as is in and on the land known as Devil's Hole, Death Valley National Monument, or may be necessary for the needs and purposes of maintaining the pool and the desert pupfish population therein, and that such rights be declared to have a priority date of January 22, 1952."

Later that month, the United States and Spring Meadows, Inc. entered into a stipulation, which stated, among other things, that pumping from three wells would cease after 9 September 1971 and that legal actions would be stayed until final settlement or end of the agreement. The water level continued to decline, however, and again exposed the rock shelf in 1972. In June 1972, the United States amended its suit to include a request for a preliminary injunction to stop pumping of additional wells.

The court granted a preliminary injunction on 5 June 1973, enjoining Spring Meadows, Inc. from pumping that would lower the water level in Devil's Hole more than 0.91 m (3.0 ft) below the copper-washer datum. The court also appointed a special master to set pumping limits and monitor the water level. Judge Thomas Foley concluded that unappropriated waters were withdrawn on the date of reservation of Devil's Hole to the extent necessary for the requirements and purposes of the reservation, including preservation of the pool of water and the Devil's Hole pupfish. In April 1974, the district court made the injunction permanent and limited pumping to that which would maintain a mean daily water level of 0.91 m (3.0 feet) below the copper washer (*United States v. Cappaert,* 375 F. Supp. 456 [Nev. 1974]).

Cappaert Enterprises (successor to Spring Meadows, Inc.) and the state of Nevada appealed the decision. The Court of Appeals for the Ninth Circuit affirmed the lower court decision, holding that the doctrine of implied reservation applied to groundwater as well as surface water and that the engineer's action on water-right applications of Cappaert Enterprises did not prevent the United States from seeking resolution in federal court (*United States v. Cappaert,* 508 F.2d 313 [1974]).

Cappaert Enterprises appealed to the U.S. Supreme Court (*Cappaert et al. v. United States et al.,* Civil No. 74-1107, June 7, 1976; 426 U.S. 128; 48 L.Ed.2d 523; 96 S.Ct. 2062 [hereafter referred to as *Cappaert*]). The Court reviewed it "to consider the scope of the implied-reservation-of-water-rights doctrine" (p. 533). The Supreme Court affirmed the lower court's decision, and Chief Justice Warren Burger delivered the opinion for a unanimous Court. With regard to the federal reserved-water-rights doctrine, the Court said the following (p. 534–538):

> This Court has long held that when the Federal Government withdraws its land from the public domain and reserves it for a federal purpose, the Government, by implication, reserves appurtenant water then unappropriated to the extent needed to accomplish the purpose of the reservation. In so doing the United States acquires a reserved right in unappropriated water which vests on the date of the reservation and is superior to the rights of future appropriators.
>
> In determining whether there is a federally reserved water right implicit in a federal reservation of public land, the issue is whether the Government intended to reserve unappropriated and thus available water. Intent is inferred if the previously unappropriated waters are necessary to accomplish the purposes for which the reservation was created.
>
> The implied-reservation-of-water doctrine, however, reserves only that amount of water necessary to fulfill the purpose of the reservation, no more.
>
> ... [S]ince the implied-reservation-of-water doctrine is based on the necessity of water for the purpose of the federal reservation, we hold that the United States can protect its water from subsequent diversion, whether the diversion is of surface or groundwater.
>
> Federal water rights are not dependent upon state law or state procedures and they need not be adjudicated only in state courts.

The Court, therefore, held the following:

1. When the United States reserved Devil's Hole, it also gained rights in unappropriated waters.
2. The district court had appropriately restricted pumping to allow for the minimum water level to protect the scientific interest that Congress identified in the pool as the habitat of the pupfish.
3. The United States could protect its right from later appropriations, be they surface or groundwater.
4. The United States does not have to perfect its water rights according to state law and can litigate its water-rights claims in federal court.

On 24 March 1978, the U.S. district court issued a permanent injunction to limit the pumping of groundwater when required to achieve and maintain at Devil's Hole a daily mean water level of 0.82 m (2.7 ft) below the copper washer. This injunction provided for a water level 0.09 m (0.3 ft) higher than that defined in either the Supreme Court decision or the earlier permanent injunction of the district court. Justification for this higher water level came from NPS-supported studies conducted during the litigation (Deacon and Williams 1991).

Lessons Learned

The Importance of Research and Monitoring

How Research Affected Court Decisions The district court probably would have rendered a different decision had there been inadequate data to characterize the water level in Devil's Hole. Scientific testimony on pupfish habitat requirements and the importance of rock-shelf inundation for feeding and propagation formed the basis of the 1973 district court ruling. By forbidding groundwater pumping from particular wells and allowing the water level at Devil's Hole to rise to 0.91 m (3.0 ft) below the on-site datum, the preliminary injunction intended to maintain water coverage of most of the rock shelf.

The Supreme Court recognized the government's entitlement to a reservation of water because of the express language proclaiming that protection of the pupfish was a purpose of the reservation. In determining the quantity of the water right, the Supreme Court granted continuing jurisdiction to the district court, which held a hearing in 1978 to consider biological and hydrological data before it permanently set a water level.

After the preliminary injunction, granted in 1973, the district court used research data on pupfish life history and physical habitat to raise the minimum acceptable water level to 0.82 m (2.7 ft) below the datum. The court revised the level after determining that the data (1) showed a high positive correlation between primary productivity (algal growth) and pupfish population size, (2) showed that the new water level would result in primary productivity rates capable of supporting a minimally viable pupfish population during the winter months (estimated at 200 individuals), and (3) identified the percentage of both shelf coverage and available habitat volume associated with different water levels. Researchers developed two independent lines of reasoning from this information and presented them to the court. Both suggested that the 0.82-m (2.7-ft) mark should be adopted as the standard for a minimum water level to protect the pupfish.

How Research Affected Management Decisions Although the uniqueness of the biological resources at Devil's Hole was well known by 1950, little was known about groundwater flow paths, locations of recharge areas, or the connection between Devil's Hole and other surface and ground waters. When Devil's Hole was added to Death Valley National Monument, direct control of the spring seemed adequate protection for the pupfish.

Awareness of Devil's Hole's unique physiographic and biological resources grew during the late 1950s and early 1960s. Research on the hydrology and biology of the Ash Meadows area also increased, primarily as a result of scientific curiosity and management concerns. Data were collected on the water levels in the pool and nearby wells, the discharge of area springs, and pupfish habitat requirements.

Extensive commercial well development in the late 1960s outside Death Valley affected the water level in Devil's Hole. NPS managers thus concluded that direct control of the pool was not enough to ensure protection of the pupfish. Knowledge of the groundwater system increased. Correlations between withdrawal of water at various wells and lowered water levels in the pool showed that hydrological actions outside the park boundaries affected park resources. NPS saw a clear and present threat to the entire rare and endangered biotic community of Ash Meadows. Additional research and management actions were needed to ensure species survival.

When NPS management actions taken through the engineer proved to be of short-lived benefit, the federal government once again considered the direct-control option: a "Pupfish National Park and Wildlife Refuge." This option, however, appeared to have high risk and require a long, uphill battle. The survival of the pupfish could not await the inevitable delays and compromises. Research had already identified the spawning and feeding needs of the pupfish, and improved hydrological predictions showed that vital habitat (the rock shelf) would be lost in a very short time.

When applications were filed with the state for additional pumping, the federal government protested the applications under state law because the direct-control option was not feasible. The engineer denied the protests because Devil's Hole had no state-permitted water right, even though both biological and hydrological facts showed the need for protection. An appeal of the engineer's decision to the state court of appeals would not likely yield a different result because the court could review only issues of compliance with state statutes, not the facts of the decision. The one option remaining to the federal government, and the one taken, was filing a complaint in federal court. Although the consensus among federal agencies was that water development threatened all water-dependent resources in the Amargosa Basin, the reservation of Devil's Hole to protect an endangered species offered the best legal and factual basis for litigating the federal reserved water right.

It might appear that another option existed: negotiating a settlement with the developer. This option, however, had no practical significance until after NPS filed a complaint in federal court and the developer could see merit in negotiating. As described above, initially a settlement based on available hydrological information was, in fact, negotiated. This settlement would have shut down certain wells and monitored others with the possibility of reducing production from them. However, after a few months, when the water level at Devil's Hole continued to drop because of pumping at wells that no one even considered monitoring, hydrogeologists knew that groundwater withdrawals over a much larger area could affect the water level at Devil's Hole. This information jeopardized the feasibility of the entire irrigation development. As a result, the settlement unraveled, renewing litigation in federal court.

How Cappaert *Affected Federal Reserved Water Rights*

Cappaert is not a panacea for protecting all water rights in NPS units, nor does it answer all legal questions surrounding a federal reserved water right for a given NPS unit. It does, however, provide a sound basis upon which to assert NPS federal reserved water rights in court. Because of *Cappaert* and, not trivially the research underpinning *Cappaert,* NPS has made significant progress in protecting its water rights and the resources dependent on them. Lacking the precedence of *Cappaert,* NPS would face far greater uncertainty in court.

Cappaert set a standard that can protect the rights of the United States against impairment by peripheral water development. The Supreme Court decision was the first federal reserved-water-right case involving an NPS unit and the first to focus on groundwater. Even though groundwater withdrawals affected the water level at Devil's Hole, the decision awarded the United States a water right to surface water only (water level in the pool). Nevertheless, the decision affects groundwater because groundwater withdrawals are enjoined when they affect the water level of the surface pool at Devil's Hole.

Ongoing and Future Concerns

Current Threats

Even though the removal of groundwater in the Ash Meadows area by the Cappaert family is no longer a threat to Devil's Hole (most of the lands next to Devil's Hole are now under federal control through several land purchases from the Cappaerts and their successors in interest), the story does not end there. Droughts, prolonged groundwater withdrawals at distant locations, and movement of blocks of the Earth's crust may affect the pool level. Recent proposals to appropriate large amounts of groundwater for expanding development in southern and central Nevada pose an additional threat to Devil's Hole and to water and water-dependent resources at Ash Meadows, Death Valley, Lake Mead National Recreation Area, and Great Basin National Park. As shown in the *Cappaert* case, research and further assessment of potential impacts to these resources are essential. The goal of NPS, armed now with a precedent-setting ruling for protecting its water rights, is to carry out resource protection strategies that incorporate research, resource management, and state and federal laws to protect its water rights from injury.

Scientists now believe that the water in Devil's Hole has traveled far. Development of the region's water resources will likely intercept some of this water at distant locations, which could, over time, again reduce the water level at Devil's Hole and other NPS springs. Hydrogeological research conducted by USGS (Winograd and Thordarson 1975; Harrill et al. 1988; Dettinger 1989) continues to show that regional groundwater-flow systems connect east-central and southern Nevada to water sources in Death Valley and Ash Meadows, including the pool at Devil's Hole.

Because of this research, NPS is taking an expanded view of the area in which water development can affect the water sources and water rights of NPS. Since 1988 NPS has opposed dozens of water-right applications in basins thought to be potential sources of recharge to springs in Death Valley, Devil's Hole, and Lake Mead. These basins are located throughout the eastern third and southern half of the state.

The water development projects that concern NPS include mining, municipal, industrial, and resort developments. For example, in 1989 the Las Vegas Valley Water District (LVVWD) filed water-right applications to divert about 986,792,000+ m^3 (then reduced to 234,363,100 m^3) of groundwater per year from regional groundwater-flow systems of east-central and southern Nevada. Some of the proposed diversions from the Ash Meadows subsystem (18,502,350 m^3 per year) are located about 80 to 130 km east and northeast of Devil's Hole. These diversions could pose a significant threat to Devil's Hole and the rights established in *Cappaert*. Based on the rates of annual groundwater recharge to individual basins in the Ash Meadows subsystem (Harrill et al. 1988), the groundwater recharge rate of the entire subsystem is about 43,172,150 m^3 per year. About half of this water (17,000 a.-ft) discharges through springs in Ash Meadows (Winograd and Thordarson 1975). The remainder discharges through underflow, evapotranspiration, and bare-soil evaporation. If LVVWD diverts 18,502,350 m^3 per year from the subsystem, the pool level at Devil's Hole is expected to decline, although the decline may not occur for tens or even hundreds of years. Continuing current groundwater diversions may accelerate or compound this decline. LVVWD's proposed groundwater diversions in regional groundwater-flow systems near the Ash Meadows subsystem will capture groundwater destined for Ash Meadows.

The current understanding as expressed in available scientific literature is inadequate to ensure that appropriation and diversion of large quantities of groundwater in the Ash Meadows subsystem (and in nearby regional groundwater-flow systems) will not affect the senior water rights, water resources, and water-related resources associated with Devil's Hole. Analyzing the combined effects of withdrawals from regional groundwater-flow systems will require data and sophisticated analytical models that will far exceed the scope and complexity of anything used in *Cappaert*.

Changing Goals for Research and Monitoring

The goal of monitoring the water level in Devil's Hole today is quite different from that before *Cappaert*. In the 1960s and early 1970s, the goal was to gather baseline data on aquifer systems and groundwater recharge characteristics to determine possible sources of recharge waters and causes of water-level decline. Early aquifer tests, measurements of spring discharges, and well-water levels near Devil's Hole led researchers to conclude that the pool level was sensitive to outside physical stresses, such as groundwater pumping.

In the *Cappaert* case, researchers found strong correlations between fluctuations in the pool level and the frequency or timing of nearby groundwater pumping. The data revealed that, during periods of intensive pumping, the pool level declined in a relatively short time, indicating a direct and immediate response to pumping-caused drawdown. The pool level also rose in a relatively short time in response to an end in pumping.

Because of these correlations, scientists easily concluded that pool levels changed in direct response to nearby pumping. The court could therefore determine, based on scientific fact, that injury to the property of the United States (i.e., water rights and resources) had occurred. Even though NPS and its experts had to show injury clearly, the evidence (data from hydrological and biological monitoring and expert analysis) was available and enabled the court to set a minimum threshold for the water level, which, if exceeded, would constitute injury. Once the court set the limit of the U.S. right (0.82 m below the copper-washer datum), it established a standard by which injury caused by any means could be measured.

With the gradual water-level decline recently detected and the prospect of additional water development, the role of research takes on heightened importance to assure that the mean daily water level in Devil's Hole does not fall below the court-mandated level. NPS relies upon its monitoring of the water level at Devil's Hole (a task it assumed from USGS in 1989) as a basic component in its resource protection program. Nevertheless, NPS needs a better understanding of groundwater flow paths and flow rates to predict the complex interactions and long-term effects that will likely result from distant water development.

Groundwater monitoring is now a condition for approval of water-rights applications. This monitoring will help NPS in gathering new data and in advancing research. Scientists can use the results to better understand how pumping of groundwater in the Ash Meadows area might also affect the groundwater at Death Valley and Devil's Hole. In developing monitoring plans, applicants agree that if water levels or spring flows near the diversion decline more than a set amount, they will reduce, or stop if necessary, withdrawing water until the water levels or spring flows recover.

The monitoring plans incorporate methods to ensure scientific validity and unbiased data collection. The monitoring is designed to identify and arrest potential impacts to groundwater flowing toward Death Valley and Devil's Hole, long before NPS detects impacts in the units themselves. Monitoring allows NPS to advance the scientific basis of future protective actions and to detect potential impacts on NPS rights. If potential impacts are detected, NPS has standing (both legal and scientific) to request relief through state administrative procedures or the state or federal court systems.

NPS has also prepared a plan of investigation to determine the probable cause(s) of the gradual decline in the pool level and has carried out parts of this plan. Because the recent drought and distant groundwater withdrawals over a long period may have caused groundwater levels, and thus the pool level, to decline, NPS funded an investigation to check climatological data for the Death Valley area. NPS also plans to do the following: (1) compile water-right and water-use data in the Death Valley region, (2) construct a conceptual model of the groundwater flow systems, and (3) have USGS conduct studies to assess possible movement of blocks of the Earth's crust. The NPS plan also includes periodic workshops for scientists and managers to (1) discuss the hydrology and geology of Devil's Hole and the surrounding area and (2) share ideas about how to determine the cause(s) of the recent gradual water-level decline. These efforts should help NPS to evaluate potential cause(s) of the decline and construct predictive hydrological models. The program will incorporate information from existing monitoring activities and provide an early-warning system to protect NPS water rights from injury. The successful application of such a program will depend upon aid and cooperation from state, federal, and local agencies.

Protection of important resources such as Devil's Hole requires caution in the development and use of water in the Death Valley area. By making groundwater monitoring a condition for approval of development permits, the state allows legitimate enterprise to progress unimpeded while simultaneously advancing the state of hydrological science through improvements in predicting system response. Monitoring and research (modeling) provide decision makers with an improved set of facts and the information scientists and administrators need to protect resources in the face of water development.

Costs and Values of Research and Monitoring

Litigation is time-consuming and expensive. Costs include salaries and travel expenses for NPS staff and lawyers from the Office of the Solicitor and U.S. Department of Justice, costs of technical or historical research to develop evidence, expert-witness salaries and travel expenses, and costs of exhibits and other courtroom presentation aids. We could not obtain information on the total cost of the court action for Devil's Hole but believe it is a six-digit figure. By comparison, the cost of a recent Colorado case involving federal reserved water rights for national forests is two orders of magnitude higher (U.S. Department of Justice, pers. comm., 1991).

The immediate value of the court action was the protection of an endangered species. One cannot estimate the value of the scientific and social benefits of maintaining the pupfish population and the ecosystem that it represents. However, the U.S. Congress did indicate the value when it passed the Endangered Species Act in 1973, giving broad protection, often at high cost, to endangered species.

As a precedent for federal reserved water rights, *Cappaert* established criteria by which the Supreme Court would decide on the entitlement and quantification of reserved water rights for protection of resources. The Court set the tone for future reserved-water-rights cases by relying upon research results to establish a water level to protect the attributes identified in the proclamation reserving Devil's Hole.

In the years following *Cappaert,* the precedent set at Devil's Hole protected NPS resources and probably lowered the costs of water-rights adjudications. Research and monitoring were integral to these savings and more by fostering negotiations and settlements outside of court.

The protection process for the pupfish took 28 years, from reservation of Devil's Hole as a national monument to the completion of a recovery plan (Table 8.1). The need to protect the pupfish, however, did not disappear with the court decision; on the contrary, it has increased with advances in water development technology.

Only if it is adequately administered does a water right have value. To assure adequate administration and protection, NPS now spends up to $21,000 annually monitoring the water level at Devil's Hole to detect impacts from other water users. The situation now, however, is quite unlike the

Table 8.1. Chronology of events in protecting the Devil's Hole desert pupfish, Death Valley National Monument, 1952–1980.

Year	Action
1952	Devil's Hole was established as a detached unit of Death Valley National Monument.
1956	Monitoring of the pool's water level in Devil's Hole began.
1962	A copper-washer datum was established for monitoring the pool's water level.
1963	USGS (Worts 1963) completed a hydrological study requested by NPS.
1967	Spring Meadows, Inc. began intensive groundwater development.
1969	The water level in Devil's Hole began to decline.
1970	The Desert Pupfish Task Force was formed and outlined steps to save the desert pupfish. NPS asked the Nevada state engineer to prohibit further groundwater pumping by Spring Meadows, Inc. NPS asked the Desert Research Institute to conduct a preliminary hydrological study. NPS filed protests and requested a delay in action on the water-right applications by Spring Meadows, Inc. NPS asked USGS to conduct a detailed hydrological study. The Nevada state engineer approved the water-right applications.
1971	A complaint for injunctive relief was filed in U.S. district court for the district of Nevada to prevent Spring Meadows, Inc. from pumping specific wells. A stipulation for continuance providing for the cessation of pumping from wells was filed in U.S. district court.
1972	The complaint for injunctive relief was amended, adding more wells to the list of those to be enjoined.
1973	The U.S. district court granted a preliminary injunction, which enjoined Spring Meadows, Inc. from pumping that would lower the pool level in Devil's Hole more than 0.91 m (3.0 ft) below the established datum. A special master was appointed to set pumping limits and monitor the water level of Devil's Hole. The appellate court granted a water level of 1.01 m (3.3 ft), pending a permanent injunction by the district court.

Table 8.1. *Continued.*

Year	Action
1974	USGS published an open-file report (Dudley and Larson 1976) on the effects of irrigation pumping on pupfish habitat. The Pupfish Recovery Team was formed. The U.S. district court entered a permanent injunction limiting the pumping to maintain a mean daily pool level of 0.91 m (3.0 ft) below the established datum and issued an order continuing the services of the special master. The appellate court upheld the injunction entered by the district court.
1976	The U.S. Supreme Court affirmed the district and appellate court decisions and stated that the United States acquired water rights in unappropriated waters by reserving Devil's Hole.
1978	The U.S. district court issued a permanent injunction to limit the pumping of underground waters from wells to achieve and maintain a daily mean pool level of 0.82 m (2.7 ft) below the established datum.
1980	The recovery plan for the pupfish was completed (Devil's Hole Pupfish Recovery Team 1980).

earlier one with the Cappaerts. Pumping in the complex carbonate aquifer system at a point 100 or more miles away could lower the water level at Devil's Hole after several years. If, to show injury and protect the pool, NPS relied only upon a drop in water level recorded by its Devil's Hole monitoring program, then the pupfish could be extirpated. The effects, unlike those produced by the Cappaerts, could be irreversible. Because of long distances and time lags, the effects on Devil's Hole due to shutting off the wells would be delayed, just as were the effects from the outset of pumping—tens to hundreds of years. Protection thus requires more than the monitoring program at the pool.

The cost of monitoring, as a condition of water-permit approval, is high. Permit holders have largely funded data collection outside of NPS units. The federal government funds data collection and analysis at Devil's Hole. However, the possible cost associated with not monitoring—the loss of an endangered species—is incalculable.

Note

1. On 31 October 1994, the designation was changed to Death Valley National Park through the Desert Protection Act.

Literature Cited

Cook, S. F., and C. D. Williams. 1982. The status and future of Ash Meadows, Nye County, Nevada. Unpubl. report to the attorney general. Office of the Attorney General, State of Nevada, Carson City.

Deacon, J. E., and C. D. Williams. 1991. Ash Meadows and the legacy of the Devil's Hole pupfish. P. 69–91 in W. L. Minckley and J. E. Deacon, eds. Battle against extinction: native fish management in the American West, vol. 5. University of Arizona Press, Tucson.

Dettinger, M. D. 1989. Distribution of carbonate-rock aquifers in southern Nevada and the potential for their development: summary of findings, 1985–88. Program for the study and testing of carbonate-rock aquifers in eastern and southern Nevada: summary report 1. U.S. Geological Survey and University of Nevada, Desert Research Institute. State of Nevada and Las Vegas Valley Water District, Carson City. 37 p.

Devil's Hole Pupfish Recovery Team. 1980. Devil's Hole pupfish recovery plan. U.S. Fish and Wildlife Service, Portland, Oregon. 46 p.

Dudley, W. W., Jr., and J. D. Larson. 1976. Effect of irrigation pumping on desert pupfish habitats in Ash Meadows, Nye County, Nevada. U.S. Geological Survey Professional Paper 927. 52 p.

Fiero, G. W., and G. B. Maxey. 1970. Hydrogeology of the Devil's Hole area, Ash Meadows, Nevada. Center for Water Resources Research, Desert Research Institute, University of Nevada, Reno. 25 p.

Gilbert, C. H. 1893. Report on the fishes of the Death Valley Expedition, collected in southern California and Nevada in 1891, with descriptions of new species. North American Fauna 7:229–234.

Harrill, J. R., J. S. Gates, and J. M. Thomas. 1988. Major ground-water flow systems in the Great Basin region of Nevada, Utah, and adjacent states: U.S. Geological Survey Hydrologic Investigations Atlas HA-694-C.

Hershler, R., and D. W. Sada. 1987. Springsnails (Gastropopoda: Hydrobiidae) of Ash Meadows, Amargosa Basin, California-Nevada. Proceedings of the Biological Society of Washington 100(4):776–843.

Hoffman, R. J. 1988. Chronology of diving activities and underground surveys in Devil's Hole and Devil's Hole Cave, Nye County, Nevada, 1950–86. U.S. Geological Survey Open-file Report 88-93. 12 p.

Johnson, L., and J. Johnson, eds. 1987. Escape from Death Valley: as told by William Lewis Manley and other '49ers. University of Nevada Press, Reno. 213 p.

Kilroy, K. C. 1991. Ground-water conditions in Amargosa Desert, Nevada-California, 1952–87. U.S. Geological Survey Water Resources Investigation Report 89-4101. 93 p.

U.S. Department of the Interior. 1970. A task force report on let's save the desert pupfish. Washington, D.C. 8 p.

Wales, J. M. 1930. Biometrical studies of some races of cyprinodont fishes from the Death Valley region with description of *Cyprinodon diabolis* n. sp. Copeia 3:61–70.

Winograd, I. J., and W. Thordarson. 1975. Hydrogeologic and hydrochemical framework, south-central Great Basin, Nevada-California, with special reference to the Nevada test site. P. C1–C126 in Hydrology of Nuclear Test Sites. U.S. Geological Survey Professional Paper 712-C.

Worts, G. F., Jr. 1963. Effect of ground-water development on the pool level in Devil's Hole, Death Valley National Monument, Nye County, Nevada. U.S. Geological Survey, Water Resources Division, Carson City, Nevada. 27 p.

9

Urban Encroachment at Saguaro National Monument

William W. Shaw

Few, if any, protected natural areas are unaffected by the pressures to develop and use nearby lands for resource extraction, agriculture, towns, or cities. "External threats" have been acknowledged as a critical issue facing parks throughout the world (McNeeley and Miller 1983). A survey of U.S. national-park superintendents found that 66% of 203 parks in the United States had problems with incompatible uses on adjacent lands (National Parks and Conservation Association 1979). In that survey, the most frequently cited conflicting uses of neighboring lands were residential and commercial developments.

Urbanization affects parks in a variety of ways, including water and air pollution, diversion and depletion of water sources, loss of habitat, and displacement of biota. Wildlife and wildlife habitats are especially vulnerable to urbanization. Half of all wildlife problems cited in the National Parks and Conservation Association survey were caused by residential and commercial development through habitat loss, pollution, loss of water sources, and direct impacts such as vehicle collisions, increased hunting or poaching, and predation or harassment of wildlife by domestic or feral pets.

Furthermore, as urbanization continues outside parks, these areas become increasingly isolated from other natural areas. "Insularization" of parks due to urban encroachment can have a variety of effects on an area's biota ranging from restrictions on immigration, emigration, and dispersal, to long-term consequences on speciation through the isolation of small gene pools (MacArthur and Wilson 1967; Wilson and Willis 1975; Miller and Harris 1979; International Union for the Conservation of Nature 1980; Wilcox 1980; Boecklin and Simberloff 1986; Newmark 1987). Although the principles of island biogeography offer valuable insights, few studies have documented these processes or established baseline data to monitor the long-term impacts of urban encroachment on wildlife in national parks near expanding cities.

In general, urbanization tends to isolate adjacent parks and protected areas. However, the extent of this isolation can be mitigated by sensitive urban design (Shaw et al. 1986). A movement is growing to incorporate wildlife conservation into the planning and design of metropolitan areas (Leedy et al. 1978; Shaw and Mangun 1984; Stenberg and Shaw 1986). Urban wildlife conservation programs are dependent on basic information about wildlife, habitats, and human activities and attitudes about wildlife resources. This paper describes several recent and ongoing studies that are attempting to produce this kind of information, especially as it pertains to the management of a protected area near a rapidly growing city.

Urbanization and Saguaro National Monument

Saguaro National Monument[1] was established in Tucson, Arizona, in 1933 to protect the unique saguaro-paloverde biotic community. With the addition of a second unit in 1961 and boundary expansions in 1976 and 1991, the monument now totals more than 35,200 ha and includes plant communities ranging from desertscrub and mixed cactus associations at its lower altitudes (659.6 m) to montane fir and pine forests at its peaks (2,639.6 m).

In 1933, when Saguaro was established, the population of Tucson was only 35,000 and the monument was separated from the city by 15 miles of dirt roads. By 1986 the population of the metropolitan area was well over 600,000 and was expected to continue to grow 3% per year into the next century (Tucson Trends 1986).

Rapid expansion of the metropolitan area is resulting in major changes in the lands adjacent to the monument. Homes have been built within a few feet of the monument's border, and a 2,830-ha planned community may be constructed adjacent to the eastern portion of the monument.

Careful planning is needed to minimize the degradation of the monument's natural resources. The conservation of wildlife and other resources will require coordination among monument personnel, community planners, and developers, as well as new and better information about the monument's wildlife and the impacts of urbanization on these resources.

Research Approach

Urbanization is a complex process that can affect wildlife resources in a myriad of ways. The studies at Saguaro acknowledge this complexity by focusing on several kinds of research questions and coordinating and integrating a variety of distinct but related investigations. These studies fall into three general groups: land-use/land-cover studies, specific wildlife studies, and studies that address the human dimensions of this issue. The breadth of these subjects necessitated a variety of research expertise, and the investigations described below involved several faculty members and graduate students.

Land-Use/Land-Cover Studies

An essential baseline for many kinds of studies is a description of the land use and land covers within the monument and on surrounding urban or potentially urban areas. Using a variety of sources, including previous studies, monument and county records, aerial photographs, and field inventories, my colleagues and I developed a computerized database for the monument and a 3.2-km strip of adjacent private lands. The database included the following information: Public Land Survey section lines; agricultural and mining areas; Saguaro National Monument boundaries, including the Tucson and Rincon Mountain Districts; vegetation communities; houses within 2 miles of the monument; classified land-ownership parcels; major roads in and around the monument; selected hiking trails inside the monument; trail needs based on the eastern Pima County trail plan; 100-ft topographic contour lines; riparian corridors; official zoning designations of the Pima County Planning Department; and experimental plots for the saguaro biomonitoring project.

The database used PC ARC/INFO (Environmental Systems Research Institute, Inc. 1990), which is a geographic information system (GIS) that includes utilities for entering, editing, analyzing, and plotting geographically based information. The system was selected because it is versatile and widely used and because the database can be readily transferred to many other GISs. We also used a second GIS program, GRASS (U.S. Army Construction Engineering Research Laboratory 1989), which has certain advantages in the display and manipulation of data and is less complicated to use than ARC/INFO. Natural-resource management agencies also use GRASS (Shaw et al. 1991).

One of the most valuable attributes of GIS programs is the ease with which they can be modified and expanded as new or better information becomes available. The land-use/land-cover information provides the foundation for a GIS database that park personnel can use to make management decisions.

Wildlife Studies

Urban encroachment can affect many kinds of wildlife. Saguaro National Monument supports a wide variety of wild animals, including 74 mammal species (Doll et al. 1989), 58 species of reptiles and amphibians (Lowe and Holm 1991), and 187 bird species (Southwest Parks and Monuments Association 1985). We used the following criteria to identify species for initial research efforts: (1) scarcity or low reproductive capability, (2) vulnerability to human activities, (3) dependence on resources outside the monument, (4) potential for displacing native wildlife, (5) vulnerability to predation or disturbance by dogs and cats, and (6) high public interest or visibility.

Based on these criteria, we developed research proposals for the following species or issues:

1. Cavity-nesting birds. European starlings (*Sturnus vulgaris*) and house sparrows (*Passer domesticus*) are exotic species whose populations are closely associated with human developments. Both of these species nest in cavities and may displace native cavity-nesting birds.

2. Desert tortoise. Saguaro National Monument provides habitat for the desert tortoise (*Gopherus agassizii*). Researchers believe that this species is declining in numbers throughout its range; California and Nevada populations have been listed as threatened by the U.S. Fish and Wildlife Service. Desert tortoises are believed to be extremely vulnerable to urbanization.

3. Mule deer and collared peccary. Mule deer (*Odocoileus hemionus*) and collared peccary (*Dicotyles tajacu*) have large home ranges; populations at Saguaro are known to range beyond monument boundaries into adjacent private lands. These species also attract a high level of public interest and concern.

Human Dimensions

Urbanization is a sociopolitical process. Some of the most important questions about the effects of urban growth on wildlife involve the behavior and attitudes of humans. For this reason, our studies at Saguaro included two elements that involve humans and their pets:

1. Residential survey. We conducted a survey of 500 households within 1.6 km of Saguaro to generate information about historical sightings of wildlife, pet ownerships and behavior, and homeowner attitudes and perceptions about wildlife resources.

2. Domestic dogs and cats. As the lands adjacent to parks are developed for human occupancy, the populations of domestic dogs and cats inevitably increase. Very little is known about the size and behavior of these populations in the Tucson area or about their possible impacts on ground-nesting birds, reptiles, and small mammals in Saguaro National Monument.

Results

Presented below are the results of our completed studies.

Land-Use/Land-Cover Studies

A comprehensive land-use/land-cover database has many applications. One important use for this information is as a baseline for monitoring changes that may affect the monument. Comparing changes in zoning, land-ownership, and housing patterns over time can provide valuable insights for park managers and community planners concerned with minimizing the impacts of urban expansion on the monument's resources. Furthermore, by overlaying two or more layers of land-use information, managers can readily identify critical land-use areas. For example, near Saguaro, riparian communities are especially important in providing habitat and movement corridors for many kinds of wildlife (Shaw et al. 1991). Housing pattern data superimposed on vegetative community maps reveal existing and potential fragmented habitats (Fig. 9.1). Finally, this GIS land-use database is extremely valuable in facilitating specific wildlife studies, as explained below.

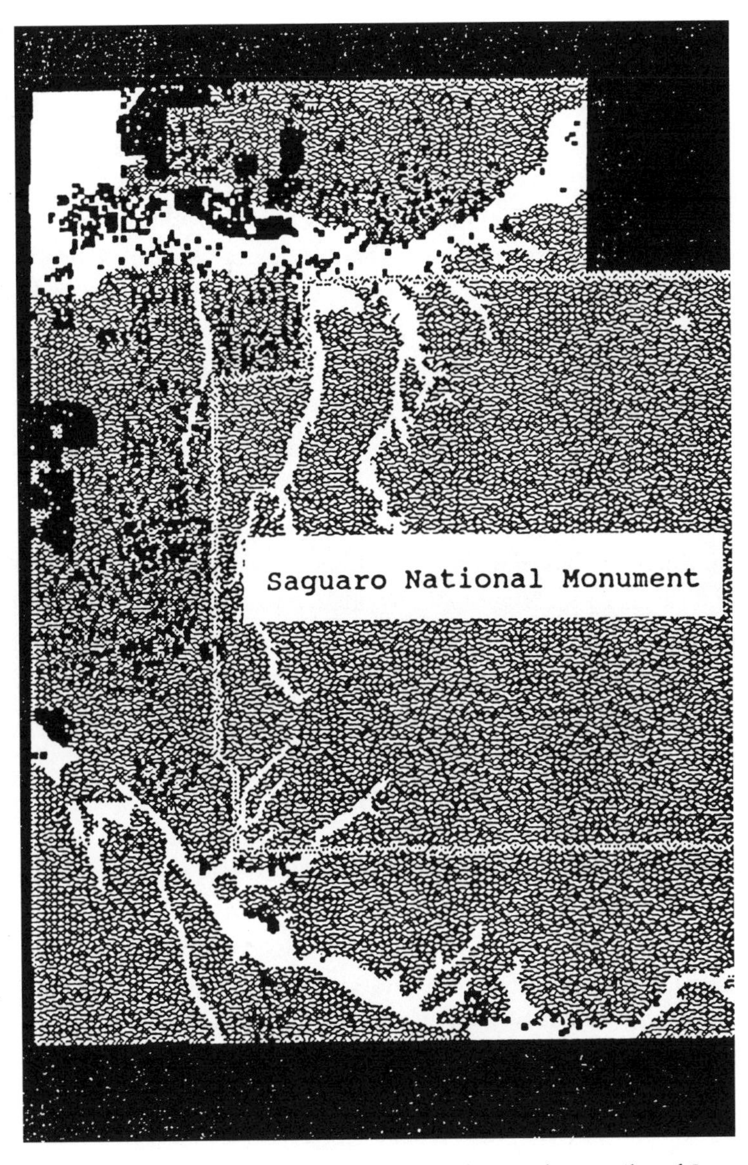

Figure 9.1 Distribution of housing units (black cubes) within 2 miles of Saguaro National Monument, Rincon Mountain District (white areas are riparian communities).

Wildlife Studies

Cavity-Nesting Birds Bibles and Mannan (Mannan and Bibles 1990; Bibles 1991) studied cavity-nesting birds in six 10-ha plots with similar paloverde/cacti/mixed-scrub vegetation. All plots contained saguaro cacti (*Carnegiea gigantea*) with cavities. Two of the plots were located within the monument more than 1 km from the nearest human residence. The other plots were selected to represent likely areas of future development (i.e., natural areas near residences or golf courses). After monitoring cavity use for one breeding season, Bibles and Mannan manipulated the availability of cavities on some plots by plugging cavities that were either unused or used by exotic species.

These researchers found no evidence of competition for nest cavities between exotic and native cavity-nesting birds. They concluded that urbanization along the monument's borders will undoubtedly result in increased numbers of starlings and house sparrows. However, because of the abundance of cavities in a saguaro cactus community and temporal differences in breeding behavior, these exotic birds probably will not displace native species in the monument. In areas where saguaros are scarce or have been reduced by development, however, competition for nest cavities may occur. Periodic monitoring of the study plots in the monument will be useful in assessing populations of exotic birds as urbanization continues.

Desert Tortoise Little was known about the distribution and status of tortoise populations at Saguaro. For this reason, an initial research effort involved a parkwide study. We found that tortoises were distributed throughout most of the monument below about 1,370 m. Although the highest densities of tortoises were in rocky slopes and desert washes, we found tortoises throughout the monument's desert communities (Goldsmith and Shaw 1988).

Researchers believe that tortoises are quite vulnerable to urbanization. Potential threats include loss of habitat, capture by humans, and vulnerability to automobiles and pets. Because tortoises that reside in Saguaro may wander across monument boundaries, an important question for assessing the impacts of urban encroachment on these animals involves their home range. Twelve tortoises were equipped with radio transmitters and monitored twice weekly during active periods and at an interval of 10–14 days during hibernation (Goldsmith and Shaw 1988). To estimate how far outside the monument a tortoise might travel while traversing its home range, we measured the maximum cross section of each tortoise's home range. Adult home ranges averaged 467 m and ranged from 250 m to 960 m. Based on this information, we believe that tortoise populations within 1 km of the monument border will be adversely affected by urbanization along the boundary.

Mule Deer and Collared Peccary Saguaro National Monument supports populations of mule deer and collared peccary, but little was known about the extent to which these wide-ranging animals depend on resources outside the monument. To study the habitats used by these animals, researchers equipped 10 deer and 5 peccaries with radio-collars and monitored their movements (Bellantoni and Krausman 1991).

Several findings in this study provide useful insights for the park managers. For example, these investigators (Bellantoni and Krausman 1992) found that mule deer appear to alter their home range on a seasonal basis that coincides with the availability of free water in the monument (Fig. 9.2). During drought periods, deer frequently entered the residential neighborhoods near the monument, but when water was available they remained primarily within the monument boundaries. If urbanization causes deer to abandon external sources of water, the "effective size" of the monument (in terms of deer habitat) will be reduced.

Unlike deer, the collared peccary herds of Saguaro use the low-density suburbs near the monument extensively throughout the year. Four of the five herds monitored spent substantial time in the private lands near the monument (Fig. 9.3). Indeed, many of the monument's neighbors attract collared peccaries to their homes by feeding them (Shaw et al. 1992). Of serious concern to park managers is the potential for problems with panhandling wildlife that have become habituated to human activities.

Human Dimensions

Residential Survey Using the mail survey methodology of Dillman (1978), we mailed a 12-page questionnaire to a random sample of 500 households within 1.6 km of the monument (Shaw et al. 1992). After we deducted new arrivals, departures, and nondeliverable questionnaires, the adjusted

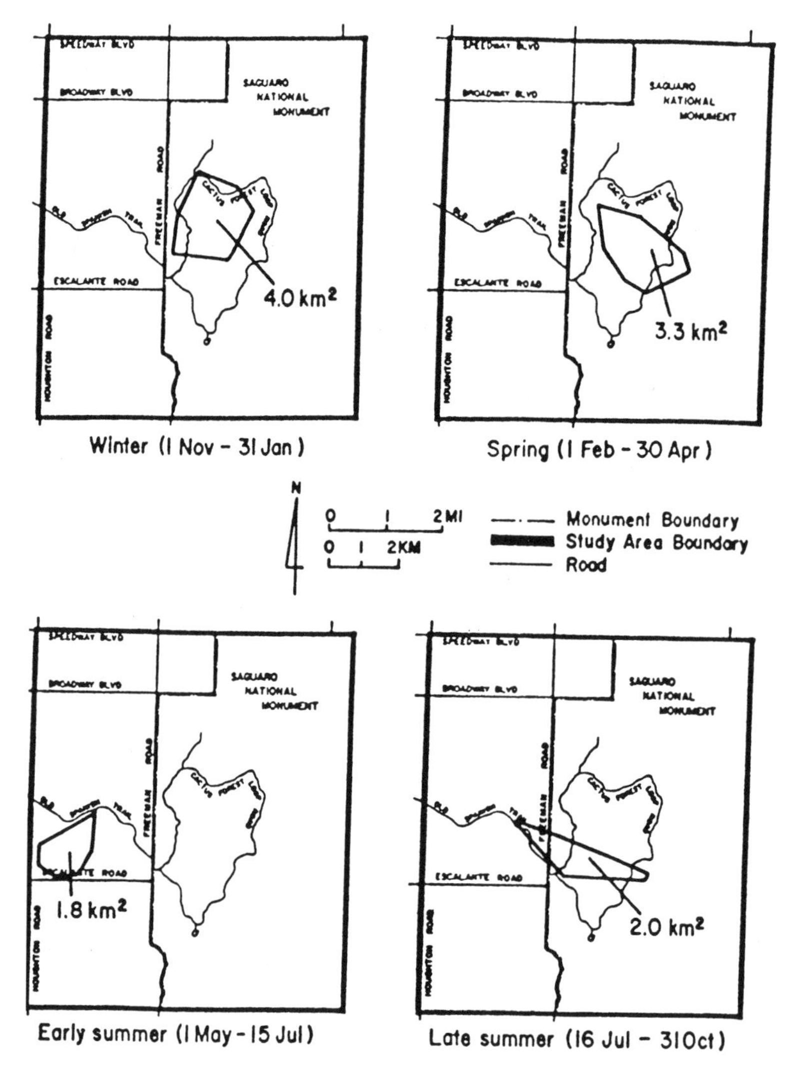

Figure 9.2 Minimum-convex-polygon estimates of seasonal home-range size (km^2) for one desert mule-deer male (no. 341), Saguaro National Monument, winter 1988–late summer 1989 (from Bellantoni and Krausman 1991).

sample size was 460. A total of 393 households responded, a response rate of 85%.

The survey covered many relevant issues for park managers, including wildlife-related attitudes, perceptions, and behaviors. Questions dealing with

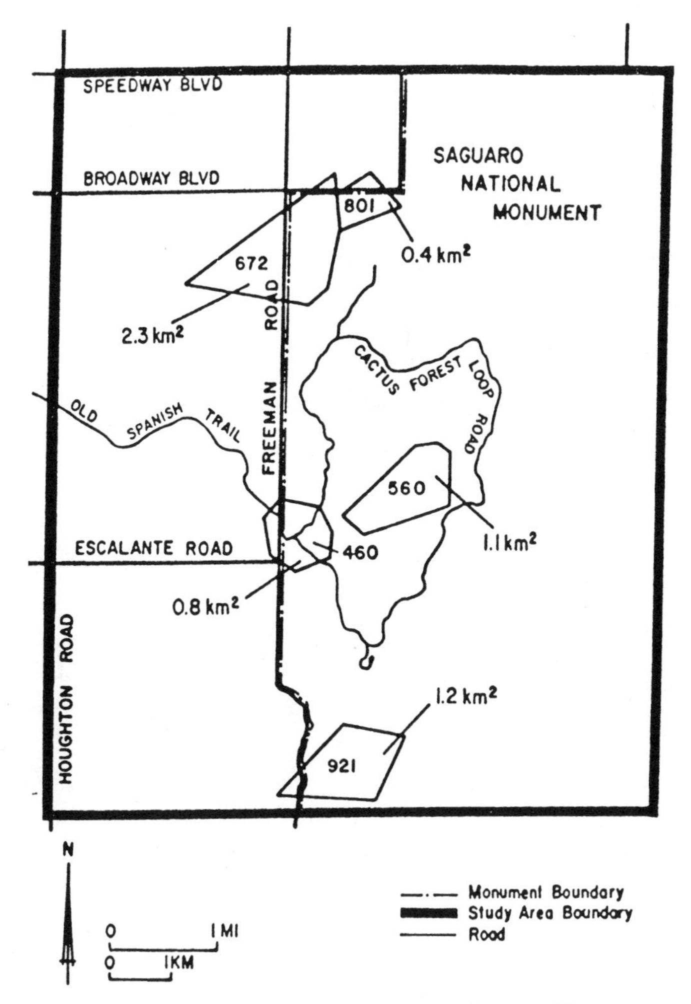

Figure 9.3 Minimum-convex-polygon estimates of seasonal home-range size (km²) for five collared peccary herds, Saguaro National Monument, 1988–1989 (from Bellantoni and Krausman 1991).

observations of wildlife revealed that at current development levels, many species of wildlife use the low-density residential lands near the monument (Fig. 9.4). Furthermore, interactions with some kinds of wildlife are encouraged by many residents. Sixty percent of the households in this sample feed wild birds, and 31% feed other kinds of wildlife, including collared peccary (12%) and coyotes (11%).

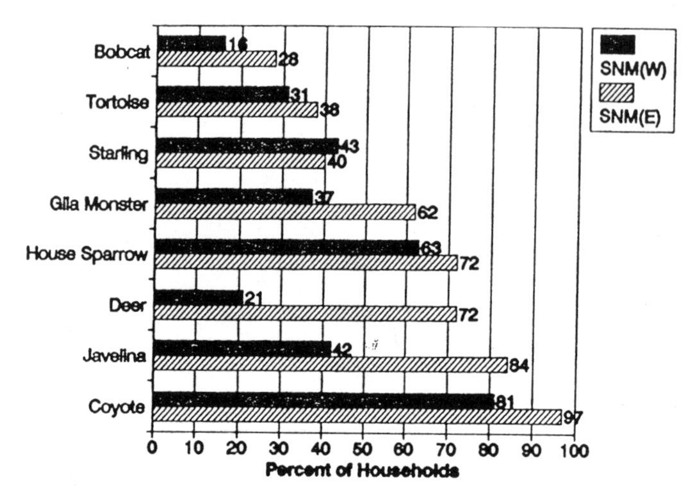

Figure 9.4 Percent of households within 1.6 km (1 mi) of Saguaro National Monument (SNM) that have observed selected species of wildlife on their property within the past 12 months (survey conducted in 1991).

Other questions dealt with the neighbors' use of Saguaro and with their attitudes toward the monument and alternative kinds of developments in the area (Fig. 9.5).

Domestic Dogs and Cats Several methods were used to study the populations of dogs and cats near Saguaro and their potential effects on the wildlife populations in this monument. The residential survey included several questions that dealt with pet ownership. The average household within 1.6 km of Saguaro has 1.5 dogs and 0.9 cats. From this data, we estimated that at existing levels of development, 2,471 dogs and 1,728 cats reside within 1.6 km of the monument (Goldsmith et al. 1991).

We collected 113 prey items from cat owners and found that domestic cats consumed a variety of small animals, including mammals, birds, lizards, and snakes. Sixty-two percent of the prey items were mammals, 27% birds, and 11% reptiles.

To study the movements of domestic cats, we equipped 16 cats from 11 homes with radio-collars and monitored their behavior. Our study, as well as others, documented the role of cats (both domestic and feral) as predators on wildlife. However, we found no evidence of feral cats in or near

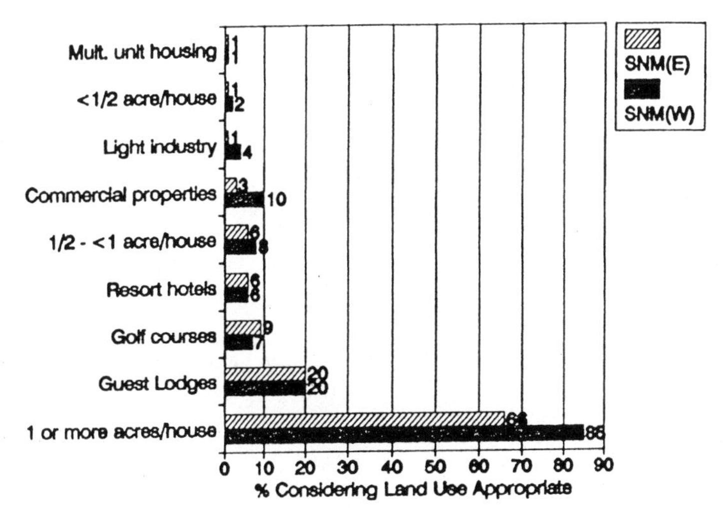

Figure 9.5 Attitudes of Saguaro National Monument (SNM) neighbors toward alternative land uses in their neighborhood.

Saguaro, and the domestic cats we studied seldom left the immediate vicinity of their homes. One possible explanation for this behavior is the large population of coyotes along the monument border. Groups of eight or more coyotes were frequently observed in these neighborhoods. One of our collared cats wandered away from its home and was killed by a predator, and many of the respondents in the residential survey reported losing cats to coyotes. Analysis of the activities of cats suggests that even with increased urbanization, the impacts of cats on Saguaro wildlife will be limited to the immediate vicinity of homes (Fig. 9.6).

Although groups of dogs were frequently seen by monument staff, these were believed to be domestic pets that return to their owners' homes. We found no evidence of feral populations of dogs. However, groups of domestic dogs are known to harass and kill wildlife, and Saguaro staff reported that a group of nine dogs threatened a hiker in the monument.

Management Implications

The fundamental premise for research in national parks is the idea that improved information will enable park managers to do a better job as stewards

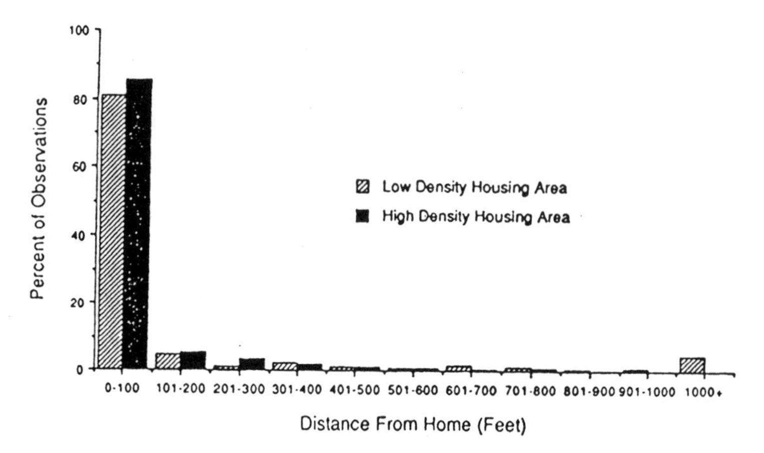

Figure 9.6 Observations of radio-collared cats in terms of distance from cats' homes (from Goldsmith et al. 1991).

of these public trusts. Obviously, one essential type of information concerns the status and condition of natural resources within a protected area. However, as the pressures for change and development on lands adjacent to parks increase, park managers must expand their research perspective and ask questions about these external pressures and their implications for park resources. Furthermore, the phenomenon of urban encroachment can never be fully described in physical and biological terms. For this reason, research must look beyond natural resources and consider the sociological, political, and economic ramifications of urban growth.

Research focusing on adjacent lands and their inhabitants can assist park managers in several ways. The first and most obvious value of this research is to assist managers in assessing and mitigating the impacts of urban pressures on a protected area. For example, at Saguaro, studies of mule deer movements across the monument boundary suggest that the monument's deer herd is dependent on outside sources of water during drought periods. If park managers are to maintain the deer populations that reside primarily within the monument, they must ensure that critical areas outside the monument remain undeveloped or provide new sources of water inside or near the monument. Similarly, studies of desert tortoises revealed that tortoises within the monument may occupy home ranges up to 1 km across. Therefore, development adjacent to the monument may affect tortoise populations well within the protected area. Park managers can use this information in working with developers and city planners to ensure appropriate setbacks or low-density zoning on lands near the monument's tortoise populations.

A second value of these studies is that they provide the park manager with leverage in the arena of competing political and economic interests that characterize urbanizing lands. For example, the results of the deer and tortoise research cited above came as no surprise to biologists and park managers. Mitigation strategies based on these studies, however, may have economic implications for private landowners. Even with the philosophical support of local politicians, meaningful protective measures are far easier to justify when scientific research documents their need.

These investigations have also assisted park managers in their relationships with park neighbors. By understanding the sociodemographic characteristics as well as the attitudes and beliefs of these people, park managers are better able to enlist local support for their programs. At Saguaro, free-running dogs are a serious concern of park managers. The residential survey not only provided estimates of the number of dogs in the region but also revealed that there is general support for the enforcement of leash laws, even among dog owners. Conversely, the study of residents combined with a study of cat movements revealed that unlike many other areas, domestic and feral cats are not a major threat to the monument's wildlife populations.

Finally, a comprehensive, research-supported, knowledge base has been invaluable in supporting efforts to enhance and protect the natural heritage preserved in Saguaro National Monument. Armed with research information, a coalition of local and national interests combining developers, neighborhood associations, local and national politicians, and local and national conservation groups has been successful in obtaining congressional legislation to authorize expansion of the monument's boundaries. This expansion will enhance Saguaro by encompassing superlative areas adjacent to the monument that are threatened by imminent development. The defensible knowledge base was a major factor in reaching the broad-based consensus that supported this expansion and ultimately led to congressional authorization in 1991.

Conclusions

The boundary of any protected area is an imperfect barrier to the flow of water, air, nutrients, energy, plants, and animals. The dynamics of this interchange are most pronounced when adjacent lands are developed for intensive human use, such as in cities. Urban encroachment represents a threat to the natural-resource-protection objectives of parks and reserves throughout the world.

To address the issues created by urbanization of adjacent lands, park managers need a variety of information. Basic information about the status of park resources and a system for monitoring changes in these resources are essential. In addition, park managers need information about the behavior and attitudes of their human neighbors.

By bringing together specific but related studies of land uses, land covers, wildlife movements, and human activities, investigators at Saguaro National Monument have developed a multidisciplinary database for evaluating and monitoring the effects of urbanization.

Acknowledgments

The studies described in this paper were funded by the National Park Service, Southwest Parks and Conservation Association, Pima County, and the city of Tucson.

Note

1. On 14 October 1994, the designation was changed to Saguaro National Park.

Literature Cited

Bellantoni, E. S., and P. R. Krausman. 1991. Habitat use by desert mule deer and collared peccary in an urban environment. Technical Report 42. Cooperative National Park Resource Studies Unit, University of Arizona, Tucson. 39 p.

———. 1992. Movements of mule deer in collared peccaries in Saguaro National Monument. P. 181–193 in C. P. Stone and E. S. Bellantoni, eds. Proceedings of the Symposium on Research in Saguaro National Monument, 23–

24 January 1991. National Park Service, Rincon Institute, and Southwest Parks and Monuments Association.

Bibles, B. D. 1991. Is there competition between exotic and native cavity-nesting birds in the Sonoran Desert?: an experiment. Unpubl. M.S. thesis, University of Arizona, Tucson. 30 p.

Boecklin, W. J., and D. Simberloff. 1986. Area-based extinction models in conservation. P. 247–276 in D. K. Elliott, ed. Dynamics of Extinctions. John Wiley & Sons, New York.

Dillman, D. A. 1978. Mail and telephone surveys: the total design method. John Wiley & Sons, New York. 325 p.

Doll, M., R. Ellis, R. Hayes, R. Sidner, H. McCrystal, R. Davis, and R. Hall. 1989. A checklist of the herpetofauna and mammals of Saguaro National Monument. Southwest Parks and Monuments Association. 2 p.

Environmental Systems Research Institute, Inc. 1990. PR ARC/INFO Version 3.4D. Redlands, California.

Goldsmith, A., and W. W. Shaw. 1988. Final report on desert tortoise ecology. Southwest Parks and Monuments Association. 13 p.

Goldsmith, A., W. W. Shaw, and J. Schelhas. 1991. The impacts of domestic dogs and cats on the wildlife of Saguaro National Monument. Unpubl. report., National Park Service, Western Region. 27 p.

International Union for the Conservation of Nature. 1980. World conservation strategy. Gland, Switzerland. 44 p.

Leedy, D. L., R. M. Maestro, and T. M. Franklin. 1978. Planning for wildlife in cities and suburbs. Report 331. Planning Advisory Service, American Society for Planning, Chicago. 64 p.

Lowe, C. H., and P. H. Holm. 1991. The amphibians and reptiles at Saguaro National Monument, Arizona. University of Arizona, Cooperative Park Studies Unit, Technical Report 37. 20 p.

MacArthur, R. H., and E. O. Wilson. 1967. An equilibrium theory of insular avifaunas with special reference to the California Channel Islands. Condor 76: 370–389.

Mannan, R. W., and B. D. Bibles. 1990. Impacts of exotic cavity-nesting birds in Saguaro National Monument. Unpubl. report in Saguaro National Monument files. 29 p.

McNeeley, J. A., and K. R. Miller. 1983. IUCN, national parks and protected areas: priorities for action. Environmental Conservation 10:13–21.

Miller, R. I., and L. D. Harris. 1979. Predicting species changes in isolated wildlife preserves. P. 79–82 in R. M. Linn, ed. Vol. 1 of Proceedings of the 1st Conference on Science Research in the National Parks, New Orleans, Louisiana, 9–12 November 1976. National Park Service Transactions and Proceedings Series 5.

National Parks and Conservation Association. 1979. NPCA adjacent lands survey: no park is an island. National Parks Conservation Magazine 53(3):4–9.

Newmark, W. D. 1987. A land-bridge island perspective on mammalian extinctions in western North American parks. Nature 325:430–432.

Shaw, W. W., J. Burns, and K. Stenberg. 1986. Wildlife habitats in Tucson: a strategy for conservation. University of Arizona, School of Renewable Natural Resources. 17 p.

Shaw, W. W., J. Davis, S. Drake, and A. Goldsmith. 1991. A Geographic Information System data base for the natural resources of Saguaro National Monument and vicinity. Unpubl. report in Saguaro National Monument files. 10 p.

Shaw, W. W., A. Goldsmith, and J. Schelhas. 1992. Studies of urbanization and the wildlife resources of Saguaro National Monument. P. 173–180 in C. P. Stone and E. S. Bellantoni, eds. Proceedings of the Symposium on Research in Saguaro National Monument, 23–24 January 1991. National Park Service, Rincon Institute, and Southwest Parks and Monuments Association.

Shaw, W. W., and W. R. Mangun. 1984. Nonconsumptive use of wildlife in the United States. U.S. Fish and Wildlife Service Resource Publication 154. 20 p.

Southwest Parks and Monuments Association. 1985. A checklist of the birds of Saguaro National Monument. 2 p.

Stenberg, K., and W. W. Shaw, eds. 1986. Integrating wildlife conservation and new residential developments. Proceedings of national conference, 20–22 January 1986. University of Arizona, School of Renewable Natural Resources. 203 p.

Tucson Trends. 1986. Tucson, Arizona: Valley National Bank and Tucson Newspapers, Inc. 96 p.

U.S. Army Construction Engineering Research Laboratory. 1989. GRASS 4.0. Geographical Information System software.

Wilcox, B. 1980. Insular ecology and conservation. P. 95–117 in M. E. Soule and B. A. Wilcox, eds. Conservation Biology and Evolutionary-Ecological Perspective. Sinaur Associates, Inc., Sunderland, Massachusetts.

Wilson, E. O., and E. O. Willis. 1975. Applied biogeography. P. 522–534 in M. L. Cody and J. M. Diamond, eds. Ecology and Evolution of Communities. Belknap Press, Cambridge, Massachusetts.

10

Karst Hydrogeological Research at Mammoth Cave National Park

E. Calvin Alexander, Jr.

The National Park Service (NPS) mandate to protect and preserve the physical, biological, and historical resources placed in its care necessitates a close collaboration between research and resource management. Each unit in the national park system requires complex, interdisciplinary approaches to be successfully managed. Mammoth Cave National Park is a prime example of how complicated this process can become. At Mammoth Cave, NPS manages both the intricate surface ecosystems and the underground Mammoth Cave system. This network of cave passages and underground rivers supports its own rich and diverse flora and fauna. These biota are intimately interconnected in complex ecological relationships and are coupled with conditions on the land surface. The management of such resources is dependent on knowledge of the biological and physical aspects of both the surface and subsurface environments. The research to gain this knowledge is, in turn, dependent on the park managers to protect and preserve this unique and fascinating resource.

Mammoth Cave National Park, the Mammoth Cave system, and the Central Kentucky Karst include the longest cave and best documented karst system in the world. Strong traditions of world-class, long-term research, exploration, and monitoring at the park exist within several scientific disciplines. International recognition of Mammoth Cave's unique size and importance led to the park being designated a World Heritage Site in 1981, the Mammoth Cave region being designated an International Biosphere Reserve in 1990, and the underground rivers of the region being designated Outstanding Resource Waters by the Commonwealth of Kentucky.

Geological Setting of the Park

Mammoth Cave National Park lies within a 2,500-km^2 region known as the Central Kentucky Karst, which is at the axis of the easternmost extremity of the Illinois Basin. This region contains more than 720 km of mapped cave passage. Some long-time workers in the area believe that at least 1,600 km of cave will be humanly passable when all such passages are discovered and mapped. More than 520 km of passage have been mapped within the Mammoth Cave system. The park contains 21,332 ha, approximately 74% of the amount originally authorized. The town of Park City is south of Mammoth Cave National Park, and the towns of Cave City, Horse Cave, and Munfordville are east of it. All of the towns lie along Interstate 65.

The Green River flows west through the park, dividing it into "the north side of the river" and "the south side of the river." The north side of the river is mostly hilly topography characterized by surface drainage with some karst features. The surface is primarily underlain by clastic rocks, though limestone is exposed in some of the valley floors. The south side of the river, in the Chester Cuesta, is composed of steep-sided ridges of limestone capped with sandstone. The intervening valleys are limestone-floored, dry, and typically pocked with numerous sinkholes. There is no surface drainage except after extreme precipitation events, but even then, the flow is rapidly diverted underground. Part of the park area to the east and south is bounded by the Chester Escarpment, which is about 60 m high. Beyond the Chester Escarpment, to the east and south, is the Sinkhole Plain, an area characterized by numerous sinkholes and no surface drainage—consequences of the extensive underground drainage network (Palmer 1981).

Palmer (1981) summarized the stratigraphy of the Mammoth Cave region. The Sinkhole Plain is bounded on the south by the less soluble, lower third of the St. Louis Limestone. This lower third, plus the subjacent Salem-Warsaw Limestone, composes the Glasgow Upland (Quinlan 1970; Quinlan et al. 1983). The ridges in the Chester Upland are capped by the Big Clifty Formation (chiefly sandstone) and flanked by the Girkin Formation (mostly limestone) and the Ste. Genevieve Limestone. Most of the Sinkhole Plain is underlain by the upper two-thirds of the St. Louis Limestone. The Mammoth Cave system and other caves in the region are formed in the Girkin Formation, Ste. Genevieve Limestone, and upper third of the St. Louis Limestone. In general, surface water flows across the lower third of the St. Louis Limestone and sinks into small depressions. The water then flows through the aquifer to the Green River (the regional base level). Regionally, the rock units dip toward the north and northwest at an average of approximately 9 m/km.

Quinlan and Ray (1981, 1989) used dye tracing and mapping of the potentiometric surface to delineate 27 groundwater basins and 7 subbasins south of the Green River. The Mammoth Cave system occupies all or a portion of six groundwater basins. Most of Turnhole Spring groundwater basin, one of the larger basins in the region, lies outside the park boundaries but flows through part of the Mammoth Cave system and discharges into the Green River within the park. The groundwater basins north of the Green River are just now being characterized.

Research and Resource Management

Despite their obvious interdependence, cooperation between research and resource management at Mammoth Cave National Park has historically varied from close to contentious. When conflicts arose, it was usually because research is a process of discovering facts and relationships, whereas management is a process of decision making based on objectives, values, and priorities. Personality clashes between individual managers and scientists also played important parts in the conflicts. Finally, within the Mammoth Cave region, an atmosphere of conflict, suspicion, and competition between the residents and park personnel affects the efforts of every superintendent and scientist.

The success of a superintendent or scientist working in the Mammoth Cave system often depends on how well and how quickly regional conflicts are recognized and addressed. Conflicts arise because of five factors: (1) the local population's bad feelings about the way the park was created, (2) the organizational structure of NPS, (3) the involvement of the concessionaires and their interests in the park, (4) the dynamic nature of the natural resources, and (5) the growing knowledge base generated by scientific research.

The origins of some of the conflicts that mold much of the karst hydrogeological research are discussed below as necessary background to a review of the research itself. Despite these pitfalls and conflicts, much scientific work has been and continues to be accomplished at the park, and the management implications of that work are being incorporated into park operations.

This chapter examines the discovery, documentation, and scientific understanding of the Mammoth Cave system and its karst aquifer. It also evaluates the varying roles of NPS management[1] in the ongoing research process. In preparing for this chapter, my colleagues and I reviewed the published results of scientific research, written plans for future research projects, and historical accounts of Mammoth Cave National Park. We also conducted interviews during the summer of 1991 with about two dozen individuals who are or have been active at the park. (See Alexander [1993] for an expanded account of this review.)

The Creation of the Park

Mammoth Cave was discovered by European settlers in 1798 or 1799. The cave has been shown as a commercial tourist attraction since 1816, beginning with various private owners. For more than a century, Mammoth Cave and surrounding show caves were operated as private businesses, while generations of people lived on and farmed the land overlying the entire cave system.

In 1924 the Mammoth Cave National Park Association (MCNPA), a private subscription organization, was founded by Kentuckians interested in creating a national park (Goode 1986). In 1926 the U.S. Congress authorized the establishment of Mammoth Cave National Park (44 Stat. 635). The maximum authorized area was to be 70,617 a. (28,578 ha), provided that at least 45,309 a. (18,336 ha) including and surrounding Mammoth Cave were donated in fee-simple title to the federal government before the park was established (National Park Service 1983). The impression of the local residents at the time was that "the government" (rather than the private citizens of MCNPA) was buying this land at "force out" prices (Murray and Brucker 1979).

The land acquisition did not proceed as quickly as MCNPA had expected. In 1928 the Commonwealth of Kentucky created the Kentucky National Park Commission and granted this commission the right of eminent domain.

Initially, the commission had very limited success in county courts in its attempts to acquire land, and it lost many cases. In 1934 NPS assumed the obligation to acquire the necessary land. NPS moved the cases into federal courts and successfully used eminent domain to complete the land acquisition (Sides 1991). Mammoth Cave National Park was formally dedicated on 18 September 1946.

The use of eminent domain created major local animosity toward the national park concept. Local landowners who had "beaten" park proponents in county courts had to fight NPS in federal court—and lost. The final result was that approximately 500 families (Goode 1986)—more than 2,000 people—were displaced from the area that became Mammoth Cave National Park. Many current residents are related to those original families. The intervening years may have dimmed the resulting animosity toward and distrust of "the government" in general, and NPS in particular, but their memory is recalled whenever NPS does something unpopular (Smith 1967; Meloy 1979; Murray and Brucker 1979).

NPS Organizational Structure

Several organizational characteristics of NPS are critical to how the resources of the national park system are managed. According to Smith (1968) and Everhart (1983), the basic organizational structure introduced when NPS was established in 1916 is fundamentally unchanged today. In NPS, responsibility for administration is vested with rangers, who advance up the line of command, whereas staff functions are filled by naturalists, engineers, scientists, and other specialists, who make recommendations about management. Rangers typically hold degrees in a variety of subjects and have skills in a variety of activities. In contrast, staff officers include scientists and interpreters who have professional training and experience in specific disciplines, but rarely at higher than the B.S. or B.A. level.

Smith (1968) cited four reasons for the initial adoption of such a line-and-staff system of administrative management and control. First, before NPS was established, park units were controlled by the military. Some of the park personnel, who were then in the military, transferred to NPS. Second, when NPS was organized, its leaders found that veterans of the Spanish-American War and World War I were among those who were the most qualified to administer what were then remote areas. Third, protection by a quasi-military ranger force was a major responsibility in the earliest days of NPS because the parks were remote and difficult to reach. Fourth, NPS was established in an era when the regulatory function of the federal government was in vogue.

To ensure more loyalty to NPS than to the individual park units, NPS adopted the military's philosophy of transferring key personnel every 3 to 4 years. Such transfers have been and continue to be the key to promotion. Everhart (1983, p. 56) stressed,

> Undeniably, the threat of crime has caused a substantial shift of emphasis toward the direction of law enforcement. Rangers now receive 400 hours of intensive law enforcement training at the Federal Law Enforcement Training Center, plus periodic refresher courses.... At a conference of Park Superintendents, the moderator of a session on law enforcement noted that "hardly anything has been more talked about" within the Park Service, concluding that "our people are undertrained in resource management and overtrained in law enforcement. A tremendous imbalance has been created."

Science in NPS is performed by staff officers with no command authority. Science in NPS in general and at Mammoth Cave in particular is supervised by the superintendent, who is typically a line officer and not a scientist. One important result of this organizational structure is that, if a conflict arises between management and science, the management command structure takes precedence. Exceptions to this chain of command are extremely rare.

Concessionaire Interests at the Park

National Park Concessions, Inc. (NPC), the main concessionaire at Mammoth Cave and four other national parks, developed directly out of the Kentucky National Park Commission. The individuals who were active in the commission helped to form NPC. In this sense, the concessionaire predates the park. NPC's corporate headquarters were within the park until about 1980, when they were moved to Cave City.

NPC, like concessionaires operating in other national park units, is essentially a protected monopoly. (Public Law 89-249, Section 5, governs the preferred renewal of contracts with the concessionaires.) NPC has become entrenched in the Mammoth Cave region with successive generations of employees. It maintains deliberate ties to the regional community and its congressman. NPC policy has been to employ residents of each county containing or surrounding the park. These local people, in effect, created a network to protest strongly any actions by park management that conflicted with NPC interests. Successive park superintendents have said that their toughest battles were waged against NPC.

NPC interests and activities have often opposed NPS efforts to manage and protect the resource. However, because of the protected monopoly within the park, NPC has enormous powers, allowing it to take actions that the park superintendent may not be able to control. Some of these actions have been described by Frome (1992, p. 174–175).

Dynamic Resources

Occasional visits to Mammoth Cave might lead one to believe that it never changes. However, regular visits and an intimate knowledge of the resource lead one to realize that it constantly changes and that the cave and park are dynamic, not static, resources. The resource is being simultaneously created and destroyed by the same processes. Cave passages and speleothems are created by flowing water, which can also hasten their destruction.

The Growing Knowledge Base

Scientific studies and systematic cave-mapping projects created a growing base of knowledge that documented the dynamic resources of the park and cave system. As that knowledge base grew in size and sophistication, it became clear that

1. the dynamic resource is changing at a greater rate than initially thought,
2. the resource is much larger than originally thought and is an integral part of a system extending well beyond the boundaries of the park,
3. the resource is more threatened and more fragile than originally thought.

This growing knowledge base has also revealed that both the Mammoth Cave system and the park are regularly affected by a variety of human activities occurring in the region but outside park boundaries. This discovery destroyed the myth that Mammoth Cave could be managed as an isolated, independent unit. From the manager's perspective, scientists and explorers continually increase the complexity of management as new information is discovered.

Scientific Research Modes

Hydrogeological research[2] at Mammoth Cave is divided into four modes: (1) individual scientists and their students; (2) self-directed volunteer groups, such as the Cave Research Foundation (CRF) and others; (3) an NPS research geologist and other NPS employees; and (4) various other government and private agencies. These four modes overlap, and individual projects and scientists commonly fit into two or more categories simultaneously. However, each research mode has different characteristics and products.

Individual Scientists

Individual, non-NPS scientists (primarily academics and often members of CRF) have made long-term career commitments to study the resources of the Mammoth Cave system. Academic scientists often obtain funding and funnel graduate students into appropriate research projects in the park and region to continue the professor's long-term research program. Examples of this include the following:

1. the hydrogeological, geomorphological, and mineralogical work in the Central Kentucky Karst by William White and his students and associates (1960–present), which culminated in two dissertations and in the White and White (1989) book summarizing karst geomorphology and hydrology of the park;

2. the geological, stratigraphic, speleogenetic, and hydrogeological research in the Mammoth Cave system by Arthur and Margaret Palmer, 1960–present (Palmer 1981; 1989a,b);

3. the geomorphic and stratigraphic studies by Pohl (1955) and Quinlan (1970);

4. the hydrogeological research in the Mammoth Cave system by Ralph Ewers and his students (Meiman et al. 1988; Recker et al. 1988; Ryan 1992).

Volunteer Organizations

For decades, a major part of the scientific research, the bulk of the resource inventory, and most of the cave exploration and mapping in the Mammoth Cave system has been done by volunteers. This volunteer pool is international in origin and contains individuals from many different professions. Their efforts are largely self-motivated, self-directed, and self-financed. The quality and professionalism of this volunteer work is excellent.

The lead organization in the volunteer effort is CRF. CRF has been primarily responsible for (1) logistic, financial, and structural support for much of the scientific effort in the system; (2) systematic cave exploration and mapping, chiefly within the park; and (3) coordination with other volunteer groups in integrating the segments of Mammoth Cave into the extensive system it is currently known to be. Other volunteer organizations involved with the Mammoth Cave region include the Central Kentucky Karst Coalition, the Fisher Ridge Project, the North Shore Project, and various chapters of the National Speleological Society. In addition, there have been numerous individual, independent volunteers.

NPS Scientists

A research geologist, James F. Quinlan, was hired by Mammoth Cave National Park in July 1973. The research geologist coordinated all research activities in long-term studies of regional karst geomorphology, speleology, park and regional hydrology, geochemistry, petrology, mineralogy, sedimentology, cave climatology, and paleontology. He also performed geological studies and established monitoring systems for physical resources, including water-drainage patterns. He was to receive general administrative supervision from the superintendent and maintain close ties with the regional chief scientist, but was to have wide latitude for professional, independent judgment and action. His tenure at Mammoth Cave was scientifically very productive and is reviewed herein.

A new research structure was established at the park in the fall of 1988, and the research geologist resigned from NPS in 1989. The research position was then assigned to the regional chief scientist. In October 1990, park managers asked the Southeast Regional Office to establish a hydrologist position at the park. In November 1990, the office approved this position in lieu of a research scientist, and the position was filled in January 1991.

Other Agencies

State and federal government agencies and private industry have also performed research in the Mammoth Cave system. Examples of this include the first dye trace in the region, performed by Anderson (1925) of Louisville Gas and Electric Co.; geological mapping efforts by the U.S. Geological Survey (USGS); and contract hydrogeological studies by USGS (Cushman et al. 1965). Most of the USGS efforts, however, were part of larger state and federal programs and involved the park only because it was in the study area. These studies provided useful information on Mammoth Cave and the Mammoth Cave system. An example of these USGS efforts is Plebuch et al. (1985).

Summary

Long-term research and resource monitoring at Mammoth Cave National Park have primarily been accomplished through the personal dedication of individuals, be they independent CRF members or NPS employees. These professionals represent a significant reservoir of knowledge and expertise that are available to NPS. The following overview is limited to the principal investigators. Space limitations prevent the summary from being all-inclusive.

Review of Research

The exploration, mapping, and scientific study of the Central Kentucky Karst, including the Mammoth Cave system, have been at the leading edge of U.S. speleology and karst hydrogeology for almost two centuries. Cave exploration, mapping, and scientific observations in the area began in the 1800s. Scientific research began in the 1920s and continues to the present.

The research from the 1950s to 1991 may be divided into three periods: (1) work done from the 1950s to 1973, before the NPS research geologist was hired; (2) 1973 to 1989, when he worked at the park; and (3) 1989 to 1991 (when this chapter was written), after he left the park.

1950s–1973

The first speleological research plan for the Mammoth Cave region was outlined by CRF in 1960 (Cave Research Foundation 1960). It included exploration, cartography, geology, hydrology, and geochemistry.

Various aspects of the hydrological research were performed by Will White, his graduate student, Jack Hess, and graduate students recruited by CRF from other institutions. These studies were conducted primarily from 1957 to 1974 and can be traced through CRF's Annual Reports (1960–1973); see also Deike and White (1969). Most of the work described by White and White (1989) was completed during this period. The reviews by Quinlan (1970) and White et al. (1970) are useful syntheses of the information obtained in the late 1960s.

Fieldwork was completed for three significant doctoral dissertations: Deike (1967), a study of cave-passage development, especially as potentially influenced by vertical or near-vertical joints and fractures; Hess (1974), a study of changes in water quality in springs and surface streams as related to storm events and seasonal distribution of rains (Hess and White 1988, 1989, 1993); and Miotke (1975), a study of regional geomorphology, water quality, and groundwater movement.

White (1969) contributed the beginnings of an American synthesis concerning the kinds and characteristics of karst aquifers, an aquifer type little understood at that time. S. G. Wells, F.-D. Miotke, and H. Papenberg (Miotke and Palmer 1972) performed the first successful, modern dye traces in the region and also provided the first demonstration of the hydrological relationship between the Sinkhole Plain and the development of the Mammoth Cave system.

The results of the work done from the 1950s to 1973 provided the foundation for the research conducted after 1973.

1973–1989

Results As measured by the number of significant hydrogeological discoveries made, their practical applications in the Mammoth Cave region and carbonate terranes elsewhere in the world, and the number of publications written (more than 120), which have been widely cited, the 16-year period from 1973 to 1989 was extremely productive. A summary of the research geologist's work is given by Quinlan and Ewers (1989).

The research geologist made six significant accomplishments on behalf of NPS. First, he demonstrated that heavy-metal-laden sewage in an underground river beneath the nearby town of Horse Cave flows 7 km via an underground distributary (like the mouth of a major river) to 46 springs at 16 locations along an 8-km reach of the Green River, rather than to just 1 or 2 springs (Quinlan and Rowe 1977; Quinlan and Ewers 1981a, 1985; Quinlan et al. 1983; Quinlan 1990). Recognizing the possibility of distributary flow and understanding how to test for it are important for spill response and for designing groundwater-monitoring systems in limestone and dolomite (carbonate rocks).

Second, he delineated, by dye tracing and potentiometric mapping, 27 groundwater basins south of the Green River (Quinlan and Ray 1981, 1989) and more than 20 basins north of the Green River (J. F. Quinlan, unpubl. data). This work is based on more than 450 tracer tests with fluorescent dyes, mapping of the potentiometric surface (water table) from about 2,000 water-level measurements in wells within a 3,000-km^2 area, and mapping of approximately 100 km of cave passage. These published maps also show inferred groundwater flow paths, springs, and cave passages.

The maps are routinely used by NPS, the U.S. Environmental Protection Agency (EPA), state agencies, and consultants to identify catchment areas that might adversely affect the water quality of any of the springs and cave streams shown. They are also routinely used to predict not only the dispersal routes of hazardous materials that might be discharged into the ground or spilled accidentally, but also the water supplies that might be polluted by such releases. For example, the maps show that Mammoth Cave and its unique, fragile biota could be adversely affected by anything that pollutes water in any of several groundwater basins.

The 1981 map was used in the design of a $13-million, regional sewage system to service the park and adjacent towns and to protect park resources. The two maps (Quinlan and Ray 1981, 1989) demonstrate that troughs in the potentiometric surface correspond to zones of maximum groundwater flow; the coincidence of major underground rivers with such troughs in four areas strengthened that conclusion. These maps set a new standard for the mapping and management of groundwater drainage basins in karst regions.

Third, he determined that waste water from Park City (which has no sewers) drains into springs along the Green River within the park. He also recognized the mechanism by which much of this sewage sometimes drains, via high-level overflow routes, into portions of Mammoth Cave visited by tourists (Quinlan et al. 1983; Quinlan and Ewers 1989).

Fourth, he documented the sequential evolution and capture of groundwater basins (Quinlan and Ewers 1981b).

Fifth, he synthesized observations in the Central Kentucky Karst and other karst terranes and used them to develop new methods to monitor groundwater quality in aquifers (Quinlan and Ewers 1985; Quinlan 1989, 1990). He showed that springs, cave streams, and wells that tracer tests have identified as drainage sites for a specific area are the most logical, efficient, reliable, and economical places to monitor for pollutants in carbonate terranes.

Sixth, he developed a strategy for responding to spills of hazardous materials along Interstate 65 and the nearby railroad right-of-way (Quinlan 1986). This strategy is applicable anywhere in the park area and nationally.

This period was marked by the completion of Ralph Ewers's dissertation on the origin and development of caves (Ewers 1982; summarized by Dreybrodt 1988, p. 223–229) and by the development of new techniques for monitoring karst aquifers. During this time, Palmer (1981, 1989a,b) also made definitive syntheses of cave and landscape development in the Mammoth Cave area that are relevant to its paleohydrology.

The personalities, research interests, and talents of James Quinlan and Ralph Ewers were complementary. They often collaborated and jointly published research results and guidebooks. One of their collaborative projects was the design, development, testing, and installation of a unique, state-of-the-art Karst Water Instrumentation System (KWIS) to continuously monitor stage, velocity, specific conductivity, and temperature of surface and cave streams that drain into the park and a spring along the Green River. Meiman et al. (1988) and Ewers (1994) combined these digitally recorded data with data from a rain-gauge network and soil-moisture sensors and interpreted them through computer-aided analysis.

Collection and interpretation of such data are the only reliable ways to determine a karst aquifer's reaction to storm events and to develop valid computer models of groundwater flow within the aquifer. This technique is the one most likely to yield maximum understanding of an unconfined carbonate aquifer and its response to storm-related transport of pollutants. It has been emulated by many researchers worldwide. The KWIS project was funded for 6 years by several cooperative agreements between NPS and Eastern Kentucky University. Four KWIS units were operated and at least four more were scheduled for construction, but the superintendent unilaterally terminated this research just as it was coming to fruition.

Quinlan summarized his own research plan as being in four phases.

1. Phase I: dye tracing to (*a*) delineate groundwater basins north and south of the Green River and (*b*) refine the results of such dye-tracing. Stream caves outside the park[3] would be mapped to understand how their plumbing systems functioned.

2. Phase II: regular chemical analysis of water quality at selected springs and cave streams to (*a*) determine the natural variations that occur; (*b*) interpret the chemical and mixing processes that occur in the aquifer between where water and pollutants enter the ground and where they are discharged at springs, which would allow researchers to predict the dilution of spilled materials; and (*c*) calibrate instrumentation used for monitoring water quality and thus gain greater reliability for interpretation of water-quality data and the dynamics of storm events and pollutant dispersal.

3. Phase III: specific geophysical techniques to (*a*) locate conduits from the surface (Lange and Quinlan 1988), (*b*) find those that are inaccessible to human inspection and mapping, and (*c*) identify the best locations to place environmental monitoring instruments in the cave rivers.

4. Phase IV: computer-aided mathematical analysis and interpretation of data to (*a*) develop a model describing flow in karst aquifers and (*b*) use that model to understand hypothetical aquifer behavior and plan for spill response. (At that time, no one had succeeded in successfully modeling flow in karst aquifers; see Huntoon [1995].)

Significance of Research Many of the research projects conducted between 1973 and 1989 are summarized in the superintendent's Annual Research Reports and in the CRF Annual Reports. The North Shore Task Force, initially a group of recreational cavers, expanded its interests to scientific research (George 1989).

The significance of the NPS-sponsored research has been as multifaceted as the research itself. The research has proven its value in the protection and management of the Mammoth Cave system and has won national and international recognition. The principles of groundwater movement in a karst terrane that Quinlan and Ewers discovered during this time are being applied to the protection of many other karst aquifers worldwide (Quinlan and Ewers 1985; Quinlan 1989, 1990; American Society for Testing and Materials 1995). This applicability is also true for the methods developed to conduct the research: those methods serve as guides for many other karst researchers (Quinlan 1982; Meiman et al. 1988; Ewers 1994).

Quinlan and Ewers were awarded the Geological Society of America's 1986 E. B. Burwell, Jr., Memorial Award for one of their papers (Quinlan and Ewers 1985). This award is given annually for "a published paper of distinction which advances knowledge concerning the principles and practice of engineering geology." Because approximately 20% of the United States is underlain by karst terrane, the paper influenced research efforts not only in Mammoth Cave and NPS, but also nationally (Quinlan 1989, 1990) and internationally. The groundwater-monitoring concepts espoused in this paper have been and are being used in regulations adopted by EPA and numerous states. They have also been incorporated into national standards for groundwater monitoring (American Society for Testing and Materials 1995).

The research geologist's work has also contributed to scientific knowledge. Five concepts of groundwater movement, previously recognized in other karst terranes, are now described more fully in the Mammoth Cave region than anywhere else. These concepts are (1) distributary flow; (2) shunting of water by high-level overflow routes, often into adjacent groundwater drainage basins; (3) shared headwaters of groundwater drainage basins; (4) location of all major stream caves in troughs on the potentiometric surface; and (5) association of all major troughs with axes of trunk drainage in the subsurface (Quinlan and Ewers 1989). His research also led to the development of a new hydrogeological technique: the use of optical brighteners and heavy metals in spring water as a prospecting tool to search for effluent from a sewage-treatment plant (Quinlan and Rowe 1977).

In addition to conducting research, the research geologist was instrumental in causing a shift in research policy. The park superintendent in 1974 held that the research geologist should not conduct research outside park boundaries, even though the research geologist's position description specified that he was to study park and regional hydrogeology. Resolution of the issue by the regional solicitor gave the research geologist permission to perform his studies without regard to park boundaries.

1989–1991

Results The previous 16 years of karst hydrogeological research was largely conceived and implemented by a Ph.D.-level research scientist with broad international experience. In the fall of 1988, the superintendent established a new research structure, under his direction. This structure was initially titled the Office of Science and Resource Management. The Southeast Regional Office formally approved the new structure in January 1989 and elevated it to full division status in January 1991. The division's activities are planned and executed by individuals with B.S. or M.S. degrees and substantially less experience in karst hydrogeology than the previous research scientist possessed.

Other scientists are continuing their research programs in the park and region. Much of this work is not yet published, but several summaries are included in the superintendent's Annual Research Reports and CRF's Annual Reports.

A water-quality monitoring program was initiated in the park in 1990 (Meiman 1990a,b; National Park Service 1990). This program represents the primary thrust of the current karst hydrogeological research funded by NPS. The NPS program was designed to monitor trends in base-flow and event-related water quality. This project's scope includes the identification of (1) existing base-flow ("chronic") and event-related ("acute") water-quality problems in the Green River drainage basin and (2) potential pollution sources and problems. The researchers' goals include determining the level of compliance with government water-quality standards and collecting data to determine current water-quality effects on the biological, aesthetic, and recreational resources of Mammoth Cave.

The primary groundwater basins on the north side of the river have been given the highest priority for water-quality measurement. Researchers will create a generalized aquifer-vulnerability map and will evaluate and modify sampling methods as needed. They will also develop methods to determine whether specific high bacterial "events" are due to sewage. The researchers will determine sample parameter(s) and analytical method(s) to reliably identify public health threats. The park staff hopes to monitor herbicides using the same sampling sites and methodologies.

Significance of Research Although the present water-quality monitoring project seems to implement previous long-range plans, it has serious shortcomings. The methodology, sampling intervals, and analytical standards may not meet current "industry" standards set by EPA, nor do they reflect the current standards of professional karst research. Monthly sampling intervals cannot provide the information necessary to make detailed, accurate, or even representative observations of this exceptionally dynamic aquifer. These intervals can miss significant events on a regular basis. Hence, the generalized studies will not produce the specific products required for resource management and protection (e.g., aquifer-vulnerability maps).

It is too early to rate the effectiveness of the current park research program. Although much of this research focuses on immediate management concerns, useful results are beginning to appear at appropriate professional meetings (Meiman 1990b, 1992; Ryan 1992). However, the change in emphasis from long-term research conceived by highly qualified scientists to short-term research conceived largely by the park superintendent and two scientists with less experience raises concerns about the long-term significance of the current research.

Academic and CRF researchers largely conduct studies without knowledge of NPS needs or priorities. It remains to be seen if NPS, CRF, and other researchers can cooperate and share information.

The Impact of Research on NPS Management

The reactions of superintendents to the research performed in the Mammoth Cave region have depended upon their individual skills, philosophy, and management style. Although no superintendent to date has been a scientist, two superintendents, 1976–1979 and 1979–1985, maximized the results of this research by permitting and encouraging a regional research partnership. The research on dye tracing showed the need for a regional sewage-treatment project to protect groundwater resources, and hence the caves, of both the park and the region. These results led to the creation of the Caveland Sanitation Authority, a regional sewage-treatment agency that serves Horse Cave, Cave City, Park City, and Mammoth Cave National Park. The sewage-treatment system for Horse Cave and Cave City was completed in 1989. The systems for the park and Park City are under construction. Details of this cost-sharing partnership may be found in Mikulak (1988).

The successes of these two superintendents prove that productive partnerships between management and research are possible. In both cases, the superintendents did not try to manage the research. Rather, they participated in setting objectives from the NPS perspective, created supportive policies, and allowed the scientists to manage the details of the research. These superintendents communicated with the research community about pending decisions and sought advice. Although they did not always follow that advice, the perception among scientists was that the managers considered their advice and valued it. The superintendents used the scientific results as a basis for innovative management of the resource by seeking the best available scientific information from their own and outside sources. They welcomed information and used it to make better management decisions. Finally, they forged effective partnerships by enlisting the participation of scientists and other skilled professionals in their management teams. The successes of these superintendents increased the prestige and value of the park at the local, regional, national, and international levels.

The attempt to relocate the park headquarters buildings and visitor center onto Joppa Ridge is an illustration of the use of scientific research results in complex management decisions. Since 1966 visitors have been concentrated at the visitor center near the Historic Entrance to the Mammoth Cave system, directly above the cave. According to the General Management Plan (National Park Service 1983, p. vi),

> From 1965 to 1975, the Historic Entrance area was heavily congested with cars and people throughout the summer and on peak travel days in spring and fall. In an attempt to relieve congestion occurring at that time, the 1976 Master Plan evaluated several alternative solutions. Based on existing conditions and available information, a preferred alternative was selected that proposed developing a staging area at the periphery of the park near Union City [on Joppa Ridge]. In concept, this staging area would concentrate parking and basic visitor services in a less fragile area of the park away from the entrances to the primary cave system.

In 1979 the discovery of Logsdon/Hawkins River beneath Joppa Ridge showed that a major cave system lies beneath the proposed staging area. Relocation of the staging area to Joppa Ridge would have moved it into the upstream portion of the groundwater basin. Pollution associated with the staging area would have then affected a larger section of the cave system. In light of the difficulties identified by exploration and research, the concept of a staging area on Joppa Ridge was removed from the 1983 General Management Plan. The new cave discoveries and a drop in visitation were cited as the reasons for the change.

Some superintendents have been indifferent or even hostile to research performed by various groups at Mammoth Cave. Without productive partnerships, research was sometimes done in spite of park management. Under these regimes, the research results may not have been used as fully as the researchers would have preferred.

Several superintendents were neutral to hostile toward exploration and mapping efforts. Park managers wanted to maintain the pretense that the cave does not extend beyond the park boundaries. This misconception allowed park managers to avoid confronting the controversial land-acquisition issue for several decades. This pretense is no longer viable, however. The current Resource Management Plan (National Park Service 1990) proposes development of specific strategies to protect external cave resources. NPS management now acknowledges that many kilometers of the Mammoth Cave system and several entrances are in unacquired lands that lie within the originally authorized boundary. The present Land Protection Plan, however, does not address this issue. The Southeast Regional Office, citing the policy of the Secretary of the Interior to address only "perceived inholdings," deleted the park's proposal to include this issue in the Land Protection Plan. Although the enabling legislation for Mammoth Cave National Park indicates that the legislative intent was to protect "all the caves," several cave systems that are now connected in the Mammoth Cave system lie outside the authorized boundary.

The Impact of NPS Management on Research

Some impacts of NPS management on long-term scientific research and monitoring have been positive; others have been negative. One positive impact was the original decision to hire a research geologist at Mammoth Cave. His outstanding research results are now applied internationally, as well as locally. His working relationships with the park superintendents, however, ranged from harmonious and supportive to indifferent and even hostile. His tenure finally ended when one superintendent took control of the research program and began making changes contrary to scientific advice.

Funding for hydrogeological research at Mammoth Cave during the research geologist's tenure came principally from NPS headquarters in Washington and its regional office in Atlanta. This stipend was substantially supplemented by funds he solicited from EPA, the Kentucky Water Resources Research Institute, local banks, the regional planning agency, and various companies. Although the park furnished him with an office and administrative support, no park funds were ever programmed to support hydrogeological research during his tenure from 1973 to 1989.

Perhaps the most important lesson to be learned from the history of NPS-supported research at Mammoth Cave is that research at almost any park should be supervised by the regional chief scientist, not the park superintendent. Because of their lack of scientific expertise, many superintendents are unqualified to supervise research. Although some superintendents have been extremely supportive of research programs, they, unfortunately, are in the minority.

Conclusions

Karst hydrogeological research has a long and distinguished history at Mammoth Cave National Park. Research and resource monitoring in the last 30 years have been particularly productive and have demonstrated that Mammoth Cave, the Mammoth Cave system, and their regional neighbors are linked together in a symbiotic relationship. Mammoth Cave National Park and the Central Kentucky Karst contain the longest cave system and the best documented, conduit-flow, karst groundwater system in the world. Information and scientific concepts developed during the study of the Mammoth Cave system form the basis for karst monitoring and regulation throughout the rest of the United States. The unique values identified by such investigations have resulted in Mammoth Cave being designated a World

Heritage Site in 1981 and in the Mammoth Cave region being designated an International Biosphere Reserve in 1990.

Long-term scientific research and resource monitoring at Mammoth Cave National Park have been accomplished by four different groups. First, several university scientists have dedicated major portions of their research careers to the study of the Central Kentucky Karst. These efforts have been successful both professionally and scientifically. Second, several self-directed, volunteer organizations have assumed primary responsibility for the exploration and mapping of the Mammoth Cave system. The decades-long efforts of these individuals resulted in the discovery and integration of many individual caves into the longest system in the world. These same groups have also supported scientific research and resource-monitoring efforts at many levels. Third, an NPS research geologist worked at Mammoth Cave National Park from 1973 through 1989. His efforts were highly successful and have been the basis of several fundamental changes in the environmental management of the Mammoth Cave system. Fourth, a few private, state, and federal agencies have performed important studies in the area.

NPS management support of and response to this research have varied. A few superintendents have supported scientific research and successfully incorporated the growing knowledge base into management decisions that affected the park and region. Park managers decided to hire the research geologist and thereby initiated a very productive period of research. One superintendent successfully initiated a regional partnership between the park and its neighbors to deal with a major threat to the groundwater quality in the region, cave, and park.

Other superintendents, however, have grudgingly used research findings or ignored them entirely. Much of the research and resource monitoring in the park was done independently, without NPS support or funding. The NPS research geologist was actively opposed by two park superintendents. A bitter clash of wills with one superintendent caused the research geologist to resign in 1989.

Scientific research and long-term resource monitoring have profoundly affected the management and managers of Mammoth Cave National Park. Research has totally changed the boundaries of the management tasks facing the superintendent. In the 1960s, the superintendent could believe that the rest of the region had no effect on the park. The superintendent could manage in isolation what some people assumed to be a time-worn show cave in central Kentucky. In the 1990s, however, the superintendent faces the daunting task of administering a national park that has the following attributes and management challenges: (1) it is a world-class resource that is part of the longest and best-documented cave system in the world; (2) it contains unique, rare, and threatened biota; and (3) it is being profoundly affected by NPS activities, concessionaire operations, visitors, and human activities in the region outside the park.

Scientific research has shown that many of the routine technical and regulatory tools that are used to manage resources are ineffective or even counterproductive in karst regions. The results of scientific research often seem to change the rules, making discoveries faster than managers can assimilate them. Nevertheless, only new, comprehensive, long-term research and resource monitoring can provide NPS management with the tools necessary to "protect and preserve" Mammoth Cave National Park.

Acknowledgments

I am solely responsible for the conclusions and opinions expressed in this article. Elizabeth Estes contributed material on the location and setting of Mammoth Cave National Park, the historical perspective of the park, and the review of historical research activities in the Mammoth Cave region. I gratefully acknowledge her major contributions.

Thanks are owed to the following individuals who helped in this project: Judy Austin, William Austin, Sarah Bishop, Jeff Bradybaugh, Roger Brucker, Robert Deskins, Dominic Dottavio, Kip Duchon, Ralph Ewers, David Foster, James Goodbar, Nicholas Gunn, Gary Hendrix, Randall Kelley, Roger McClure, Joseph Meiman, David Mihalic, James Nieland, Arthur Palmer, James Quinlan, Martin Ryan, Stanley Sides, Gordon Smith, and Richard Zopf. This manuscript has particularly benefited from extensive, detailed reviews of Alexander (1993) by park staff and Roger Brucker. The following individuals have also reviewed portions of the manuscript and offered extremely helpful suggestions: Barbara Albright, Tom Poulson, and George Shaw. I also thank Gerry Estes of Image Database Software for computer support.

This research was partially supported by NPS under Purchase Order No. PXB120-1-0178.

Notes

1. NPS "management" in this chapter is used as both a singular and a collective noun. It includes the individual managers, the policies and practices of NPS and individual managers, and the actions that result.

2. The archaeological and biological researchers at Mammoth Cave National Park are at the forefront of their respective subdisciplines, but their work is beyond the scope of this review of karst hydrogeology.

3. Numerous field assistants over a 6-year period mapped more than 100 km of passage in about 20 caves. The mapping of the Hidden River Complex, which proved critical to the understanding of pollutant dispersal, was described in a series of articles in the February 1990 issue of the *Kentucky Caver*.

Literature Cited

Alexander, E. C., Jr. 1993. The evolving relationship between Mammoth Cave National Park and its hydrogeologic symbionts. P. 11–56 in D. L. Foster, ed. Proceedings, 1991 National Cave Management Symposium, Bowling Green, Kentucky, October 23–26. American Cave Conservation Association, Horse Cave, Kentucky.

American Society for Testing and Materials. 1995. Standard guide for the design of ground-water monitoring systems in karst and fractured-rock aquifers. Standard D 5717. 7 p.

Anderson, R. B. 1925. An investigation of a proposed dam site in the vicinity of Mammoth Cave, Kentucky. Unpubl. report, Louisville Gas and Electric Company. 14 p.

Cave Research Foundation. 1960. Speleological research opportunities at Mammoth Cave National Park. P. 320–322 in R. A. Watson, ed., 1981 [reprint]. The Cave Research Foundation, Origins and the First Twelve Years, 1957–1968. Mammoth Cave, Kentucky.

Cushman, R. V., R. A. Krieger, and J. A. McCabe. 1965. Present and future water supply for Mammoth Cave National Park, Kentucky. U.S. Geological Survey Water-Supply Paper 1475-Q. 47 p.

Deike, G. H. 1967. The development of caverns of the Mammoth Cave region. Unpubl. Ph.D. dissert., Pennsylvania State University, University Park. 398 p.

Deike, G. H., III, and W. B. White. 1969. Sinuosity in limestone solution conduits. American Journal of Science 267:230–241.

Dreybrodt, W. 1988. Processes in karst systems: physics, chemistry, and geology. Springer-Verlag, Berlin. 288 p.

Everhart, W. C. 1983. The National Park Service. Westview Press, Boulder, Colorado. 197 p.

Ewers, R. O. 1982. Cavern development in the dimensions of length and breadth. Unpubl. Ph.D. dissert., McMaster University, Hamilton, Ontario. 398 p.

———. 1994. The use of digital data-loggers in defining groundwater transport mechanisms in karst aquifers. P. 47–65 in Comptes Rendus Colloque Internationale de Karstologie à Luxembourg, 1992. Publications du Service Géologique du Luxembourg, vol. 27.

Frome, M. 1992. Regreening the national parks. University of Arizona Press, Tucson. 289 p.

George, A. I. 1989. Caves and drainage north of the Green River. P. 189–221 in W. B. White and E. L. White, eds. Karst Hydrology: Concepts from the Mammoth Cave Area. Van Nostrand Reinhold, New York.

Goode, C. E. 1986. World wonder saved: how Mammoth Cave became a national park. Mammoth Cave National Park Association, Mammoth Cave, Kentucky. 92 p.

Hess, J. W. 1974. Hydrochemical investigations of the Central Kentucky Karst aquifer system. Unpubl. Ph.D. dissert., Pennsylvania State University, University Park. 219 p.

Hess, J. W., and W. B. White. 1988. Storm response of the karstic carbonate aquifer of south-central Kentucky. Journal of Hydrology 99:235–252.

———. 1989. Water budget and physical hydrology. P. 105–126 in W. B. White and E. L. White, eds. Karst Hydrology: Concepts from the Mammoth Cave Area. Van Nostrand Reinhold, New York.

———. 1993. Groundwater geochemistry in the Central Kentucky Karst aquifer. Applied Geochemistry 8:189–204.

Huntoon, P. W. 1995. Is it appropriate to apply porous media groundwater circulation models to karstic aquifers? P. 339–358 in A. I. El-Kadi, ed. Groundwater Models for Resources Analysis and Management. Lewis, Boca Raton, Florida.

Lange, A. L., and J. F. Quinlan. 1988. Mapping caves from the surface of karst terranes by the natural potential method. P. 369–390 in Proceedings of the 2nd Environmental Problems in Karst Terranes and Their Solutions Conference, Nashville, Tennessee. National Water Well Association, Dublin, Ohio.

Meiman, J. 1990a. Mammoth Cave National Park water quality monitoring program, March 1990. Mammoth Cave National Park, Mammoth Cave, Kentucky. 40 p.

———. 1990b. Mammoth Cave National Park water quality monitoring program, preliminary results of 1990: year one. P. 1–38 in Proceedings of Mammoth Cave National Park's 1st Annual Science Conference: Karst Hydrology, Mammoth Cave National Park, December 17–18, 1990. Mammoth Cave, Kentucky.

———. 1992. The effects of recharge basin land-use practices on water quality at Mammoth Cave National Park, Kentucky. P. 697–713 in J. F. Quinlan and A. Stanley, eds. Proceedings of the 3rd Conference on Hydrogeology, Ecology Monitoring and Management of Ground Water in Karst Terranes, December 4–6, 1991, Nashville, Tennessee. National Ground Water Association, Dublin, Ohio.

Meiman, J., R. O. Ewers, and J. F. Quinlan. 1988. Investigation of flood pulse movement through a maturely karstified aquifer at Mammoth Cave National Park: a new approach. P. 227–263 in Proceedings of the 2nd Environmental Problems in Karst Terranes and Their Solutions Conference, Nashville, Tennessee. National Water Well Association, Dublin, Ohio.

Meloy, H. 1979. Outline of Mammoth Cave history. Journal of Spelean History 13(1):28–33.

Mikulak, R. J. 1988. Wastewater management in cave country: an unlikely success story. P. 315–331 in Proceedings of the 2nd Environmental Problems in Karst Terranes and Their Solutions Conference, Nashville, Tennessee. National Water Well Association, Dublin, Ohio.

Miotke, F.-D. 1975. Der karst im zentralen Kentucky bei Mammoth Cave. Habilitation dissertation. Jahrbuch der Geographischen Gesellschaft zu Hannover, Jahrbuch 1973. Hannover, Germany. 355 p.

Miotke, F.-D., and A. N. Palmer. 1972. Genetic relationship between caves and landforms in the Mammoth Cave National Park area. Böhler Verlag, Würzburg, Germany. 69 p.

Murray, R. K., and R. W. Brucker. 1979. Trapped! G. P. Putnam & Sons, Inc., New York. 335 p.

National Park Service. 1983. General management plan for Mammoth Cave National Park. Mammoth Cave, Kentucky. 71 p.

———. 1990. Resource management plan. Mammoth Cave, Kentucky.

Palmer, A. N. 1981. A geological guide to Mammoth Cave National Park. Zephyrus Press, Teaneck, New Jersey. 210 p.

———. 1989a. Stratigraphic and structural control of cave development and groundwater flow in the Mammoth Cave region. P. 293–316 in W. B. White and E. L. White, eds. Karst Hydrology: Concepts from the Mammoth Cave Area. Van Nostrand Reinhold, New York.

———. 1989b. Geomorphic history of the Mammoth Cave system. P. 317–337 in W. B. White and E. L. White, eds. Karst Hydrology: Concepts from the Mammoth Cave Area. Van Nostrand Reinhold, New York.

Plebuch, R. O., R. J. Faust, and M. J. Townsend. 1985. Potentiometric surface and water quality in the principal aquifer, Mississippian Plateaus region, Kentucky. U.S. Geological Survey Water Resources Investigation Report 84-4102. 45 p.

Pohl, E. R. 1955. Vertical shafts in limestone caves. National Speleological Society Occasional Paper 2. 24 p.

Quinlan, J. F. 1970. Central Kentucky Karst. Réunion Internationale Karstologie en Languedoc-Provence, 1968, Actes: Mediterraneé Etudes et Travaux 7:235–253.

———. 1982. Groundwater basin delineation with dye-tracing, potentiometric surface mapping, and cave mapping, Mammoth Cave region, Kentucky, U.S.A. Beiträge zur Geologie der Schweiz-Hydrologie 28:177–189.

———. 1986. Recommended procedure for evaluating the effects of spills of hazardous materials on ground water quality in karst terranes. P. 183–196 in Proceedings of the Environmental Problems in Karst Terranes and Their Solutions Conference, Bowling Green, Kentucky. National Water Well Association, Dublin, Ohio.

———. 1989. Ground-water monitoring in karst terranes: recommended protocols and implicit assumptions. U.S. Environmental Protection Agency, Environmental Monitoring Lab, Las Vegas. EPA/600/X-89/00050. 78 p.

———. 1990. Special problems of ground-water monitoring in karst terranes. P. 278–304 in D. M. Nielsen and A. I. Johnson, eds. Ground Water and Vadose Zone Monitoring. American Society for Testing and Materials Special Technical Paper 1053.

Quinlan, J. F., and R. O. Ewers. 1981a. Hydrogeology of the Mammoth Cave region, Kentucky. P. 457–95 in T. G. Roberts, ed. Field Trip Guidebook, vol. 3, Annual Meeting of the Geological Society of America, Cincinnati, Ohio, November 5–8, 1981. American Geological Institute, Falls Church, Virginia.

———. 1981b. Preliminary speculations on the evolution of groundwater basins in the Mammoth Cave region, Kentucky. P. 496–501 in T. G. Roberts, ed. Field Trip Guidebook, vol. 3, Annual Meeting of the Geological Society of America, Cincinnati, Ohio, November 5–8, 1981. American Geological Institute, Falls Church, Virginia.

———. 1985. Groundwater flow in limestone terranes: strategy, rationale and procedure for reliable, efficient monitoring of groundwater quality in karst areas. P. 197–234 in Proceedings of the 5th National Symposium and Exposition on Aquifer Restoration and Ground Water Monitoring, Columbus, Ohio. National Water Well Association, Worthington, Ohio.

———. 1989. Subsurface drainage in the Mammoth Cave area. P. 65–104 in W. B. White and E. L. White, eds. Karst Hydrology: Concepts from the Mammoth Cave Area. Van Nostrand Reinhold, New York.

Quinlan, J. F., and J. A. Ray. 1981. Groundwater basins in the Mammoth Cave region, Kentucky, showing springs, major caves, flow routes, and potentiometric surface. Friends of the Karst Occasional Publication 1. 1 plate.

———. 1989. Groundwater basins in the Mammoth Cave region, Kentucky, showing springs, major caves, flow routes, and potentiometric surface. Friends of the Karst Occasional Publication 2. 1 plate.

Quinlan, J. F., J. A. Ray, R. L. Powell, and N. C. Krothe. 1983. Groundwater hydrology and geomorphology of the Mammoth Cave region, Kentucky, and of the Mitchell Plain, Indiana. P. 1–85 in R. H. Shaver and J. A. Sunderman, eds. Field Trips in Midwestern Geology, vol. 2. Geological Society of America and Indiana Geological Survey.

Quinlan, J. F., and D. R. Rowe. 1977. Hydrology and water quality in the Central Kentucky Karst: Phase I. University of Kentucky, Water Resources Research Institute, Research Report 101. 93 p. [Reprinted with corrections as National Park Service, Uplands Field Research Laboratory, Management Report 12.]

Recker, S. A., R. O. Ewers, and J. F. Quinlan. 1988. Seepage velocities in a conduit-adjacent porosity system of a karst aquifer and their influence on the movement of contaminants. P. 265–287 in Proceedings of the 2nd Environmental Problems in Karst Terranes and Their Solutions Conference, Nashville, Tennessee. National Water Well Association, Dublin, Ohio.

Ryan, M. 1992. Development of a flow-through filter fluorometer for use in quantitative dye tracing at Mammoth Cave National Park, Kentucky. P. 243–261 in J. F. Quinlan and A. Stanley, eds. Proceedings of the 3rd Conference on Hydrogeology, Ecology Monitoring and Management of Ground Waster in Karst Terranes, December 4–6, 1991, Nashville, Tennessee. National Ground Water Association, Dublin, Ohio.

Sides, S. 1991. Chronology of the history of Mammoth Cave [course notes for "Exploration of Mammoth Cave"]. P. 9–50 in Karst Field Studies at Mammoth Cave. Center for Cave and Karst Studies, Western Kentucky University, Bowling Green.

Smith, P. M. 1967. Some problems and opportunities at Mammoth Cave National Park. National Parks Magazine 41(233):14–19.

———. 1968. New approaches to National Park Service administration and management. National Parks Magazine 42(245):14–18.

White, W. B. 1969. Conceptual models for carbonate aquifers. Ground Water 7(3): 15–21.

White, W. B., R. A. Watson, E. R. Pohl, and R. Brucker. 1970. The Central Kentucky Karst. Geographical Review 60(1):88–115.

White, W. B., and E. L. White, eds. 1989. Karst hydrology: concepts from the Mammoth Cave area. Van Nostrand Reinhold, New York. 346 p.

11

Air Quality in Grand Canyon

Christine L. Shaver and William C. Malm

Good visibility is critical to maintaining the scenic integrity of the panoramic vistas at Grand Canyon National Park in northern Arizona. The Colorado River began cutting the Grand Canyon roughly 8 million years ago, exposing rock almost 2 billion years old and creating a canyon that is now about 450 km long, 1.6 km deep, and as much as 30 km wide. The importance of being able to *see* clearly the features, forms, and colors that compose the Grand Canyon is apparent in the legislative history of the park: "The sides of the gorge are wonderfully shelved and terraced and countless spires rise within the enormous chasm, sometimes almost to the rim's level. The walls and cliffs are carved into a million graceful and fantastic shapes and the many colored strata of the rocks through which the river has shaped its course have made the canyon a lure for the foremost painters of American landscape" (Senate Report 1082, 64th Congress, 2nd Session [1917]).

When Grand Canyon National Park was established by Congress in 1919 "for the benefit and enjoyment of the people" (40 Statutes-at-Large 1175), the National Park Service (NPS) was handed the responsibility of protecting park resources and values unimpaired for the enjoyment of future generations (16 U.S. Code 1, 1a-1). Almost 4 million people from around the world visit the Grand Canyon annually. Visitor surveys document that good visibility is extremely important to visitors. Results have shown that visitors are aware of haze in the atmosphere that diminishes the view and that visitors' enjoyment of the view and the park decreases as the amount of haze increases (Bell et al. 1985).

The need and responsibility for protecting the view at the Grand Canyon date back to 1919, but the legal tools necessary to effect protection were lacking until Congress amended the Clean Air Act in 1977 (42 U.S. Code 7401, et seq.). Specific statutory mechanisms were established to prevent significant air-quality deterioration in areas where air quality was better than the national ambient air-quality standards set to protect public health and welfare. The most stringent degree of air-quality protection was afforded to certain national parks and wilderness areas, including Grand Canyon, that were designated as "Class I" areas. Federal land managers were given an affirmative responsibility to protect air-quality-related values, including visibility, of Class I areas from the adverse effects of air pollution. Congress also established a national visibility goal of remedying any existing and preventing any future manmade visibility impairment in Class I areas where visibility was an important value (42 U.S. Code 7465, 7469A).

As directed by Congress, the Environmental Protection Agency (EPA) published regulations to implement the visibility protection requirements of the Clean Air Act. EPA's rules outlined a phased approach to make reasonable progress toward the national goal (45 *Federal Register* 80084 [1980]; 40 Code of Federal Regulations 51.301, et seq.). The first phase included the following: (1) monitoring visibility in Class I areas, (2) assessing potential impacts of proposed sources of air pollution to ensure that their emissions would not adversely affect visibility, and (3) remedying existing visibility impairment by installing the best available retrofit technology if the impairment can be reasonably attributed to a specific source(s) of air pollution. EPA indicated that regulatory requirements would be developed in a future, second phase to deal with "regional haze"—visibility impairment caused by multiple sources in numerous areas typically far from Class I areas.

Visibility Monitoring and Research Program

Routine Long-term Monitoring

Accepting the responsibility assigned to the federal land manager in the 1977 Clean Air Act amendments and recognizing the opportunity to address visibility problems that could be documented, NPS began monitoring visibility at Grand Canyon in 1978. At first, researchers used photography to document visibility conditions. Later they used equipment to monitor concentrations of fine particulate matter in the air and an instrument to measure contrast between the terrain and sky, from which they could estimate standard visual range (a teleradiometer). Finally, they used an instrument capable of directly measuring the amount of light that is scattered or absorbed as it passes through the atmosphere (i.e., light extinction), thereby affecting the clarity of the observer's view (a transmissometer). Since then, routine monitoring at the Grand Canyon has included optical, aerosol, and photographic components. NPS has operated similar visibility-monitoring stations, measuring all or some of these components, at more than 50 other areas throughout the country.

The monitoring data have been used to characterize visibility conditions and to conduct research on the causes of visibility impairment. Compared to national parks in other parts of the country, Grand Canyon enjoys relatively good visibility, but this visibility is very sensitive to even small changes in fine-particle concentrations. "Rayleigh" visibility, indicative of natural conditions when only natural gases in the atmosphere limit visibility to a theoretical 392 km, occurs less than 10% of the time at Grand Canyon. Visibility is reduced to one-half the Rayleigh condition 50% of the time. Fine particles, 2.5 microns or less in diameter, are responsible for the vast majority of visibility degradation. The total fine-particle mass concentration at Grand Canyon averages 3–4 $\mu g/m^3$. On the average, this fine-particle mass is made up of sulfates (36%), organics (25%), nitrates (7%), soil (20%), and light-absorbing carbon (5%). Sulfates are responsible for a disproportionate amount of visibility impairment (40–60%) because their hygroscopic tendencies allow them to grow to a size that is most effective in scattering light (National Park Service 1989a,b; Malm 1992).

Researchers have analyzed monitoring data in conjunction with meteorological data to examine the origins of air masses that carry high concentrations of visibility-reducing aerosols to Grand Canyon and other national parks. These source-receptor modeling techniques show that high-sulfate episodes at Grand Canyon most likely result from long-range transport of pollution from southern California, southern Arizona, and Mexico (Bresch et al. 1987; National Park Service 1989a,b; Malm 1992). However, EPA's regulatory framework did not provide a mechanism for requiring mitigation of the many pollution sources contributing to this regional haze.

On the other hand, photographic monitoring since 1978 had detected a common winter haze condition over the Colorado Plateau that appeared to be related to the buildup of local pollution sources. The terrain surrounding the lower Colorado River rises about 800 m above the water's surface. Air pollutants can be trapped by a persistent thermal inversion below the height of the surrounding terrain during stagnation events that last for about 1 week, resulting in a distinct visible surface layer. Photographic monitoring sites, in particular at Bryce Canyon National Park, had recorded this haze layer as much as 80% of the time during winter months (Malm et al. 1985; Chinkin et al. 1987). The region affected by this haze layer is largely undeveloped and sparsely populated. It includes a few small urban areas and industrial facilities and several large, coal-fired, electric generating stations. One of these stations is the 2,250 MW Navajo Generating Station (NGS) near Page, Arizona, which operates without sulfur dioxide control equipment.

Certification of Existing Visibility Impairment

In November 1985, the U.S. Department of the Interior's assistant secretary for Fish, Wildlife, and Parks certified to EPA that visibility in all NPS Class I areas, including Grand Canyon, was impaired because of regional haze. This blanket certification was based on data collected through the NPS long-term air-monitoring program. The assistant secretary also identified a few areas where the impairment might be due to one or more specific sources. Photographic evidence showed that local pollution sources—rather than long-range transport—could contribute significantly to the winter haze over the Colorado Plateau. NGS was initially identified as the suspected source of winter visibility impairment at Bryce Canyon.

Exploratory Study of Winter Haze

Researchers conducted a special exploratory study in January 1986 to determine the horizontal and vertical extent of the haze commonly seen during the winter at Bryce Canyon and to identify the major constituents of the haze. This study showed that the haze extended into Grand Canyon and Canyonlands National Park. Aircraft-based measurements indicated that the air above the inversion layer was essentially particle-free, whereas below the layer, the amount of light extinction was 2 to 5 times higher than clean air values. At the south end of Lake Powell, the particles were largely composed of sulfates and organics (Fitz et al. 1987). In March 1986, the November 1985 certification was amended to identify NGS as a potential contributor to visibility impairment at Grand Canyon and Canyonlands (U.S. Department of the Interior 1985; National Park Service 1986).

Winter Haze Intensive Tracer Experiment

Overview of Study Plan Shortly after the winter 1986 study was completed, planning began for a more comprehensive effort for the winter of 1987 to address questions about the nature and sources of the winter haze conditions. The overall objective of the Winter Haze Intensive Tracer Experiment (WHITEX) was to assess the feasibility of attributing emissions from a single point source to visibility impairment in prespecified geographic regions. Specifically, researchers would evaluate various receptor and deterministic models and compare their ability to link NGS emissions to visibility impairment at Grand Canyon, Canyonlands National Park, and Glen Canyon National Recreation Area (Mathai 1989; National Park Service 1989a,b).

WHITEX was planned and conducted by NPS and other participants in a cooperative industry-government effort known as SCENES (Subregional Cooperative Electric Utility, NPS, EPA, and Department of Defense Study, which was involved in a 5-year study of visibility throughout the Southwest), as well as numerous private contractors and academic institutions. It ran for 6 weeks during January and February 1987. Extensive monitoring was conducted throughout the Colorado Plateau area, with major monitoring sites established at Grand Canyon, Canyonlands, and Glen Canyon (Fig. 11.1). The intensive monitoring network included photography (still and time-lapse), meteorological equipment (meteorological stations and upper air soundings), particle and tracer samplers, transmissometers, and aircraft equipped with monitors. Eighteen optical, chemical, physical, and meteorological variables were studied at 12 sites. More than 21,000 samples were collected, and 136,000 measured constituents or concentrations were over minimum detectable limits.

A unique tracer (deuterated methane, CD_4) was injected into the stacks of NGS so emissions from the power plant could be tracked to assess NGS's con-

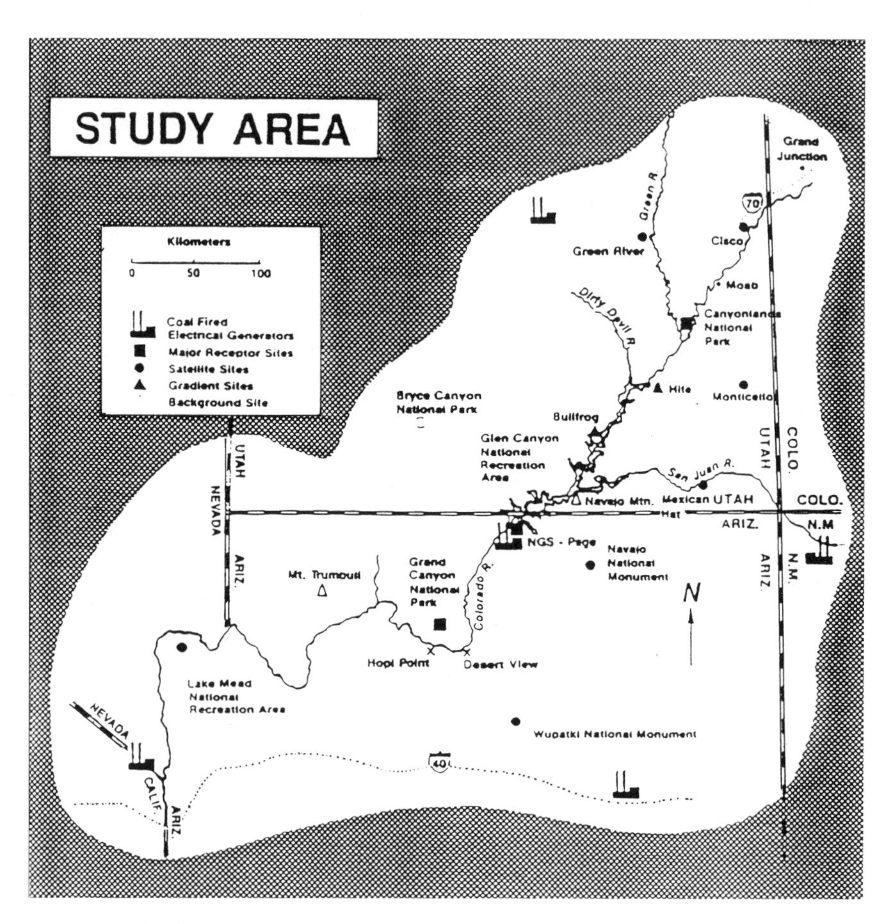

Figure 11.1 Study area.

tribution to visibility impairment in various locations. Because of the high cost of analyzing tracer data, the study planners decided to analyze the data for all monitoring sites only during the period of worst visibility (11–12 February 1987). During that time, the tracer released from NGS was detected throughout the study region, but the highest concentrations of CD_4 were found at Hopi Point in Grand Canyon and in Page, Arizona. The researchers therefore decided to analyze a larger subset of tracer data, for three periods of worst visibility, at these two sites and to conduct more detailed source-attribution analyses primarily at these locations. NPS had the lead responsibility for data analysis and preparation of a draft report.

Overview of Source-Attribution Techniques To assess the contribution of NGS emissions to visibility impairment, researchers first analyzed the monitoring data to determine the kinds of primary and secondary aerosols contained in the ambient air at various receptor sites. They then examined the contribution of each aerosol to visibility reduction. Fine particles were responsible for most of the visibility reduction, and sulfates were the largest contributor to extinction. During the poorest visibility episode, sulfates constituted 97% of the non-Rayleigh light-extinction budget.

The most difficult part of the analysis was assessing the relative contribution of primary and secondary aerosols associated with NGS. Although the tracer data were useful in tracking the path taken by the NGS plume, several issues confounded NPS's ability to calculate NGS's contribution to visibility impairment directly from measured tracer concentrations at receptor sites. First, the tracer was being used to track a secondary pollutant. Although conversion and deposition rates for sulfur dioxide and sulfate could be estimated, the associated behavior of tracer in the atmosphere was unknown. Second, during some periods, a constant ratio between the release of the tracer and emission of sulfur dioxide from the NGS stack was not maintained. Finally, unique tracers were not used to track emissions from other sources that might influence the study area, including other coal-fired power plants, raising colinearity issues.

In light of these confounding factors, researchers conducted several qualitative and quantitative analyses and compared and reconciled the results to build confidence in the results of the source-apportionment analysis. These techniques included the following:

1. Emissions. Researchers compared source strength of NGS with other sources in the region and the locations of NGS and other regional emissions with those of key receptor sites.

2. Trajectory and streakline analysis. Researchers evaluated whether the predicted presence of NGS or other source emissions is coincident with elevated sulfur concentrations or is due to random processes.

3. Spatial and temporal patterns. Researchers examined qualitatively and quantitatively the spatial patterns in visibility-reducing aerosol concentrations as a function of time through empirical orthogonal function (EOF) analysis.

4. Climatology. Researchers analyzed the synoptic climatology and meteorology to understand the origin of stagnation periods and why pollutants were transported along various pathways.

5. Deterministic windfield modeling. Researchers studied model simulations to understand how pollutants can be transported along various pathways and to build conceptual models of physiochemical processes associated with aerosol concentrations and visibility impairment.

6. Tracer mass balance regression (TMBR). Researchers used the variation of sulfur and natural or artificial tracers as a function of time to apportion visibility-reducing aerosols among various pollution sources.

7. Differential mass balance (DMB). Researchers used the ambient concentration of a unique tracer, CD_4, to estimate dispersion and used the deterministic model to calculate conversion and deposition from estimated plume age.

8. Chemical mass balance. Researchers applied the chemical mass-balance formula to estimate source contributions of primary aerosols and to set an upper limit on NGS contributions.

Each technique had advantages and disadvantages, which precluded an exact quantification of NGS's contribution to visibility impairment. However, after preliminary analyses were conducted, NPS scientists and regulatory specialists concurred that the quantitative methods (TMBR and DMB), in conjunction with the qualitative analyses, could be used to estimate NGS's contribution to haze at the two study sites.

Overview of WHITEX Findings NPS used DMB and TMBR to attribute quantitatively a portion of the sulfate aerosol concentrations at Grand Canyon and Page to NGS. The results from both analyses were quite similar and very well correlated. On the days for which CD_4 data were analyzed, the TMBR best estimate of NGS's average contribution to sulfate at Hopi Point was 73 (±4)%, whereas the DMB calculation yielded 68 (±3.5)%. TMBR was also used to calculate the fraction of light extinction caused by NGS sulfate. At Grand Canyon, NPS concluded that the mean fraction due to NGS sulfate was 42 (±13)%. During the highest extinction episode, the fraction due to NGS sulfate was 59 (±18)%.

The qualitative analyses were used to assess whether the quantitative findings were reasonable. The examination of emissions inventory for the region confirmed that NGS could be expected to contribute significantly to air-quality problems during the winter at Grand Canyon because (1) NGS is the largest coal-fired power plant in the six-state region around the Colorado Plateau; (2) NGS, located only 20 km from the boundary of Grand Canyon, is the largest sulfur dioxide source within hundreds of kilometers of Grand Canyon; (3) NGS and Grand Canyon are in the same air basin, and the height of the stack and effective plume rise, compared with the elevation of the canyon rim, would funnel NGS emissions directly into the canyon.

The trajectory and streakline analysis showed a strong correlation between predicted "hits" of the NGS plume and all periods with elevated sulfate. Researchers applied a statistical, multiresponse permutation procedure to these results and found that the probability that the association between NGS plume "hits" and elevated sulfate was due to random processes was less than 5%.

The EOF analysis of spatial and temporal trends, using data from all monitoring sites in the study region, showed that a sulfate concentration field that is strongest at Page and decreases as one moves radially out from NGS explains 70% of the variance in the field. This spatial concentration gradient is predominant under stagnant meteorological conditions and suggests that NGS emissions are a significant contributor to ambient sulfate concentrations.

Through the meteorological analysis, researchers examined multiyear synoptic climatology (1980–1984) to identify categories that would likely produce stagnant air masses. They found that such conducive conditions occurred 65% of the time during the months from November through March. Most of these events lasted at least 3–5 days; one-third to one-half lasted 6–14 days. A 9-day stagnation event, such as that which occurred during the WHITEX study and led to the highest sulfate concentrations in the region, would be expected to occur 16% of the time during the winter months.

A separate meteorological analysis, a deterministic windfield model, was conducted over the 11–12 February period and confirmed the conceptual models suggested by the TMBR, DMB, and EOF analyses. For example, the release of particles at NGS plume height into modeled wind fields resulted in the transport of those particles into the Grand Canyon region. The model also confirmed that transport of emissions from areas to the southwest and east of Grand Canyon would not occur during this synoptic regime, and transport from the north would require 2 days or longer.

Finally, although the chemical mass-balance analysis was not used to attribute secondary aerosols to respective sources, attribution of primary aerosols suggested that the only two sources associated with sulfur dioxide emissions were coal-fired power plants and copper smelters, results that are consistent with the TMBR analysis.

Based on the results of these quantitative and qualitative analyses, NPS concluded that sulfur dioxide emissions from NGS had contributed significantly to visibility impairment at Grand Canyon during the study period, causing 30–50% of the visibility reduction on the average and 40–80% during episodes. NPS also concluded that meteorological conditions during WHITEX represented typical conditions for the Grand Canyon area during the winter months (November through March).

Research Meets the Regulatory Process

Legal Criteria for Regulatory Relief

Once NPS certified in 1985 and 1986 that visibility was impaired at Grand Canyon, Canyonlands, and Bryce Canyon and that it might be caused by NGS, certain regulatory mechanisms were set in motion to guide the interpretation and use of data collected during the WHITEX study. Under EPA's regulations, after a certification is made, the affected state (or EPA if the state has not adopted a visibility protection plan) must identify each stationary facility that may "reasonably be anticipated to cause or contribute" to visibility impairment. If the impairment is found to be "reasonably attributable . . . by visual observation or any other technique that State deems appropriate," then an analysis of "best available retrofit technology" (BART) must be conducted for the facilities (40 Code of Federal Regulations 51.301[s], 51.302[c]).

BART analyses for fossil-fuel-fired electric generating facilities greater than 750 MW must be conducted according to EPA guidelines (U.S. Environmental Protection Agency 1980). The BART analysis must consider the following: the costs of compliance, the energy and non-air-quality environmental impacts, any existing pollution-control technology in use at the facility, the remaining useful life of the source, and the degree of improvement in visibility that would reasonably result from application of controls (40 Code of Federal Regulations 51.301[c] [4] [iii]). An affected coal-fired power plant may be exempted from BART requirements only if it demonstrates that it does not "by itself or in combination with other sources emit any air pollutant which may reasonably be anticipated to cause or contribute to significant impairment of visibility in any class I area" (40 Code of Federal Regulations 51.303).

Initial Application of Regulations to WHITEX Results

In April 1989, NPS transmitted a draft report on WHITEX to EPA (National Park Service 1989a). EPA sponsored a public workshop in late April to provide a forum for NPS scientists to explain the results, as well as an opportunity for other concerned parties to express their views on the scientific and legal adequacy of the NPS findings. EPA also hired an independent consultant to review the WHITEX report. The Salt River Project (SRP), operator and major owner of NGS, and other owners (Los Angeles Department of Water and Power, Arizona Public Service Co., Nevada Power Co., and Tucson Electric Power Co.) challenged the scientific credibility of the NPS report, emphasizing the uncertainties associated with the methods used to quantify NGS's contribution to visibility impairment. Attorneys representing the NGS participants questioned EPA's legal authority to impose emission-control requirements on NGS because of its alleged contribution to a "regional haze" problem. The Environmental Defense Fund, on the other hand, argued that the findings from WHITEX satisfied the regulatory criteria of "reasonable attribution."

On 5 September 1989, EPA announced that it was preliminarily attributing several episodes of winter visibility impairment at Grand Canyon to emissions from NGS (54 *Federal Register* 36948 [1989]). EPA indicated that it had reviewed the draft NPS report on WHITEX, concurred with the findings, and solicited comments on the merits of the preliminary attribution. EPA further indicated that it would publish a rulemaking proposal on the need for BART by a court-ordered deadline of 1 February 1990,[1] unless, in response to comments, EPA rejected its proposed attribution finding.

Additional Studies Conducted

Because of the controversy surrounding the WHITEX results and EPA's proposed attribution finding, the Secretary of the Interior asked the National Academy of Sciences to review the WHITEX report. The secretary's interest in further scientific review stemmed from a desire to resolve a difference of opinion between two bureaus within the U.S. Department of the Interior—NPS and the Bureau of Reclamation. The Bureau of Reclamation, like the other NGS participants, disagreed with the NPS findings in the WHITEX report and EPA's proposed attribution finding. More than half of SRP's 46% ownership interest in NGS was for the use and benefit of the United States (acting through the Bureau of Reclamation) because electric power from NGS was used to pump water for the Central Arizona Project. The National Academy of Sciences, with funding from EPA, SRP, and the U.S. Department of the Interior, convened a panel of experts to review the WHITEX report and to conduct a broader study of regional haze in national parks and wilderness areas.

In addition, SRP and other NGS participants planned a second study of visibility impairment at Grand Canyon for the winter of 1990. SRP unsuccessfully attempted to secure an agreement from EPA, NPS, and the Environmental Defense Fund to postpone further regulatory action until results from their study were available. However, SRP requested and received a court order extending the rulemaking schedule for 1 year. The SRP study was conducted between January and March 1990. The study area was smaller than that used for WHITEX, but more monitors were deployed and several tracers were released sequentially from NGS in an attempt to track plume transport and age more accurately.

The National Academy of Sciences report, entitled "Haze in the Grand Canyon: An Evaluation of the Winter Haze Intensive Tracer Experiment," was published in October 1990. The report maintained that the database and data analyses in the WHITEX report were not sufficient to ascertain an exact quantification of NGS's contribution to haze at any given time. However, the report also contained a qualitative assessment of WHITEX, which supported EPA's finding that, at some times during the study period, NGS contributed significantly to haze in Grand Canyon. In addition, the report stated that the "rate of sulfur dioxide emissions from NGS is easily large enough to serve as the source of sulfur measured in the Grand Canyon NP."[2]

Partial results from the SRP study were released in a draft report in April 1991 (Sonoma Technology, Inc. 1991). The report concluded that NGS emissions were present at Hopi Point in Grand Canyon less frequently than during the WHITEX period. In addition, on days when NGS emissions were present at Hopi Point, visibility impairment associated with those emissions was substantially less than the amount calculated in the WHITEX report.

One tactic SRP used to refute the NPS findings was to emphasize NGS's annual average or seasonal average impact on visibility at Grand Canyon. This emphasis diminishes the relative significance of worst-case impacts. If emission levels are constant, the worst-case impacts are dependent on meteorology: the highest levels of air pollution occur during stagnation episodes. Averaging impacts diminished the relative frequency and relative percentage that NGS emissions contributed to visibility degradation. Whereas the NPS analyses focused on the episodic impact, SRP used the scientific data to diminish the impact of NGS. However, the results of the SRP study also indicated that NGS was responsible for a significant quantity of sulfate and haze during specific visibility-impairment episodes at Hopi Point.

EPA's BART Proposal

In February 1991, EPA issued a proposed regulation that would have required a 70% reduction in sulfur dioxide emissions from NGS based on a 30-day rolling average (56 *Federal Register* 5173 [8 February 1991]). The proposal was based on EPA's draft BART analysis and Regulatory Impact Analysis (U.S. Environmental Protection Agency 1990a, b).[3] Although the Regulatory Impact Analysis had concluded that a 90% control option would be most cost-effective, EPA proposed a lower control option because of the uncertainties associated with quantifying NGS's contribution to winter haze at Grand Canyon.

During the public comment period and at the 18–19 March 1991 public hearing on the proposed BART regulation for NGS, EPA received more than 400 comments. The vast majority favored the installation of pollution controls at NGS. Many supported a more stringent (90–95% reduction) control requirement. New analyses were also submitted that showed the likely annual impact of NGS on other Class I areas on the Colorado Plateau and depicted computer-generated simulations of the relative impact of various control options on visibility at Grand Canyon (Air Resource Specialists, Inc. 1991). Some parties, however, objected to requiring any additional controls on NGS.[4]

Negotiated Agreement on NGS Controls

After the close of the comment period at EPA's recommendation, representatives of SRP, the Grand Canyon Trust, and the Environmental Defense Fund met to discuss alternative approaches. In August 1991, the outside parties reached agreement and together recommended that EPA adopt a sulfur dioxide emission limitation for NGS of 0.10 lbs/MMBtu (approximately a 90% control level) based on a rolling annual average. The recommended 90% control requirement, coupled with an initial delay in the installation schedule and an extended compliance-monitoring requirement, was estimated to cost less than EPA's proposed 70% control option. EPA solicited comments on the negotiated agreement (56 *Federal Register* 38399 [13 August 1991]). A final regulation adopting a 0.10 lbs/MMBtu sulfur dioxide emission limit for NGS, to be phased in between 1997 and 1999, was announced by President Bush at Grand Canyon on 18 September and published in the *Federal Register* on 3 October 1991 (56 *Federal Register* 50172).

Role of Monitoring in Issue Resolution

Responding to Scientific Challenges

Monitoring conducted during WHITEX, combined with the NPS long-term monitoring data and historical meteorological database, provided a necessary foundation for a regulatory decision. The scientific integrity of the monitoring data was never challenged. Although some questions were raised about the accuracy of the tracer data, the National Academy of Sciences concluded that the presence of the unique tracer provided unambiguous evidence that emissions from NGS traveled to Grand Canyon.

The data analysis and source-attribution techniques used by NPS, on the other hand, were the subject of substantial debate within the scientific community (e.g., Markowski 1992). NPS assumptions and algorithms used in formulating the quantitative techniques, the adequacy of the database to support the numerous analyses, and NPS's objectivity in exploring alternative models were frequently questioned. However, NPS scientists were able to respond to the challenges in a timely and efficient manner because of the comprehensive nature of the database, the wide variety of monitoring equipment (particle samplers, photography, meteorological instruments, and extinction devices), and the scientific skills and dedication of NPS researchers. NPS performed additional analyses, carefully documented the assumptions and calculations used in every analysis, and discussed departures from those assumptions and their effect on the findings. NPS conducted additional monitoring inside the Grand Canyon during the winter of 1989, which proved to be very useful in filling information gaps and interpreting the WHITEX results. Photographic evidence acquired during WHITEX showed pollution trapped below the rim of Grand Canyon; however, no monitors had been located in that area during the study.

The ability to use the qualitative data to interpret and reconcile findings from the quantitative analyses proved to be invaluable. Several key pieces of evidence supported the NPS quantitative findings: the photographic documentation, particularly time-lapse photography documenting the air movements and conversion mechanisms (e.g., the presence and effect of clouds, which facilitated the rapid conversion of sulfur dioxide to sulfate particles); the windfield modeling, which showed a plume trajectory that mimicked the air pathways observed with photography; the emissions strength and proximity of NGS compared with other pollution sources; and the detection of the unique tracer at Grand Canyon.

The EPA's general acceptance of the NPS findings was also largely affected by the legal and regulatory criteria that would be used to evaluate the BART decision. Legally, EPA was not required to base its decision on absolute certainty. Rulemaking activities are not held to the standard of "beyond a reasonable doubt." In fact, and in law, EPA was not required to prove that its decision was "more likely than not" to be the right one. EPA's decision could only be overturned if it was "arbitrary and capricious"; i.e., based on little or no evidence "or contrary to law." With respect to the latter, the Clean Air Act and EPA's regulations established criteria that were relatively easy to satisfy. Was there "*reasonable* attribution?" Were emissions from NGS *reasonably anticipated* to cause or *contribute* to visibility impairment? The data that NPS scientists collected, the analyses they performed, and the conceptual model they developed to interpret and reconcile the data proved more than adequate to satisfy the regulatory tests.

Responding to Political Controversy

The collection of good data and performance of many analyses provided an impetus for the regulatory process. More was needed, however, to achieve a successful resolution. First, an internal and external support network needed to be established for both the science and the most park-protective solution. Constant consultation between NPS scientists and air-quality regulatory experts promoted an understanding and trust in the science on one hand and an appreciation of the regulatory context and responsibility on the other. In addition, regular communication with staff and management at Grand Canyon kept them apprised, ensured consistency in dealings with the public, and facilitated an informed consensus on the policy approach. NPS's ready availability and response to requests for information from EPA, congressional staff, the media, and public interest groups helped alleviate the controversy and build alliances.

The internal and external support network, combined with strategic planning, allowed NPS to anticipate issues that might be raised. NPS was able to acquire or access data to make preliminary estimates of the costs and benefits of emissions control to answer questions before they were asked.[5] The support network was also instrumental in providing the motivation to persevere in the face of intense industrial opposition. During this lengthy process, there were many opportunities to capitulate, compromise, and agree to delays. However, the support network remained intact because it was built on trust in the integrity of the data, understanding of the regulatory framework, and commitment to the goal of improving visibility at Grand Canyon as quickly as possible.

Analysis of Relative Costs and Values of Monitoring

Economic Costs and Benefits

Quantifying the cost of the monitoring and research that provided a technical foundation for the final EPA regulation requiring a reduction in NGS emissions was relatively easy. Approximately $2 million in NPS funding was devoted to WHITEX, including the cost of monitoring devices that were subsequently deployed in national parks throughout the country. EPA and industry contributed another $1 million. The subsequent SRP-sponsored study cost $13 million.

The cost of pollution control mandated by EPA's final regulation was also easily calculated. SRP estimated (in 1992 dollars) a capital cost of $430 million and a total levelized annual cost of $89.6 million to install and operate emission-control equipment to meet the new emission limitation (56 *Federal Register* 50177 [3 October 1991]). The cost to consumers, in the form of increased utility bills, could also be reasonably estimated and was substantially less than 5% (≤ 2.50/month for the average customer; 56 *Federal Register* 50184 [3 October 1991]).

Quantifying the benefits—both environmental and economic—was also possible, although not as easily calculated. Based on data collected during WHITEX and subsequent NPS monitoring and analyses, EPA determined how much visibility would improve if certain humidity and wind conditions were present during very high sulfate episodes. EPA concluded that a 90% reduction in NGS sulfur dioxide emissions could increase standard visual range up to 300% at Grand Canyon (56 *Federal Register* 50180 [13 October 1991]).

The economic benefits of improved visibility presented a more difficult case. The calculation of economic benefits associated with a nonmarketable good, such as visibility, is far from an exact science. The Clean Air Act does not require EPA to consider the economic benefits of visibility improvement in a BART analysis; EPA must consider only how much visibility may improve after BART installation (40 Code of Federal Regulations 51.301 [c] [4] [iii]). EPA, however, did estimate the economic benefits in its proposed rulemaking in response to Executive Order 12291. At that time, EPA estimated that the economic benefits exceeded the costs of control (56 *Federal Register* 5173 et seq. [8 February 1991]).

Policy Benefits

The monitoring data and analyses documenting the contribution of a single source to a visibility problem and culminating in a requirement to control pollution at that source resulted in other benefits. First, the controversy over the WHITEX results, coupled with the high public recognition and value of Grand Canyon, bred considerable media attention that raised public and congressional awareness of visibility problems in national parks and wilderness areas. The Clean Air Act Amendments of 1990 included new requirements related to visibility protection in these areas (Clean Air Act Amendments of 1990, Public Law 101-549, 104 Statute 2399, section 816, adding new section 169B [42 U.S. Code 7469B] to the Clean Air Act). Congress authorized the appropriation of $8 million/year for 5 years to expand visibility monitoring and research. Although Congress has not yet appropriated this funding, EPA's FY 1992 appropriation bill recommended that EPA use $700,000 of its available funding to develop regional modeling approaches. EPA, on its own initiative, has increased its visibility monitoring and research budget. In addition, the Arizona congressional delegation succeeded in adding a line item to EPA's FY 1991 appropriation bill requiring a tracer study of the Mohave power plant—a large, coal-fired, electric generating station operating without sulfur dioxide emissions control in Nevada, southwest of Grand Canyon.

Second, it was widely recognized that numerous sources—large and small, near and far—contributed to the observed visibility impairment. In the 1990 Clean Air Act Amendments, Congress required EPA to establish a Grand Canyon Visibility Transport Commission by 15 November 1991. The commission has 4 years to make recommendations to EPA on promulgating regulations, including long-range strategies for addressing regional haze that impairs visibility at Grand Canyon. The amendments required that all states in the "visibility transport region" (states that provide both clean and polluted air to Grand Canyon) be included in the commission. Eight western states, EPA, and affected federal land-management agencies have begun discussing visibility issues and potential approaches. The deliberations and recommendations of the commission may affect visibility throughout the Colorado Plateau and could provide a precedent for dealing with visibility impairment in other national parks.

Third, a legal challenge of EPA's decision filed by the Central Arizona Water Conservation District (CAWCD) and other water districts that obtain electricity from NGS to pump their water led to a favorable court decision. This decision will facilitate future regulatory actions to require pollution controls at sources that impair visibility in other national parks and wilderness areas. In response to this challenge, the U.S. Court of Appeals for the Ninth Circuit found the following:

> Even if the Final [EPA] Rule addresses only a small fraction of the visibility impairment at the Grand Canyon, EPA still has the statutory authority to address that portion of the visibility impairment problem which is, in fact, "reasonably attributable" to NGS. Congress mandated an extremely low triggering threshold, requiring the installment of stringent emission controls when an individual source "emits any air pollutant that may reasonably be anticipated to cause or contribute to any impairment of visibility" in a class I Federal area (*Central Arizona Water Conservation District, et al., v.* EPA, 990 F. 2d 1531, 1541 (9th Circuit); *cert. denied,* 114 S.Ct. 94 [1993]).

In addition, the court rejected the CAWCD's claim that EPA relied on "discredited elements of a seriously flawed study, failed to refute significant criticisms which undercut the basis for EPA's [visibility] improvement estimate, and refused to consider highly relevant evidence which contradicted its estimate" (*Central Arizona Water Conservation District, et al., v.* EPA, p. 27). Instead, the court concluded the following:

> In the final analysis, Petitioners simply adhere to a different interpretation of the rather disparate and equivocal scientific data in the record. While Petitioners may not be satisfied with EPA's responses, it is not EPA's duty to satisfy all of the concerns of potentially affected or aggrieved parties. EPA conducted an extensive and involved notice and comment period, and adequately met its statutory obligations of responding to significant comments and criticisms under 42 U.S.C. 7607(d)(6)(B). . . . Because Congress delegated to EPA the power to "regulate on the borders of the unknown," this court will not interfere with the agency's "reasonable interpretations of equivocal evidence." (*Public Citizen Health Research Group v. Tyson,* 796 F.2d at 150 [*Central Arizona Water Conservation District, et al., v.* EPA, p. 1546]).

The court's decision highlighted how important it is for NPS to provide EPA and state regulatory agencies with information that is adequate to support reasoned decision making, despite scientific uncertainty. The decision also showed how important it is for NPS to promote its resource protection mandates in the face of scientific uncertainty.

Conclusions

Long-term visibility monitoring at Grand Canyon and other national parks on the Colorado Plateau documented that air pollution was degrading the ability to view scenic resources. An understanding of regulatory remedies to mitigate the problem led to a more intensive study of visibility-reduction episodes, during which the contribution of a specific pollution source was assessed. Guided by regulatory criteria that reflected a standard of "reasonableness," NPS could use the data it acquired to advocate, defend, and promote a regulation requiring that NGS substantially reduce its pollution. As a result, visibility at Grand Canyon and probably other national parks on the Colorado Plateau will significantly improve. The wheels have been set in motion to deal with other air-pollution sources that degrade visibility throughout this spectacularly scenic region.

Notes

1. As a result of a lawsuit filed in 1982 (*Environmental Defense Fund, et al., v. Reilly*, No. C82-6850 RPA [N.D.Cal.]), EPA was under a court-ordered schedule to promulgate visibility protection regulations, including BART provisions, for various states that had failed to adopt visibility plans. That schedule required EPA to propose a BART regulation to address visibility impairment at Grand Canyon by 1 February 1990.

2. NPS transmitted a technical response to the issues raised in the National Academy of Sciences report to the chairman of the report committee on 30 October 1990. NPS concurred that the TMBR and DMB techniques cannot be used alone to apportion exactly the sulfate haze at Grand Canyon to NGS emissions. However, NPS attempted to document how these quantitative techniques could be used in conjunction with the more qualitative analyses to arrive at reasonable estimates of NGS's contribution.

3. A Regulatory Impact Analysis for EPA's proposed regulation was required by Executive Order 12291 because the annual effect on the economy was $100 million or more.

4. See EPA Air Docket A-89-02A and JCM Environmental Inc.'s "Summary of Public Comments on Proposed Revision to Arizona Visibility Federal Implementation Plan for Navajo Generating Station" (July 1991).

5. Some examples are as follows: Letter from N. Kaplan to J. Bunyak, NPS Air Quality Division, responding to a request for pollution-control cost estimates (10 March 1989); Chestnut, L. G., and R. D. Rowe, "Integral Vista Benefit Analysis," Cambridge, Mass. (February 1983); FERC Form No. 1: Steam Electric Generating Plant Statistics; and EIA Form 412.

Literature Cited

Air Resource Specialists, Inc. 1991. Submission to EPA Air Docket #A-89-02A: technical information regarding visibility degradation caused by the Navajo Generating Station in the Grand Canyon and other national parks and wilderness areas within 500 km of NGS. Fort Collins, Colorado (18 April 1991).

Bell, P. A., D. B. Garnand, G. E. Haas, M. J. Kiphart, R. J. Loomis, W. C. Malm, G. E. McGlothin, J. V. Molenar, D. M. Ross, S. Solmonson. 1985. Assessment of visibility impairment on visitor enjoyment and utilization of park resources. Colorado State University, Cooperative Institute for Research in the Atmosphere, Fort Collins.

Bresch, J. F., E. R. Reiter, M. A. Klitch, H. K. Iyer, W. C. Malm, and K. Gebhart. 1987. Origins of sulfur laden air at national parks in the continental United States. P. 695–708 in P. S. Bhardwaja, ed. Proceedings of the Air Pollution Control Association International Specialty Conference, Visibility Protection: Research and Policy Aspects. Air and Waste Management Association, Pittsburgh.

Chinkin, L. R., D. A. Latimer, and H. Hogo. 1987. Layered haze observed at Bryce Canyon National Park: a statistical evaluation of the phenomenon. P. 709–719 in P. S. Bhardwaja, ed. Proceedings of the Air Pollution Control Association International Specialty Conference, Visibility Protection: Research and Policy Aspects. Air and Waste Management Association, Pittsburgh.

Fitz, D. R., P. Bhardwaja, J. Sutherland, and G. M. Markowski. 1987. Characterization and extent of winter hazes in the vicinity of Lake Powell. Paper presented at the 1987 annual meeting of the American Association for Aerosol Research, September 1987, Seattle.

Malm, W. C. 1992. Characteristics and origins of haze in the continental United States. Earth Sciences Review 33:36.

Malm, W. C., R. Eldred, and T. Cahill. 1985. Visibility and particulate measurements in the western United States. In Proceedings of the 78th Annual Meeting and Exhibition of the Air Pollution Control Association, June 1985, Detroit, Michigan, paper #85-11.3. Air and Waste Management Association, Pittsburgh.

Markowski, G. R. 1992. The WHITEX study and the role of the scientific community: a critique. Journal of the Air and Waste Management Association 42:11.

Mathai, C. V., ed. 1989. Visibility and fine particles. P. 781–897 in Transactions of the Air and Waste Management Association International Specialty Conference, Estes Park, Colorado, October 1989. Air and Waste Management Association, Pittsburgh.

National Park Service. 1986. Letter from associate director for Natural Resources to EPA assistant administrator for Air and Radiation (24 March 1986). EPA Air Docket #A-89-02A. Air Quality Division, Denver.

———. 1989a. The National Park Service report on WHITEX: draft final report (7 April 1989). EPA Air Docket #A-89-02A. Air Quality Division, Denver.

———. 1989b. The National Park Service final report on WHITEX (December 1989). EPA Air Docket #A-89-02A. Air Quality Division, Denver.

Sonoma Technology, Inc. 1991. Navajo Generating Station visibility study. Draft number 2 (16 April 1991). EPA Docket Section, Air Docket #A-89-02A, Washington, D.C.

U.S. Department of the Interior. 1985. Letter from acting assistant secretary for Fish, Wildlife, and Parks to assistant administrator for Air and Radiation, EPA (15 November 1985). EPA Air Docket #A-89-02A. National Park Service, Air Quality Division, Denver.

U.S. Environmental Protection Agency. 1980. Guidelines for determining best available retrofit technology analysis for coal-fired power plants and other stationary facilities (EPA-450/3-8–009b). Research Triangle Park, North Carolina.

———. 1990a. Draft report on best available retrofit technology analysis for the Navajo Generating Station in Page, Arizona (January 1990). EPA Docket Section, Air Docket #A-89-02A. Washington, D.C.

———. 1990b. Draft regulatory impact analysis of a revision of the Federal Implementation Plan for the state of Arizona to include sulfur dioxide controls for the Navajo Generating Station (February 1990). EPA Docket Section, Air Docket #A-89-02A. Washington, D.C.

4

Protection Versus Use

12

Rare Plant Monitoring at Indiana Dunes National Lakeshore

Noel B. Pavlovic and Marlin L. Bowles

The Endangered Species Act of 1973 (U.S. Fish and Wildlife Service 1988) directs federal land-management agencies to preserve ecosystems, although in application the focus is on rare-species population protection and management. The preservation and recovery of threatened and endangered (T&E) species is an appropriate goal because their decline is often an indication of ecosystem health and should ideally redirect management toward restoration measures. As a result, park managers have monitored and researched rare plants to better understand and manage their populations. Rare plant conservation involves species inventories, surveys, monitoring, research, and recovery (Bratton and White 1981; Palmer 1987; Pavlik and Barbour 1988). Little has been published about the successes, failures, and biological knowledge gained in monitoring programs, except for a few general (Norris 1987; Thompson and Countryman 1989; Fellers and Norris 1991) and specific species accounts (Clifton and Callizo 1987; Leitner and deBecker 1987; Pavlik and Barbour 1988).

In this chapter we review the development, application, and achievements of the Indiana Dunes National Lakeshore Threatened and Endangered Plant Monitoring Program (hereafter abbreviated T&EMP). We address biological, methodological, management, and administrative considerations through specific case studies and give examples of each T&EMP method. We took the data presented in the case studies from monitoring reports (Bowles et al.

1985, 1986a; Bowles 1988, 1989–1991) and research projects (McEachern et al. 1989; McEachern 1992; Pavlovic 1994).

The Park Setting

Indiana Dunes National Lakeshore, including Indiana Dunes State Park and the satellite areas of Pinhook Bog, Heron Rookery, and Hoosier Prairie, is 5,754 ha of dune ecosystem at the southern tip of Lake Michigan (Fig. 12.1). Congress authorized the national lakeshore in 1966 to "preserve for the educational, inspirational, and recreational use of the public certain portions of the Indiana dunes and other areas of scenic, scientific, and historic interest and recreational value" (Public Law 89-761). Congress prohibited any recreational facilities that "would be incompatible with the preservation of the unique flora and fauna or the physiographic conditions now prevailing."

After more than a century of botanical exploration (Cowles 1899; Fuller 1911, 1912; Lyon 1927, 1930; Pepoon 1927; Peattie 1930; Deam 1940; Wilhelm 1980, 1990; Klick et al. 1989), researchers have identified 1,445 plant species in the park, of which 1,135 are native. This total ranks the lakeshore

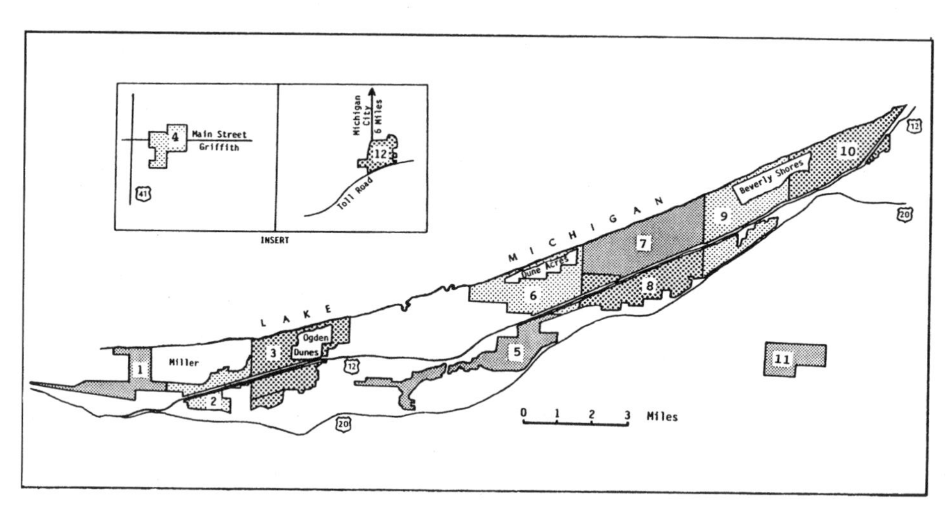

Figure 12.1 Survey unit map of the Indiana Dunes National Lakeshore (modified from Wilhelm 1990). West Unit: 1, Miller; 2, Tolleston; 3, West Beach and Inland Marsh; 4, Hoosier Prairie. East Unit: 5, Bailly; 6, Dune Acres; 7, Indiana Dunes State Park; 8, Visitor Center; 9, Keiser; 10, Tamarack; 11, Heron Rookery; 12, Pinhook Bog.

third among large (> 405 ha) national parks in species richness (Great Smoky Mountains and Grand Canyon National Parks are first and second; National Park Service 1977; Pavlovic and Cole 1994). The diversity is partly due to the variety of parent materials (sand, glacial till, and organic deposits), topographic moisture gradients, habitats, and disturbance regimes and is extraordinary given the extensive development and range of environmental insults in the region.

Twenty-six percent of Indiana's listed rare flora is found within the lakeshore. Of these 141 species, one (< 1%) is listed as extirpated (*Potamogeton pulcher*) but is now known to be extant, 56 (40%) are endangered, 36 (26%) are threatened, 38 (27%) are rare, and 10 (7%) are watch listed (nomenclature follows Swink and Wilhelm 1979). Many of the T&E plants have affinities with prairie, Atlantic coastal plain, boreal, or Eastern deciduous floras (Peattie 1922; Welch 1935; Swink and Wilhelm 1979) and are at the limits of their ranges. None of the plant species are endemic to the Indiana Dunes. Only Pitcher's thistle (*Cirsium pitcheri*) is federally listed as threatened. An additional seven species are category 2 candidate species (species that are proposed for federal listing): aromatic sumac (*Rhus aromatica arenaria*), bog bluegrass (*Poa paludigena*), butternut (*Juglans cinerea*), dune goldenrod (*Solidago racemosa gillmannii*), fame flower (*Talinum rugospermum*), Hall's sedge (*Scirpus hallii*), and Wolf's spike rush (*Eleocharis wolfii*).

Humans have greatly altered the dune ecosystem in ways that have affected rare plants. The beach and nearshore dunes have been changed by recreational pressure and shoreline development, which has altered natural erosion and replenishment processes (Hultsman 1986; Wood and Davis 1987). Ecologists have examined the roles of dune and soil formation, soil chemistry, fire regime, and microclimate on dune vegetation succession (Cowles 1899; Fuller 1911, 1912; Kurz 1923; Olson 1958; Futyma 1984; Henderson and Long 1984; Menges and Armentano 1985; Cole et al. 1992; Cole and Taylor, in press). Before the national lakeshore was established, the vegetation of the successively older, postglacial Nippising, Tolleston, Calumet, and Glenwood dunes was altered by logging, farming, sand mining, and residential and industrial development (Moore 1959; Cook and Jackson 1978; Hiebert and Pavlovic 1987; Pavlovic 1994). Fire suppression has converted savanna and prairie into more closed plant communities (Henderson and Long 1984; Taylor 1990; Cole and Taylor, in press). Researchers have shown that fire, soil characteristics, and hydrology (Klick et al. 1989) are important factors in the formation and maintenance of wetlands, including sedge meadows (Wilcox et al. 1984), mesic and wet prairie (Cole and Taylor, in press; Cole and Pavlovic n.d.), fens (Wilcox et al. 1986; Wilcox and Simonin 1987, 1988), and pannes (Hiebert et al. 1986). Ditching, draining, burning peat, and farming have greatly altered the wetlands (Wilcox et al. 1984). Atmospheric (Wilcox et al. 1985; Cole et al. 1990; Esser et al. 1991, 1992) and groundwater (Wilcox 1984, 1986) deposition of industrial pollution has contributed to the invasion of problem native and exotic species (Klick et al. 1989) and has probably altered rates of succession. The loss of most vertebrate carnivores (Whittaker et al. 1994) has indirectly affected the vegetation through increased herbivore density. Further impacts will occur when the surrounding rural land is converted to suburban and industrial landscape as the Chicago urban sprawl expands eastward.

Program Methods and Approach

The National Park Service (NPS) initiated T&EMP in 1984 under contract with the Morton Arboretum and the Natural Land Institute. Researchers inventoried, surveyed, and monitored Indiana-listed endangered and threatened vascular plants (Bacone and Hedge 1980; Fox and Hiebert 1983) during the first and second years of the contract, respectively. Thirty-six endangered and 42 threatened plant species were inventoried and monitored in 1984 and 1985; 20 endangered and 9 threatened species, however, remained undiscovered. Through a cooperative agreement with the Morton Arboretum, 64 (82%) of the initial 78 species have been remonitored since 1986. Of these, 37 (58%) have been remonitored once, 12 (19%) twice, 5 (8%) thrice, and 10 (16%) four or more times. Limited funding restricted the remonitoring to species listed by the U.S. Fish and Wildlife Service that showed declines, had few small populations, experienced management problems, or had unusual life-history behavior.

The objectives of T&EMP were to inventory, survey, map, and describe populations, including those in new areas added to the park; sample selected populations; design a monitoring system; and develop management recommendations for each species. T&EMP was designed to guide park personnel in managing T&E species through qualitative reconnaissance and quantitative monitoring of permanent plots and through assessment of habitat changes by use of permanent photographic plots, archival photographs, and records of associated species (Bowles et al. 1986b). Sampling methods for population density, genet or ramet frequency, and flower or fruit number in quadrats were designed in relation to the plant form, population size, and dispersion of each species (Bowles et al. 1986b). Researchers surveyed the monitoring plots and populations in relation to permanent features, such as roads and trails, and mapped on U.S. Geological Survey 15-minute quadrangles and on the park base map (scale 1:2,400) after 1985; (Bowles et al. 1986b). In subsequent years, rather than treating each species individually, researchers have lumped species into comparable habitat groups (e.g., savanna and prairie) because species in the same habitat often experience similar threats, management problems, and needs.

In addition to the field monitoring, researchers have relied on other sources of information and methods of data management. The plant ecologist has maintained a file of published literature on the rarest T&E species. Since 1990 investigators have mapped and digitized rare-plant populations using a Geographic Information System to help protect T&E populations from destruction through park development. While mapping specific species, the scientists make marginal notes to identify which additional species need mapping. To facilitate the tracking of numerous T&E species, the researchers are developing a dBASE rare-plant tracking system. The design is general enough to be used by any conservation group and is hierarchical, starting at the park level and proceeding to park units, individual species, and specific populations.

Since 1985 T&EMP has grown because of the rediscovery of 9 plants seen earlier in the century, discovery of 24 new species, addition of new species to the state list, and changes in species status from 1980 to 1990 (Aldrich et al., 1986; Martin 1990). Of the 24 new discoveries, 17 are still listed as endangered or threatened (Martin 1990). Of all these species, 16 have been added to the monitoring program.

In the late 1980s, park personnel initiated two projects involving demographic monitoring and experimental research (Pavlik and Barbour 1988) to address management questions concerning two federally listed plant species: Pitcher's thistle and fame flower. The goal of each project was to understand the processes that determine distribution and abundance of each species.

When classified by habitat type and disturbance regime, 131 populations of 81 species were distributed equally among habitats, and there were no differences between naturally and anthropogenically disturbed communities (Bowles et al. 1990). However, 25% of all the populations were found in anthropogenically disturbed communities, these being equally divided between disturbed natural communities and areas disturbed beyond recognition as natural communities. The majority of populations (38.9%) were in late successional disturbance regimes. Anthropogenic populations were concentrated in annual/frequent disturbance regimes.

From 1988 to 1990, 26 (56%) of the 46 monitored populations declined, 9 (20%) did not change, and 11 (24%) increased (Table 12.1). Some declines (e.g., panne spike rush [*Eleocharis geniculata*] and zigzag bladderwort [*Utricularia subulata*]) were attributed to the moderate 1988 drought. These declines were not limited to annuals that can be expected to vary more than perennials during the same period. The two woody plants, common juniper (*Juniperus communis depressa*) and dune willow (*Salix syrticola*), showed no change probably because of their longevity. A decline in the only known population of the perennial beach pea (*Lathyrus japonicus glaber*) was probably caused by grapevine encroachment and succession. Further research is warranted for species, such as gaywings (*Polygala paucifolia*), that have shown enigmatic declines. The decline in bunchberry (*Cornus canadensis*) north of Cowles Bog may represent natural fluctuations or successional change. This former white-pine swamp was logged and replaced by a red-maple/yellow-birch swamp, which is more shaded (Futyma 1984). Other species experienced declines in reproduction or production of flowering stems whereas ramet densities remained the same (stiff aster [*Aster ptarmicoides*] and creeping sedge [*Carex chordorrhiza*]). Although many of the case studies concerned declining species, about half of the monitored populations were stable or increasing. Some species, such as white (*Cypripedium candidum*) and showy ladyslippers (*C. reginae*), have benefited from prescribed burns (Bowles 1991). White ladyslipper ramets have increased from 22 to 30 under prescribed burning at Hoosier Prairie, whereas the species merely hangs on at Cowles Bog in the absence of fire and presence of cattail invasion (Table 12.1, Fen and Bog). Showy ladyslipper has responded in a similar fashion to fire.

Table 12.1. Percent changes between data of original quadrats collected in 1984 and 1985 and data collected from the same plot in 1988, 1989, 1990, or 1991. Multiple values indicate data from multiple survey units. Data were modified from Bowles (1988, 1989–1991). Asterisk indicates statistical significance (NS = nonsignificance) at $p < 0.05$.

Community/*Species* (Date of Resample)	Increase	No Change	Decline
BEACH AND DUNE COMPLEX			
Cakile edentula (1991)			35%, 66%
Euphorbia polygonifolia (1989)			90%
Juniperus communis depressa (1990)		NC	
Lathyrus japonicus glaber (1990)			*98%
Salix syrticola (1990)			NS 16%
SAND SAVANNA AND PRAIRIE			
Aralia hispida (1990)			*> 90%
Aristida tuberculosa (1988)		NC	
Aster sericeus (1990)	9%		
Carex conoidea (1989)			68%, 100%
Carex flava fertilis (1990)		NC	
Eleocharis wolfii (1989)		NC	
Habenaria flava herbiola (1989)	425%		
Hudsonia tomentosa (1988)			32%
Lycopodium clavatum (1990)	76%		
Lycopodium tristachyum (1990)	229%		
Malaxis unifolia (1990)	4%		
Melampyrum lineare (1989)			62%, 77%
Polygonella articulata (1988)			99%, 64%
Scleria pauciflora caroliniana (1989)		NC	
Selaginella rupestris (1988)			19%
Sisyrinchium angustifolium (1989)	64%		73%
Stipa avenacea (1989)	29%		26%

continued

Table 12.1. *Continued.*

Community/*Species* (Date of Resample)	Increase	No Change	Decline
WET MEADOW AND PANNE			
Aster ptarmicoides (1988)	18%	NC	
Eleocharis geniculata (1988)			18%
Habenaria hyperborea huronensis (1989)			10%
Potentilla anserina (1989)			11%
Utricularia cornuta (1989)			152%
Utricularia subulata (1989)			93%
FEN AND BOG			
Cavex chordorrhiza (1989)			91%
Carex seorsa (1990)	60%		
Cornus canadensis (1990)			30%
Cypripedium candidum (1989)	44%	NC	
Isotria verticillata (1989)			42%
Myosotis laxa (1989)		NC	
DUNE FOREST			
Chimaphila umbellata cisatlantica (1990)		NC	
Botrychium matricariaefolium (1990)			*90%
Polygala paucifolia (1990)			> 50%
Trillium cernuum macranthum (1990)	31%		

Analyses and observations suggest that four factors and their interactions affected the occurrence, distribution, and relative stability of rare plant populations: (1) frequency and intensity of anthropogenic disturbance, (2) presence or absence of natural disturbance processes, (3) intensity of stochastic environmental events, and (4) existing management regimes (Bowles et al. 1990). The case studies presented below illustrate these factors in a management perspective by concentrating on recreational, developmental, erosional, successional, anthropogenic, and enigmatic impacts.

Case Studies

This section explains in detail the internal and external threats to the T&E plants of Indiana Dunes and how important inventory and monitoring are to resource protection.

Recreational Impact

As the enabling legislation suggests, recreational activity is a central focus for the lakeshore because of the park's scenic beauty and recreational value for swimming, sunbathing, and hiking. These activities are concentrated on the beaches, foredunes, and blowouts in the nearshore environment (Hultsman 1986). In these habitats, conflicts between recreation and protection of plant species are most evident.

Sea rocket (*Cakile edentula* var. *lacustris*) grows in nearshore habitats and may be negatively affected by human use of the beaches (DeMauro et al. 1986). Sea rocket is an annual succulent herb characteristic of the Great Lakes beaches above the winter storm-wash line (Gleason 1952; Payne and Maun 1984). At Indiana Dunes, it is also scattered throughout the foredunes and along the edges of pannes especially in drought years.

Sea rocket appears to be sensitive to human trampling (see Payne and Maun 1984), as suggested by low densities of plants near the bathhouse and Wells St. Beach. Greater population reduction at the bathhouse relative to Wells St. beach is probably due to visitor density: 60,000 to 94,000 vehicles visit the 600-car West Beach parking lot from April to October, whereas Wells St. beach is a much smaller private lot.

Sea rocket populations also fluctuate greatly from year to year based on data collected from 1985 and 1988–1991. Beach censuses varied significantly over time at West Beach and Dune Acres. Although average population density was lower at West Beach than at Dune Acres in 3 out of 5 years, site differences were undetectable. These trends suggest that precipitation also affects sea rocket density. Low densities were caused by the 1988 drought, but in 1985 the densities were higher despite a more substantial drought. The absence of sea rocket from West Beach in 1990 is attributed to landward movement of sand as lake levels receded from 1987 highs. Population size is also influenced by other direct and indirect effects involving precipitation, visitation, and lake level, which determine habitat width and quality.

Data on sea rocket have provided insights into the factors important to its distribution and abundance. The relationship between high visitor use and sea rocket decline has led to the recommendation that certain beach areas be designated as low visitation areas. The Dune Acres Unit Beach may serve this purpose because access is limited to boaters and hikers. How the increased visitation from three new marinas being constructed to the southwest will affect this population remains to be seen.

The federally listed, threatened Pitcher's thistle is endemic to the sand dunes of the Great Lakes Huron, Michigan, and Superior (Harrison 1988; Loveless and Hamrick 1988). In Indiana Dunes, Pitcher's thistle formerly grew in the foredunes and blowouts but is now restricted to early and mid-successional disturbance patches in blowouts (McEachern 1992; McEachern et al. 1994; Pavlovic 1994). Before and after construction of the West Beach bathhouse in 1976, protection of the dunes through guidance and trail structures was absent. As part of the 1985 dune-impact research, a guided boardwalk/trail was constructed to reduce hiking impacts, whereas another portion was left unprotected. This experiment confirmed Pitcher's thistle's sensitivity to trampling (see also Keddy and Keddy 1984) because populations are larger in area and size in the protected areas than in unprotected areas: an average of 23 versus 5 plants per population. Trail development that directs and channels visitor use in the dunes will protect Pitcher's thistle.

Juvenile mortality of Pitcher's thistle is higher in areas having heavy public use, rabbit grazing is reducing juvenile growth at one site, and many Indiana Dunes populations have declined over the 4 years of study (McEachern 1992). The pattern of juvenile mortality would not have been apparent based on annual population counts only. Thus demographic monitoring is more explicit in assessing causes for populations trends. Four years of demographic sampling provide a rigorous benchmark for evaluating long-term Pitcher's thistle population trends.

Pitcher's thistle illustrates the complexity of managing T&E species because its viability and conservation depend on long-term protection of landscape units (Pavlovic et al. 1993; Fig. 12.2). Suitable dune habitat that is unoccu-

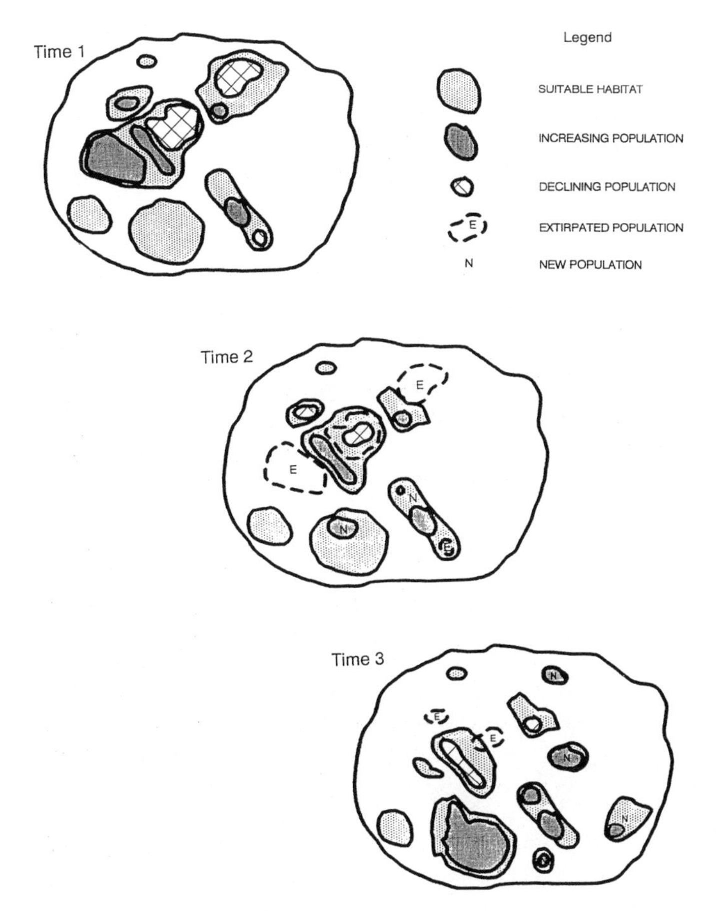

Figure 12.2 Metapopulation model for conservation of Pitcher's thistle. Sequence of three times illustrates several properties of metapopulation dynamics. Not all suitable habitat patches are occupied at one time nor are they fixed in space. Three suitable patches are unoccupied at time 1, two at time 2, and three at time 3. In the sequence, two suitable habitat patches decrease in size, one disappears, and four are created. For a species to fit a metapopulation model, subpopulations must be weakly linked so their trends and growth rates are not synchronous among subpopulations. In this model, some subpopulations and suitable habitats are declining, whereas others are increasing.

pied by Pitcher's thistle is important because it may become occupied habitat at a later time and act as a dispersal corridor to newly created suitable habitat. Park facility development in currently unoccupied habitat and isolation of single populations will lead to the eventual demise of such populations (Pavlovic et al. 1993). A workshop and a poster were developed (McEachern and Pavlovic 1991) to educate park personnel on metapopulation dynamics and conservation of Pitcher's thistle. Nevertheless, despite a mid-1980s study of anthropogenic impacts on dunes, no explicit comprehensive plan for dune protection exists at the lakeshore. Planning is underway to protect Pitcher's thistle populations and other important dune habitats (McEachern et al. 1994).

Developmental and Management Impacts

Development in national parks impacts native plants (Bratton and White 1980) through habitat loss or secondary effects. Failure of interdivisional communication concerning developments has altered several populations of rare plants.

T&EMP was initially successful because inventories identified new T&E species populations, whose presence thwarted development plans at the West (Tolleston Dunes) and East Transit Units. These sites were purchased to develop a series of parking lots to shuttle visitors to West Beach (National Park Service 1980, 1984) and Mount Baldy. The vegetation of the Tolleston Dunes was characterized as "weeds and grasses in sand-mined area north of shopping center; a few black oak" (National Park Service 1979). Inventories, however, revealed eight new rare-species populations at Tolleston (bristly sarsaparilla [*Aralia hispida*], false heather [*Hudsonia tomentosa*], beach three-awn grass [*Aristida tuberculosa*], rushlike aster [*Aster junciformis*], jack pine [*Pinus banksiana*], round-headed rush [*Juncus scirpoides*], fame flower, and Hall's sedge; Fox and Manek 1984) and two new populations at the East Unit Transit Center (few-flowered nut rush [*Scleria pauciflora*] and long sedge [*Carex folliculata*]; Bowles 1986). A more recent development plan (National Park Service 1990) proposes only to develop a trail through the Tolleston area and avoid the rare plant populations at the East Unit Transit Center.

Park managers did not consult NPS Resource Management and Research Divisions about replacing the wooden boardwalk with plastic superdeck at

Pinhook Bog. The managers thought that a boardwalk 15.2 cm wider would not have any environmental impact; however, previous rare-plant inventory and monitoring had revealed that seven T&E species abutted the boardwalk. After discovering the boardwalk replacement plans, researchers directed maintenance personnel to shift the boardwalk away from T&E species. The researchers also transplanted individuals of the following plant species within the bog: pink ladyslipper (*Cypripedium acaule*), orange-fringed orchid (*Habenaria ciliaris*), pitcher plant (*Sarracenia purpurea*), and yellow-eyed grass (*Xyris caroliniana*). It is ironic that these species were probably growing along the boardwalk because of the increased light that the trail afforded.

Bowles (1989) was unable to remonitor round-headed rush in a former roadbed near the newly constructed Douglas Environmental Education Center because the roadbed was converted to a gravel trail. Fortunately, this species grows at other sites; however, researchers do not know what component of genetic diversity was lost. Another trail was rerouted to prevent repeated trail mowing from killing the only known red baneberry plant (*Actea rubra*) in the lakeshore.

These examples illustrate the need for divisional coordination in large agencies such as NPS to ensure that the locations of rare species are considered in development planning. At Indiana Dunes National Lakeshore, a new environmental compliance form, developed by the superintendent, has alleviated this problem by ensuring that all projects are reviewed.

Beach and Bluff Erosion

Rice grass (*Oryzopsis asperifolia*) is a perennial rhizomatous grass distributed throughout the boreal regions from Newfoundland and Quebec to British Columbia, south to Pennsylvania and Indiana, and in the Rocky Mountains to New Mexico (Gleason 1952). In 1984 rice grass was known in Indiana only from a northeast-facing, mesic oak forest on a dune 45 ft above Lake Michigan, where 12 genets grew in three colonies. By 1987 eight plants were lost to bluff erosion that was exacerbated during record-high lake levels because of a hardened breakwater revetment and jetty to the east in Michigan City (Wood and Davis 1987). In addition, two of the four remaining plants had died of other causes.

This example illustrates the vulnerability of small populations to stochastic events that are exacerbated by human actions. The bluff-erosion threat to rice grass was predicted (Bowles et al. 1985). The plants could have been saved and transplanted if those responsible for T&EMP after 1985 had known of the threat and monitored lake levels.

Succession and Fire Protection

Many plant species grow in successional habitats and require disturbance processes for their persistence (Bowles et al. 1990; Pavlovic 1994). With habitat fragmentation and the suppression of fire in the dunes, such species have declined as oak invasion has proceeded.

Black oat grass (*Stipa avenacea*) is a perennial of dry woodland throughout eastern North America (Gleason 1952). Two populations in a black oak woodland and savanna were sampled in 1984 and showed lower culm density and fewer culms per genet, respectively, in more shaded habitat. The more open savanna site was last burned in 1980 (Taylor 1990); oat-grass has also declined here since 1984 as the black oak scrub has grown up. These comparisons between sites through time confirm that shading is deleterious to black-oat-grass sexual reproduction and vegetative spread, although the grass can persist vegetatively. In the dunes, fire is required to maintain open savanna habitat by arresting woody plant encroachment, although in other areas savanna may also depend on gap phase succession. This knowledge, plus the presence of other species requiring open habitats, has led to designating the Kansas Ave. area as a high-priority burn site.

Of the six club-moss species that grow at the lakeshore, two are quite rare partly because they are at the limits of their ranges (Mickel 1979; Lellinger 1985): running ground pine (*Lycopodium clavatum*] and ground cedar (*L. tristachyum*). These were located and monitored in 1984, each at individual sites and both at one site that they share. Since then, populations at the individual sites have been extirpated. For example, the one *L. clavatum* population was smothered under 7.6 cm of white-oak leaf litter from adjacent fire-suppressed dune forest!

At the last remaining site for both species, the plants persist, but fertile spore-producing strobili have never been seen. Succession is occurring at this site through even-aged invasion of white and black oak in response to fire suppression. Canopy closure and the decline of these club mosses were the impetus to initiate a research-prescribed burn in the spring of 1992. Portions of these small populations were protected from the fire to serve as controls and to ensure that this experiment did not extirpate both species. Running ground pine is also fire sensitive because its runners are above ground (Lellinger 1985). Before the prescribed burn, two very healthy clumps of running ground pine were discovered adjacent to a nearby trail; however, these were heavily damaged by illegal all-terrain-vehicle activity.

Role of Anthropogenic Habitats in Species Maintenance

As in other NPS units (White 1984), the monitoring program revealed that many successional rare plants rely on anthropogenic disturbances or habitats for their persistence (Fig. 12.2) in the absence of natural disturbance regimes (Bowles et al. 1990; Pavlovic 1994). Examples include annual (beach three-awn grass and jointweed [*Polygonella articulata*]) and perennial xeric species (false heather, sand club moss [*Selaginella rupestris*], and fame flower) and annual hydric species (round-headed rush and Hall's sedge of former roadbeds and sand-mined areas, and perennial grasses [wood millet (*Millium effusum*) and grove blue grass (*Poa alsodes*)] along trails where periodic maintenance and hiking activity mimic natural gap disturbances). Human disturbances mimic natural disturbance processes and have allowed these species to survive.

Fame flower is a small perennial succulent herb of the Midwest and Great Plains having small pink flowers that bloom once from 4 to 7 p.m. (Cochrane 1993). Fame flower is endangered in Indiana and is confined to anthropogenic sites (70% of Indiana sites), including unpaved roadbeds, trails, and roadsides in sand dunes, except at a few sites where it grows in a high-quality savanna.

Paradoxically, species such as fame flower that are confined to anthropogenic habitats are dependent on continued human disturbance (trampling, mowing, and grading) for their existence as long as natural disturbance processes are altered or eliminated from adjacent native habitat (Bowles et al. 1990). Fame flower grows in an abandoned road and trail habitat, where it receives different levels of foot trampling and off-road-vehicle activity. This has maintained high population densities where disturbance was light.

Experimental plantings of fame flower seeds in six habitats—ranging from sand blowout to black oak woods—that experienced ambient and supplemental rainfall demonstrated that exposed sand blowouts are too dry to allow seedling establishment. More vegetated conditions are favorable for successful seedling establishment, but shading from closed tree canopies and persistent oak-leaf litter causes the seedlings' demise. Thus the greatest emergence and survival occurred in the savanna gaps. Demographic monitoring alone would not necessarily reveal this information; however, such monitoring has shown that seedlings are fire intolerant but adults are resistant and capable of resprouting (N. B. Pavlovic, unpubl. data.). Reintroduction of fame flower into its native habitat will require reintroduction of fire, which will benefit not only fame flower but also many other species living in the savanna ecosystem. This information is important to deciding how to restore species such as fame flower that are now largely confined to anthropogenic disturbances or habitats.

Unexplained Population Declines

The state-listed, endangered blue hearts (*Buchnera americana*), a hemiparasitic herb with opposite sessile lanceolate leaves and a raceme of purple flowers, is known from two mesic sand prairies in Indiana. It inhabits sandy or gravelly upland woods and prairies from New York to Michigan and Illinois and south to Georgia and Texas (Gleason 1952) but is also reported from mesic and wet meadows (Ostlie 1990).

Two monitoring plots were established in 1984 in the park when the population was dense and large. Subsequent resampling showed a decline of plants, even though individuals still grew outside the plots. Since 1989 no blue hearts have been seen, although the other Indiana population outside the park has continued to emerge and flower. This decline occurred despite copious flowering and seed set: each capsule produces hundreds of seeds, and a 1986 study (R. E. Cook and F. Oviatt, unpubl. data) found that 75% of 108 ramets bloomed from 57 plants.

Rankin and Pickett (1985) suggested that fire favored blue hearts and that persistence between fires is ensured through the presence of a subterranean parasitic stage. The populations in the park have declined, however, despite the area being burned in 1986, 1988, and 1993. We hypothesize that in mesic sites abundant precipitation and groundwater or reduced grass competition when water levels are high initiate the emergence of the blue-hearts' subterranean stage as photosynthetic plants. When precipitation is high, fire may increase emergence through nutrient enrichment or removal of shade from a thick duff layer. The failure of blue hearts to reappear in 1990 and 1991, when groundwater levels were rising and high, and their reappearance in 1992 suggest there is a time lag in blue-hearts' response.

This example illustrates the potential for false positive declines as detected in a monitoring program and the need to temper population trends with biological knowledge and intuition. It also illustrates the value of integrating population monitoring with environmental sampling (i.e., groundwater). More frequent monitoring and groundwater data from the other Indiana population outside the park would have aided in interpreting the decline in the park.

Bristly sarsaparilla is an outcrossing perennial herb (Thomson 1988) of dry open woods in northeastern North America (Gleason 1952). Voss (1985) implied that this species is successional by stating that it forms large vegetative colonies in newly disturbed sandy soils and persists for some years thereafter. This species dramatically declined from 108 to 8 genets and also had a reduction in umbel number per plant (16 to 1 umbel) from 1984 to 1990. In 1991 and 1992 no plants were found in this population, which suggests extirpation. No plants have reappeared since. Perhaps this species depends upon some dynamic patch-disturbance regime that requires dispersal among patches and metapopulation dynamics for persistence.

Discussion and Conclusions

Threats and the Value of T&EMP

Pimm and Gilpin (1989) identified four threats (four horsemen of the biological apocalypse) to biological diversity: habitat loss, introductions, overexploitation, and secondary effects. T&EMP is valuable because it can provide the basis for thwarting such threats.

In some instances the threats are predictable and obvious, whereas in others they are unpredictable and difficult to understand. The threat of bluff erosion to rice grass was known before the rise in Lake Michigan level. On the other hand, researchers had no way of predicting the loss of bristly sarsaparilla and the "disappearance" of blue hearts. In the former case, the cause of the decline is still uncertain. Without T&EMP, park managers know neither what they have nor what they may have lost.

Loss of habitat and T&E species through park development must be mitigated by careful planning. The loss of a population of round-headed rush could have been prevented with adequate on-site planning and coordination. The examples of the East and West Transit centers, restoration of special plants at Pinhook Bog, and trail alteration for red baneberry show that under a favorable resource-protection environment, development plans can be altered to protect T&E plants and mitigate visitor use of the park. This favorable environment requires financial and manpower support from the agency leadership, upper-level park management, and resource personnel dedicated to the protection and preservation of park resources.

Loss of native vegetation and T&E species through the invasion of exotics and problem native species is technically more difficult to solve because it requires active management to eliminate the threatening species. At Indiana Dunes, exotic species, such as purple loosestrife (*Lythrum salicaria*), glossy buckthorn (*Rhamnus frangula*), field thistle (*Cirsium arvense*), reed canary grass (*Phalaris arundinacea*), and honeysuckles (*Lonicera* spp.), pose a serious threat to much of the vegetation (Klick et al. 1989), as do problem native invaders, such as black locust (*Robinia pseudoacacia*) and common reed (*Phragmites australis*). We are unaware of any documented T&E species that have declined because of exotic plants at Indiana Dunes, although this has undoubtedly occurred in already devastated habitats, such as wetlands. Aggressive native blackberries and grapevines may have played a role in decimating bristly sarsaparilla and beach pea. Problem exotic and native herbivores, such as deer, can also affect rare plants (Miller et al. 1992), for example, by trampling a small population of false heather and herbivory on nodding trillium (N. B. Pavlovic, pers. obs.). We also do not know of any threats to T&E species of commercial and aesthetic value. Because of the numerous beautiful orchids and carnivorous T&E plant species at the lakeshore, collection could be a problem as it was in the past (Pepoon 1927; Peattie 1930).

The indirect threats, however, are the problems that are often unpredictable, subtle, and difficult to identify (Bratton and White 1980). Is the apparent loss of bristly sarsaparilla due to blackberry invasion or to loss in genetic diversity in this small population, or is the decline an artifact of a short life span? Are such population fluctuations normal for this species? Is air pollution having a negative impact on the lakeshore's T&E species? Investigators cannot answer these questions with the monitoring data at hand, but T&EMP points the way to needed research.

Population Dynamics and T&EMP

It is well known that species fluctuate in space and time in response to environmental variation at various scales. Some species fluctuate wildly (Vandermeer 1982), whereas others appear stable for long periods. Cryptic life stages, such as those exhibited by orchids and hemiparasites, add to long-term variability in population trends and make demographic studies difficult (Menges 1991).

It is this temporal variability that makes detection of declining T&E population health tricky. Only with long-term data, such as those for blue hearts, Pitcher's thistle, fame flower, and sea rocket, can researchers obtain a benchmark for distinguishing usual fluctuations from catastrophic declines and from increasing visitor impacts. Long-term population data can also be used to assess population viability and extinction risks (Dennis et al. 1991). The blue-hearts study illustrates the value of monitoring both T&E species and the environment so population changes can be correlated with environmental trends.

Evaluation of T&EMP

For nearly a decade, the Indiana Dunes T&EMP has provided information to park managers on the protection and preservation of T&E species. This program is no exception to the rule that programs must evolve in response to new information, new needs, and perhaps better methodology in monitoring. From this evaluation, we believe that T&EMP needs revitalization.

In conjunction with T&EMP, small experiments by park managers to test species biology and management hypotheses can be valuable (Cook and Dixon 1986). The experiments, such as those planned for running ground pine, ground cedar, and white ladyslipper, can be helpful in understanding and mitigating plant declines.

T&EMP has been largely reactive rather than proactive. More funds could be saved if threats to T&E species could be anticipated and recovery instituted before the biological and managerial crisis level is reached. Given the unpredictable nature of threats, researchers will never be able to anticipate all declines. Prediction of threats requires a classification of realistic threats to populations and a good knowledge of conditions in the field, information that T&EMP provides. Although researchers cannot always predict threats, they can at least mitigate population declines before the crisis level is reached.

Although the program was minimally funded from 1987 to 1990, no single person was responsible for critical assessment of the program. We think the program has suffered in the area of follow-up actions. We provided guidance as to which species should be monitored, but species rescue and recovery were ancillary activities. Rare plant research and population mapping have continued under the direction of the senior author. The monitoring program, however, has lapsed, despite a monitoring action plan (Oviatt 1986), because of lack of funds and the failure of resource managers to defend the need to administrators. We hope these deficiencies will be corrected and the program will be institutionalized.

Despite these shortcomings, T&EMP has indicated the health of and trends in rare plant populations and has yielded insights into the ecosystem processes on which the species depend. Inventory and monitoring are vital in maintaining and protecting this component of biodiversity. Without knowledge of resources such as T&E plants, managers of national parks will fail to balance their dual mandate to preserve natural resources and provide public enjoyment. Ultimately, through the knowledge gained in T&EMP, park personnel should be able to preserve and protect T&E plants by managing ecosystem processes, rather than by intensive (and expensive) species-specific triage. Declines in T&E species are, in fact, an indicator of ecosystem health and should point the way to understanding important disturbance and ecosystem processes.

Acknowledgments

Sandy Whisler assisted in data analysis. Shelly Swisher and Sandy Whisler prepared the figures. Midwest regional chief scientist Ron Hiebert and Indiana Dunes chief scientist Richard Whitman supported the senior author in writing the paper. Kathryn McEachern kindly provided her Pitcher's-thistle population data from West Beach. We also thank the many employees and volunteers who helped the rare-plant monitoring program either by finding and mapping new species or by assisting in monitoring.

Literature Cited

Aldrich, J. R., J. A. Bacone, and M. A. Homoya. 1986. List of extirpated, endangered, threatened, and rare vascular plants in Indiana: an update. Indiana Academy of Science 95:413–419.

Bacone, J. A., and C. L. Hedge. 1980. A preliminary list of endangered and threatened vascular plants in Indiana. Proceedings of the Indiana Academy of Science for 1979 89:359–371.

Bowles, M. L. 1986. Reconnaissance for endangered or threatened plant species in the East Unit Transit Area. Unpubl. report, Indiana Dunes National Lakeshore. 3 p.

———. 1988. A report on special floristic elements at the Indiana Dunes National Lakeshore: new species monitoring and update of selected existing populations. Unpubl. report, Indiana Dunes National Lakeshore. 70 p.

———. 1989–1991. A status report on endangered and threatened plants of the Indiana Dunes National Lakeshore: monitoring of species new to the lakeshore and remonitoring of selected species. Unpubl. annual reports, Indiana Dunes National Lakeshore. 48, 97, 101 p.

Bowles, M. L., M. M. DeMauro, N. Pavlovic, and R. D. Hiebert. 1990. Effects of anthropogenic disturbances on endangered and threatened plants at the Indiana Dunes National Lakeshore. Natural Areas Journal 10:187–200.

Bowles, M. L., W. J. Hess, and M. M. DeMauro. 1985. An assessment of the monitoring program for special floristic elements at the Indiana Dunes National Lakeshore: Phase I. The endangered species. Unpubl. report, Indiana Dunes National Lakeshore. 70 p.

———. 1986a. An assessment of the monitoring program for special floristic elements at the Indiana Dunes National Lakeshore: Phase II. The threatened species. Unpubl. report, Indiana Dunes National Lakeshore. 70 p.

Bowles, M. L., W. J. Hess, M. M. DeMauro, and R. Hiebert. 1986b. Endangered plant inventory and monitoring strategies at Indiana Dunes National Lakeshore. Natural Areas Journal 6(1):18–26.

Bratton, S. P., and P. S. White. 1980. Rare plant management—after preservation what? Rhodora 82:49–75.

———. 1981. Rare and endangered plant species management: potential threats and practical problems in U.S. national parks and preserves. P. 459–474 in

H. Synge, ed. The Biological Aspects of Rare Plant Conservation. John Wiley & Sons, New York.

Clifton, G., and J. Callizo. 1987. Monitoring rare plant populations in the Knoxville area of California. P. 397–400 in T. S. Elias, ed. Conservation and Management of Rare and Endangered Plants. California Native Plant Society, Sacramento.

Cochrane, T. S. 1993. Status and distribution of *Talinum rugospermum* Holz. (Portulacaceae). Natural Areas Journal 13:33–41.

Cole, K. L., D. R. Engstrom, R. P. Futyma, and R. Stottlemyer. 1990. Past atmospheric deposition of metals in northern Indiana measured in peat core from Cowles Bog. Environmental Science and Technology 24:543–549.

Cole, K. L., K. F. Klick, and N. B. Pavlovic. 1992. Fire temperature monitoring during experimental burns at the Indiana Dunes. Natural Areas Journal 12:177–183.

Cole, K. L., and N. B. Pavlovic. n.d. Howes Prairie, a remnant of the Indiana Dunes Prairie preserved by periodic flooding. Unpubl. report, Indiana Dunes National Lakeshore Science Division. 33 p.

Cole, K. L., and R. S. Taylor. In press. The history of an Indiana Dunes sand prairie: the effects of time, water, and fire. Journal of Vegetation Science.

Cook, R. E., and P. Dixon. 1986. A review of recovery plans for threatened and endangered plant species: a report for the World Wildlife Fund. Unpubl. report, World Wildlife Fund, Washington, D.C. 88 p.

Cook, S. G., and R. S. Jackson. 1978. The Bailly area of Porter County, Indiana. Unpubl. report, Indiana Dunes National Lakeshore. 110 p.

Cowles, H. C. 1899. The ecological relations of the vegetation on the sand dunes of Lake Michigan. Botanical Gazette 27:95–117; 167–202; 281–308; 361–391.

Deam, C. C. 1940. Flora of Indiana. Indiana Department of Conservation, Indianapolis. 1236 p.

DeMauro, M. M., M. L. Bowles, W. J. Hess, R. Hiebert, and N. B. Pavlovic. 1986. The impact of anthropogenic disturbances on populations of *Cakile edentula* (sea rocket) and *Euphorbia polygonifolia* (seaside spurge) at the Indiana Dunes National Lakeshore [abstract]. P. 15 in Program for the 1st Indiana Dunes Research Conference—Indiana Dunes: A Century of Scientific Enquiry, May 1–3, 1986, Gary, Indiana. National Park Service, Science Publications Office, Atlanta.

Dennis, B., P. L. Muncholland, and J. M. Scott. 1991. Estimation of growth and extinction parameters for endangered species. Ecological Monographs 61: 115–144.

Esser, K. B., J. G. Bockheim, and P. A. Helmke. 1991. Trace element contamination of soils in the Indiana Dunes. Journal of Environmental Quality 20:492–496.

———. 1992. Mineral distribution in soils formed in the Indiana Dunes, USA. Geoderma 54:91–205.

Fellers, G. M., and V. Norris. 1991. Rare plant monitoring and management at Point Reyes. Park Science 11:20–21.

Fox, C., and R. Hiebert. 1983. State endangered, threatened, and rare plant species of the Indiana Dunes National Lakeshore. Unpubl. report, National Park Service. 6 p.

Fox, C., and W. Manek. 1984. Assessment of the West Unit Transit Center proposals: impacts on the vegetation of Tolleston Dunes. Unpubl. report, Indiana Dunes National Lakeshore Science Division. 52 p.

Fuller, G. D. 1911. Evaporation and plant succession. Botanical Gazette 52:193–208.

———. 1912. Germination and growth of the cottonwood upon the sand dunes of Lake Michigan near Chicago. Transactions of the Illinois State Academy of Science 5:137–143.

Futyma, R. P. 1984. Paleobotanical studies at Indiana Dunes National Lakeshore. Unpubl. report, National Park Service. 242 p.

Gleason, H. A. 1952. The new Britton and Brown illustrated flora of the northeastern United States and adjacent Canada. Volumes I–III. Hafner Press, New York. 482, 655, 595 p.

Harrison, W. F. 1988. Endangered and threatened wildlife and plants: determination of threatened status for *Cirsium pitcheri*. Federal Register 53(137):27137–27141.

Henderson, N. R., and J. N. Long. 1984. A comparison of stand structure and fire history in tow black oak woodlands in northwestern Indiana. Botanical Gazette 145:222–228.

Hiebert, R. D., and N. B. Pavlovic. 1987. Past land use effects on succession in razed residential sites at the Indiana Dunes: implications for management. P. 47–70 in K. L. Cole, R. D. Hiebert, and J. D. Wood, eds. Vol. 1 of Proceedings of the 1st Indiana Dunes Research Conference: Symposium on Plant Succession. National Park Service, Science Publications Office, Atlanta.

Hiebert, R. D., D. A. Wilcox, and N. B. Pavlovic. 1986. Vegetation patterns in and among pannes (calcareous interdunal ponds) at the Indiana Dunes National Lakeshore, Indiana. American Midland Naturalist 116:276–281.

Hultsman, W. 1986. Visitor use and evaluation of impact mitigation at Indiana Dunes National Lakeshore. Unpubl. report, Indiana Dunes National Lakeshore. 57 p.

Keddy, C. J., and P. A. Keddy. 1984. Reproductive biology and habitat of *Cirsium pitcheri*. Michigan Botanist 23:57–67.

Klick, K., S. O'Brien, and L. Lobik-Klick. 1989. Exotic plants of Indiana Dunes National Lakeshore: a management review of their extent and implications. Unpubl. report, National Park Service. 192 p.

Kurz, H. 1923. Hydrogen ion concentration in relation to ecological factors. Botanical Gazette 76:1–29.

Leitner, B. M., and S. deBecker. 1987. Monitoring the Geyser's Panicum (*Dicanthelium lanuginosum* var. *thermale*) at the Geysers, Sonoma County, California. P. 391–396 in T. S. Elias, ed. Conservation and Management of Rare and Endangered Plants. California Native Plant Society, Sacramento.

Lellinger, D. B. 1985. A field manual of the ferns and fern-allies of the United States and Canada. Smithsonian Institution Press, Washington, D.C. 389 p.

Loveless, M. D., and J. L. Hamrick. 1988. Genetic organization and evolutionary history in two North American species of *Cirsium*. Evolution 42(2):254–265.

Lyon, M. W., Jr. 1927. List of flowering plants and ferns in the Dunes State Park and vicinity, Porter County, Indiana. American Midland Naturalist 10:245–295.

———. 1930. List of flowering plants and ferns in the Dunes State Park and vicinity, Porter County, Indiana: supplement. American Midland Naturalist 12:33–43.

Martin, M. 1990. List of state endangered, threatened, rare and watch list plants. Indiana Department of Natural Resources, Division of Nature Preserves, Indianapolis. 13 p.

McEachern, A. K. 1992. Disturbance dynamics of Pitcher's thistle (*Cirsium pitcheri*) populations in Great Lakes sand dune landscapes. Unpubl. Ph.D. dissert., University of Wisconsin, Madison. 216 p.

McEachern, A. K., M. L. Bowles, and N. B. Pavlovic. 1994. Recovery planning for the threatened Great Lakes thistle *Cirsium pitcheri* according to a metapopulation model. In M. L. Bowles and C. Whelan, eds. Recovery and Restoration of Endangered Species. Cambridge University Press, Cambridge.

McEachern, A. K., J. A. Magnuson, and N. B. Pavlovic. 1989. Preliminary results of a study to monitor *Cirsium pitcheri* in Great Lakes National Lakeshores. Unpubl. report, Indiana Dunes National Lakeshore. 96 p.

McEachern, A. K., and N. B. Pavlovic. 1991. Metapopulation dynamics in species recovery planning: Pitcher's thistle as an example [poster abstract]. In E. Forris, ed. Proceedings of the 53rd Midwest Fish and Wildlife Conference. Iowa Department of Natural Resources, Des Moines.

Menges, E. S. 1991. The application of minimum viable population theory to plants. P. 45–61 in D. A. Falk and K. E. Holsinger, eds. Genetics and Conservation of Rare Plants. Oxford University Press, New York.

Menges, E. S., and T. V. Armentano. 1985. Successional relationships of pine stands at Indiana Dunes. Indiana Academy of Science 94:269–287.

Mickel, J. T. 1979. How to know the ferns and fern allies. Pictured Key Nature Series. Wm. C. Brown Co., Dubuque, Iowa. 229 p.

Miller, S. G., S. P. Bratton, and J. Hadidian. 1992. Impacts of white-tailed deer on endangered and threatened vascular plants. Natural Areas Journal 12:67–74.

Moore, P. A. 1959. The Calumet region: Indiana's last frontier. Indiana Historical Bureau. 685 p.

National Park Service. 1977. NPFLORA/COMMON coverage by acreage of NPS units. Unpubl. report. 5 p.

———. 1979. Assessment/review of proposals for the West Unit, Indiana Dunes National Lakeshore. Denver Service Center. 30 p.

———. 1980. General management plan, Indiana Dunes National Lakeshore. Denver Service Center. 381 p.

———. 1984. Task directive, West Unit Transit Center. Denver Service Center.

———. 1990. General management plan amendment, development concept plan, environmental assessment: West Unit, Indiana Dunes National Lakeshore. Denver Service Center. 77 p.

Norris, L. L. 1987. Status of five rare plant species in Sequoia and Kings Canyon National Parks. P. 279–282 in T. S. Elias, ed. Conservation and Management of Rare and Endangered Plants. California Native Plant Society, Sacramento.

Olson, J. S. 1958. Rates of succession and soil changes on southern Lake Michigan sand dunes. Botanical Gazette 19:125–170.

Ostlie, W. 1990. Element stewardship abstract for *Buchnera americana*—blue hearts. The Nature Conservancy. 12 p.

Oviatt, F. 1986. Action plan for endangered and threatened plant species within Indiana Dunes National Lakeshore. Unpubl. report, National Park Service. 36 p.

Palmer, M. E. 1987. A critical look at rare plant monitoring in the United States. Biological Conservation 39:113–127.

Pavlik, B. M., and M. G. Barbour. 1988. Demographic monitoring of endemic sand dune plants, Eureka Valley, California. Biological Conservation 46:217–242.

Pavlovic, N. B. 1994. Disturbance-mediated persistence of rare plants: restoration implications. P. 159–193 in M. L. Bowles and C. Whelan, eds. Recovery and Restoration of Endangered Species. Cambridge University Press, Cambridge.

Pavlovic, N. B., M. L. Bowles, S. R. Crispin, T. C. Gibson, K. Herman, R. Kavetsky, A. K. McEachern, and M. R. Penskar. 1993. Pitcher's thistle (*Cirsium pitcheri*) recovery plan. U.S. Fish and Wildlife Service, Minneapolis. 111 p.

Pavlovic, N. B., and K. L. Cole. 1994. Plants of the Indiana Dunes National Lakeshore. Indiana Dunes National Lakeshore. 38 p.

Payne, A. M., and M. A. Maun. 1984. Reproduction and survivorship of *Cakile edentula* var. *lacustris* along the Lake Huron shoreline. American Midland Naturalist 111:86–95.

Peattie, D. C. 1922. The Atlantic coastal plain element in the flora of the Great Lakes. Rhodora 24:57–70; 80–88.

———. 1930. Flora of the Indiana Dunes. Field Museum of Natural History, Chicago. 432 p.

Pepoon, H. S. 1927. An annotated flora of the Chicago area. Chicago Academy of Sciences, Chicago. 554 p.

Pimm, S. L., and M. E. Gilpin. 1989. Theoretical issues in conservation biology. P. 287–305 in J. Roughgarden, R. M. May, and S. A. Levin, eds. Perspectives in Ecological Theory. Princeton University Press, Princeton, New Jersey.

Rankin, W. T., and S. T. A. Pickett. 1985. The demography of *Buchnera americana*. ASB [Association of Southeastern Biologists] Bulletin 32:68.

Swink, F., and G. Wilhelm. 1979. Plants of the Chicago region. Morton Arboretum. 922 p.

Taylor, F. S. 1990. Reconstruction of twentieth century fire histories in black oak savannas of the Indiana Dunes National Lakeshore. Unpubl. M.S. thesis, University of Wisconsin, Madison. 123 p.

Thompson, E. H., and W. D. Countryman. 1989. Rare plant conservation in Vermont in the 1980's. Rhodora 91:110–115.

Thomson, J. D. 1988. Effects of variation in inflorescence size and floral rewards on the visitation rates of traplining pollinators of *Aralia hispida*. Evolutionary Ecology 2:65–76.

U.S. Fish and Wildlife Service. 1988. Endangered Species Act of 1973 as amended through the 100th Congress. 45 p.

Vandermeer, J. 1982. To be rare is to be chaotic. Ecology 63:1176–1178.

Voss, E. G. 1985. Michigan flora. Part II. Dicots (Saururaceae-Cornaceae). Bulletin 59. Cranbrook Institute of Science, University of Michigan, Ann Arbor.

Welch, W. H. 1935. Boreal plant relicts in Indiana. Proceedings of the Indiana Academy of Sciences 45:78–88.

White, P. S. 1984. Impacts of cultural and historical resources on natural diversity: lessons from Great Smoky Mountains National Park, Tennessee. P. 120–132 in J. L. Cooley and J. H. Cooley, eds. Natural Diversity of Forest Ecosystems. USDA, Forest Service, Institute of Ecology, Athens, Georgia.

Whittaker, J. O., J. Gibble, and E. Kjellmark. 1994. Mammals of the Indiana Dunes National Lakeshore. National Park Service Scientific Monograph NPS/NRINDU/NRSM-94/24. 130 p.

Wilcox, D. A. 1984. The effects of NaCl deicing salts on *Sphagnum recurvum* P. Beauv. Environmental and Experimental Botany 24(3):295–304.

———. 1986. The effects of deicing salts on vegetation in Pinhook Bog, Indiana. Canadian Journal of Botany 64:865–874.

Wilcox, D. A., S. I. Apfelbaum, and R. D. Hiebert. 1984. Cattail invasion of sedge meadows following hydrologic disturbance in the Cowles Bog wetland complex, Indiana Dunes National Lakeshore. Wetlands 4:115–120.

Wilcox, D. A., N. B. Pavlovic, and M. L. Mueggler. 1985. *Scirpus cyperinus*: selected ecological characteristics and its role as an invader of disturbed wetlands. Wetlands 5:87–97.

Wilcox, D. A., R. J. Shedlock, and W. H. Hendrickson. 1986. Hydrology, water chemistry, and ecological relations in the raised mound of Cowles Bog. Journal of Ecology 74:1103–1117.

Wilcox, D. A., and H. A. Simonin. 1987. A chronosequence of aquatic macrophyte communities in dune ponds. Aquatic Botany 28:227–242.

———. 1988. The stratigraphy and development of a floating peatland, Pinhook Bog, Indiana. Wetlands 8:75–91.

Wilhelm, G. S. 1980. Report on the special vegetation of the Indiana Dunes National

Lakeshore. Indiana Dunes National Lakeshore Research Program Report 80-01. 262 p.

———. 1990. Report on the special vegetation of the Indiana Dunes National Lakeshore. Indiana Dunes National Lakeshore Research Program Report 90-02. 373 p.

Wood, W. L., and S. E. Davis. 1987. A perspective on the present and future conditions of the Indiana Dunes National Lakeshore coastline. P. 34–46 in W. A. Wilcox, R. D. Hiebert, and J. D. Wood, Jr., eds. Proceedings of the 1st Indiana Dunes Research Conference: Symposium on Shoreline Processes. National Park Service, Science Publications Office, Atlanta.

13

Wilderness Research and Management in the Sierra Nevada National Parks

Jan W. van Wagtendonk and David J. Parsons

The concept of managing wilderness had its beginning in the Sierra Nevada of California in 1890 when Sequoia and Yosemite National Parks were established. Expansion of Sequoia in 1926 and establishment of Kings Canyon National Park in 1940 reinforced the importance of wilderness. Management has evolved from army troops patrolling for trespassers to the sophisticated use of computers and the application of scientific data in decision making. The success of the efforts to manage these spectacular areas has been largely dependent on the availability of data from long-term research studies.

The Sierra Nevada form a chain of mountains stretching over 300 km along the eastern edge of California. Elevations range from near sea level on the western slope to nearly 4,000 m at the crest. Yosemite National Park lies in the central Sierra Nevada, whereas Sequoia and Kings Canyon National Parks occupy the southern end of the range. The wilderness areas in the parks are characterized by rugged glaciated mountains incised by steep river canyons. The vegetation consists of dense conifer forests, interspersed with open expanses of granite and alpine meadows above tree line.

For many years after the establishment of the parks, users of the wilderness were few. The remote high-elevation backcountry was accessible only by foot or on horseback. Interest in the environment and the availability of lightweight backpacking equipment during the 1960s and early 1970s led to

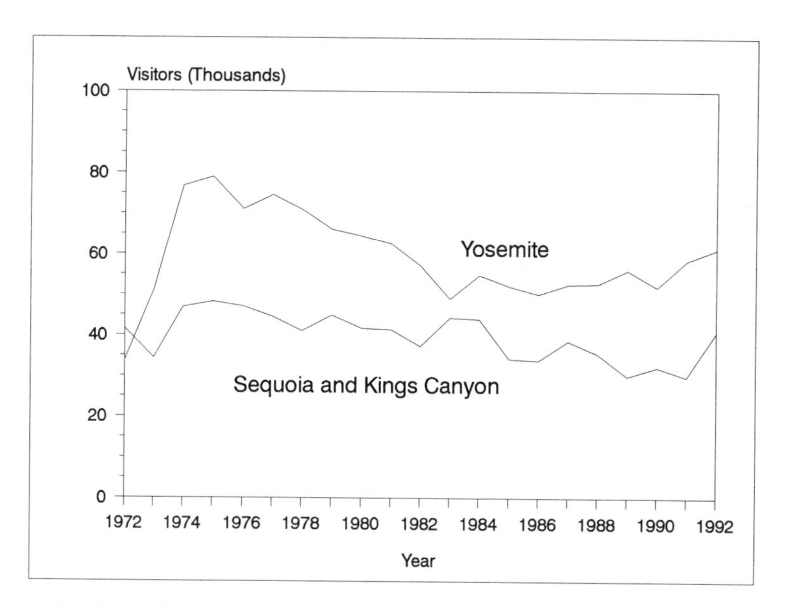

Figure 13.1 Recorded visitors in wilderness and backcountry areas of Yosemite, Sequoia, and Kings Canyon National Parks, 1972–1992.

dramatically increased use (van Wagtendonk 1981; Parsons 1983). By 1975 wilderness use had reached a peak when both Yosemite and Sequoia/Kings Canyon recorded more than 200,000 visitor nights each (Fig. 13.1). Since that time, wilderness use has decreased, leveling off at about 10% below those peak levels.

The legislative mandates establishing Sequoia and Yosemite National Parks state that they will be managed to "provide for preservation from injury of all timber, mineral deposits, natural curiosities, or wonders within said reservation, and their retention in their natural condition." Preservation of wilderness values was specifically mentioned as the primary purpose for establishing Kings Canyon National Park.

In 1984 the California Wilderness Act set aside 95% of Yosemite and 85% of Sequoia/Kings Canyon as parts of the National Wilderness Preservation System. Additional backcountry areas in each park that were not designated by the act are managed as de facto wilderness. The act described wilderness areas as lands that retain their primeval character and that must be managed to preserve their natural conditions. House Report 98-582 accompanying the act required the National Park Service (NPS) to monitor and assess impacts in areas within and adjacent to the wilderness.

NPS policies state that the conditions and long-term trends of wilderness resources will be monitored to identify the needs for and results of management actions (National Park Service 1988). Monitoring programs are required not only to assess physical and biological resources, but also to identify impacts of human activities on wilderness resources and experiences. Research is recognized as one of the statutory purposes of wilderness and is permitted as long as resource values are preserved (Graber 1988; Parsons and Graber 1991).

The wilderness management programs in the Sierra Nevada parks have largely been based on findings from research conducted by park scientists. From the earliest observations to modern computer simulations, research results have been applied to solve a diversity of issues from overgrazing to overused campsites.

Research and Monitoring Program

Muir (1894) was the first to record resource impacts in the Sierra Nevada when he described the effects of sheep grazing on mountain meadows. Qualitative assessments of wilderness conditions were also available in Sierra Club base-camp trip logs from the early 1900s. Scientific record keeping did not begin until the 1930s and, as was often the case, was initiated only after park managers believed that excessive impacts were occurring.

Early Studies

Scientific collection of wilderness impact data began in Sequoia National Park during the 1930s. Backcountry use in the Sierra Nevada was unrestricted during these early years; visitors could go wherever and do whatever they wanted. Park managers responded by sending out investigators to collect data on the impacts and make recommendations.

Sumner (1936, 1942) inventoried meadows in Sequoia and Yosemite National Parks and noted overgrazing and trampling by recreational stock as the most common forms of direct impact. More insidious were the indirect impacts associated with increased access through roads and trails. As access improved, more visitors came to the mountains. Soon additional facilities were developed to accommodate the visitors. These visitors widened trails, trampled meadows, denuded campsites, and left trash throughout the backcountry.

Sumner (1947, 1948) repeated his studies a few years later and found no improvement in meadow and trail conditions. Armstrong (1942) also studied grazing impacts in meadows in Kings Canyon National Park but concluded that they would recover if stock were managed carefully. Black (1952) and Rutter and Black (1953) addressed grazing, camping, trails, and stock impacts in Sequoia/Kings Canyon. They were guardedly optimistic about the prospects for recovery but recommended aggressive management. These reports resulted in an increased emphasis on wilderness management. Grazing was restricted in some areas, and ranger patrols were established to inform and regulate visitors.

As a result of visual observations of the effects of overgrazing, Sharsmith (1959, 1961) conducted studies of meadows in all three parks. He concluded that the meadows were in serious decline and that restrictions on use would have to be enforced. Thede et al. (1963) and Briggs (1965) completed exhaustive reviews of previous studies and conducted extensive field surveys of the backcountry areas of the parks. Their reports formed the basis for managing those areas. Grazing capacities were established, and use was managed with permits and adjusted annually with data from grazing reports. As a result, impacts associated with stock use were reduced for a period of time.

Meanwhile, impacts by the visitors themselves were locally concentrated at campsites and along trails, streams, and lakeshores (Parsons and DeBenedetti 1979). The threat of wildfires from campfires set in flammable areas or left to burn unattended was a serious concern. The collection of fuel stripped trees of all dead wood, especially at high-elevation sites. Humans were also affecting wildlife species by stocking and taking fish, trapping furbearers, and preempting bighorn-sheep range. Based on data that researchers collected on these impacts, campfire permits were issued to contact visitors and gather information on the amount and distribution of use. Unfortunately, no system for analyzing the data was instituted.

Current Research Programs

The great influx of backcountry visitors in the late 1960s and early 1970s prompted a series of new inventories and research studies. Holmes (1972), as part of a larger study of carrying capacities for the Yosemite backcountry, inventoried every trail and campsite in the park for human-caused impacts. For the first time, quantitative data were available for future comparisons. The realization that wilderness impacts far exceeded acceptable limits led to the decision to control visitor use. Based on research recommendations, managers in Yosemite decided in 1973 to restrict the number of people camping in a particular travel zone in the backcountry.

Park managers started a comprehensive research program to refine the numbers used for the zone limits. Harvey et al. (1972) conducted a study of wilderness impacts due to trampling by humans and stock and to fuel collection. This study was the first in the Sierra Nevada to scientifically document the relative trampling impacts of humans and horses.

Social impact studies were also underway in Sequoia/Kings Canyon (Kantola 1975) and Yosemite (Lee 1977; Absher and Lee 1981) to determine the effect visitors were having on each others' experience. These researchers interviewed visitors about their attitudes toward crowding, resource impacts, and satisfaction. They also observed the same visitors in the backcountry under varying social and environmental conditions. The conclusion was that wilderness enjoyment was affected more by human behavior and resource condition than by measures of crowding, such as total number of people encountered. These studies lent credence to the management decision to control use through more general restrictions, such as trailhead quotas or zone limits, rather than limit total use.

To better understand aquatic impacts, Silverman and Erman (1979) studied water quality in heavily used lakes in Kings Canyon National Park and found little adverse effect. However, Taylor and Erman (1979, 1980) found elevated levels of benthic plants and invertebrates in the same lakes. As a result of these studies, park managers began to remove campsites within 30.5 m of water.

Increasing use of the wilderness areas resulted in more frequent contacts between humans and black bears. These contacts are recorded as incidents if they result in a personal injury or property damage. Keay and van Wagtendonk (1983) found a positive linear relationship between use levels and bear incidents. Based on their results, park managers provided information about where incidents might occur using maps in visitor centers. Brochures and interpretive programs were also used to inform visitors how they might avoid incidents. Although these efforts resulted in an early decline in incidents, subsequent bear-proofing of front-country campgrounds has exacerbated the backcountry problem.

Visitor-use impact studies in Kings Canyon National Park (Parsons 1983) found that use levels and ecological impacts could be lessened by a variety of management restrictions. Parsons and DeBenedetti (1979) and Stohlgren and Parsons (1986) documented extremely slow recovery when heavily used campsites around alpine lakes in Sequoia National Park were closed to visitor use. This information led to the recognition that reducing use levels or using short-term closures were of limited value in allowing recovery. Studies at Bullfrog Lake, however, showed that campsites do recover in forested areas (Parsons and DeBenedetti 1979).

The value of campsite impact data as a basis for management decisions and future reevaluations of impacts was evident. Parsons and McLeod (1980) described field methods for assessing those impacts. Park managers used data from campsite impact studies to derive wilderness use capacities (Parsons et al. 1981). Parsons (1986) also used these data to determine that one-night camping limits and wood-fire bans were effective means of controlling impacts. As a result, park managers permanently closed some areas and restricted camping to one night only in others.

The studies in Yosemite were complementary to those in Sequoia National Park, although the former concentrated on the means for determining and implementing use limits. Van Wagtendonk (1979a) conceived a wilderness carrying-capacity model based on ecological, physical, human, and managerial components. He applied this model by analyzing wilderness resources, including ecological fragility and social density (van Wagtendonk 1986). Parsons (1986) combined the model with campsite data to determine and implement capacities for Sequoia.

Van Wagtendonk (1979b) used a computerized wilderness-simulation model to predict visitor movements based on travel times and existing use patterns. He and another researcher determined the travel times used by the model (van Wagtendonk and Benedict 1980a). Because computers were a new tool for managing wilderness, rangers in Yosemite were initially skeptical of the model. After they compared the results to their own field experience, however, they acknowledged the usefulness of the model and accepted the limits recommended by the researchers.

The use of wilderness permits to monitor and limit visitor use in the three parks began in 1972. Because use limits rely on accurate permit data, van Wagtendonk and Benedict (1980b) studied permit compliance and validity in Yosemite, and Parsons et al. (1982) compiled similar data for Sequoia/Kings Canyon. The methods for converting permit data into travel patterns were developed by van Wagtendonk (1978). Although the researchers found that limits affected the spatial and temporal distributions of use patterns, they did not affect the total amount of use (van Wagtendonk 1981). Park managers used this information to adjust use limits.

Limits were just the first step in developing a management system that would maximize visitor freedom in accordance with wilderness resource goals. The quota program was developed in Yosemite to translate zone carrying capacities and travel patterns into trailhead quotas (van Wagtendonk and Coho 1986). Park managers implemented quotas by issuing permits for only a specified number of people for each trailhead each day. Placing controls at the entry points allowed visitors to go where they wanted and stay as long as they wanted. Because trailhead quotas are relatively simple to understand and to implement, managers embraced them with little hesitancy.

Trailhead quotas are now in effect in the wilderness areas of all three parks, as well as in the adjoining Forest Service wilderness areas (van Wagtendonk et al. 1990, 1992). When supplemented with site-specific restrictions, such as one-night-only campsites or area closures, the quotas have proven to be both effective and popular. Whether enforcing use limits and trailhead quotas results in fewer impacts is still an unanswered question, however. Subsequent monitoring of resource conditions is important to determine the effectiveness of these management actions.

Monitoring Programs

Once the management systems were in place, it became necessary to monitor results and make refinements. Visitor use has been monitored through mandatory wilderness permits since 1972. Rangers check permits throughout the wilderness and record the number of people they encounter in each area each day. Data from the permits have been useful for chronicling temporal and spatial trends as well as for limiting use. Similar data are used in Sequoia/Kings Canyon to track and regulate the amount and distribution of recreational stock use. Park managers make adjustments each year based on the monitoring results.

Long-term monitoring of trail and campsite impacts has also proven to be useful for detecting change. Sydoriak (1986) implemented a comprehensive inventory and monitoring system for trail and campsite impacts in Yosemite. This system followed much of the earlier effort and allowed comparisons to be made. A computerized database allows access to the data and is used to direct mitigation programs (Stohlgren 1988). Monitoring of campsites will continue in Sequoia/Kings Canyon on a periodic basis (Parsons and Stohlgren 1987). Park managers annually review the data and adjust quotas upward or downward.

Management of the backcountry meadows has been a particular source of controversy in Sequoia/Kings Canyon. Reviews of meadow conditions (Ratliff 1985) and the history of pack-stock use and management (DeBenedetti and Parsons 1979, 1983; McClaren 1989), together with a recent inventory of all park meadows (Neumann 1990), have provided a basis for the development of an ecologically sensitive, meadow–and–stock-use management program. Park managers are using research data on the effects of different quantities and timing of herbage removal in different meadow types (Stohlgren et al. 1989) to revise standards for evaluating the impacts of stock on meadows. The management program includes systematic monitoring to evaluate meadow conditions and the trend in those conditions. Annual grazing programs, including opening dates and total allowable use, are based on the monitoring results.

Major Issues and Challenges

The history of wilderness use in the Sierra Nevada highlights the major management issues: impacts from grazing and limitations on visitors. Several decisions based on monitoring and research data have resulted in specific management actions.

Various areas have been closed to camping or grazing. The Bullfrog Lake area in Kings Canyon National Park was closed as early as 1960 (Parsons and DeBenedetti 1979). Subsequently, all areas within 6.4 km of Yosemite Valley and Tuolumne Meadows in Yosemite National Park were closed to camping. Some meadows in all three parks have been permanently closed to grazing, whereas others cannot be grazed during the early season until the soil dries enough to withstand stock use.

In Sequoia/Kings Canyon, managers have banned fires in areas with limited firewood, have established designated campsites, and have restricted camping to one or two nights in some areas to reduce impacts (Parsons 1983). In Yosemite, wood fires are prohibited above an elevation limit, and camping in designated sites is required in one limited area. Wilderness permits for overnight use were instituted in 1972 in all three parks and became mandatory the next year. Travel-zone limits were used to control use in Yosemite in 1973. Trailhead quotas based on campsite impacts were instituted in Sequoia/Kings Canyon in 1975. Quotas based on zone limits were implemented in Yosemite in 1977.

For many years, stock use has been largely restricted to maintained trails, although some historical unmaintained routes continue to be permitted. Recent restrictions include closing some areas and trails, prohibiting loose herding in all but the most dangerous situations, and not allowing stock to be tied to trees. Closures of meadows to grazing have been based on periodic monitoring and analysis of meadow conditions as well as qualitative judgments. Several meadows have been closed recently in Sequoia/Kings Canyon to preserve examples of pristine, untrampled meadow vegetation as a baseline for future comparisons.

Before 1972 there were no restrictions on party size. Then a party limit of 25 was enforced, whereas the number of allowable stock varied. Based on an analysis of wilderness permit data, managers of the Sierra Nevada parks and forests have recently limited party size to 15 and the number of stock to 25 (van Wagtendonk et al. 1992). Public comments on the decision were requested and were overwhelmingly in support of the reductions, although some groups vociferously opposed the decision.

The management actions have been unchallenged, except for the closure of certain areas to stock use and the increase from 20 to 25 head of stock in Sequoia/Kings Canyon. Stock users object to any closure as long as park managers lack quantitative standards derived from long-term data on ecosystem changes. On the other hand, hikers have argued that stock have far greater impact than humans and should be eliminated from the wilderness entirely. Both groups are requesting that managers of Sequoia and Kings Canyon write an Environmental Impact Statement before adopting a new wilderness management plan. The plan would increase the network of trails maintained for stock use and raise the stock limits from 20 to 25 animals per group.

The Role of Monitoring and Long-term Research

Both research and monitoring have played pivotal roles in the management of wilderness in the Sierra Nevada national parks. Scientific research that built upon previously recorded data was instrumental in determining how issues were perceived, approached, resolved, and followed. Without scientifically based information, it would have been impossible to make informed decisions. Because the data had been collected over a long period and supported the decisions, the decisions were more acceptable to the managers and the public. Although it would be difficult to assess the cost of not having valid data, the fact that only one decision has been contested attests to their value.

Much of the research and monitoring that was done in the Sierra Nevada was progressive for its time and addressed major problems in wilderness management. New emphases, such as the U.S. Forest Service program to limit acceptable changes, are based on measurable objectives, which must be determined by public involvement and policy reviews. Monitoring must be focused on issues that trigger management actions to ensure that objective wilderness conditions are maintained. Research and monitoring will play pivotal roles in the evaluation and refinement of management objectives. The feedback loop between management and research, if employed in a process of adaptive management, would sustain a process of program refinement.

Scientific data from long-term research and monitoring programs will become more important as a growing population, air pollution, and habitat fragmentation place increasing pressures on the preservation of wilderness ecosystems. We Americans can no longer think of these areas in isolation from the human environments that surround and affect them. We can speak of the Sierra Nevada ecosystems as specific entities, but we must realize that they are also part of the larger global ecosystem.

Literature Cited

Absher, J. D., and R. G. Lee. 1981. Density as an incomplete cause of crowding in backcountry settings. Leisure Science 4(3):231–248.

Armstrong, J. E. 1942. A study of grazing conditions in the Roaring River District, Kings Canyon National Park, with recommendations. National Park Service, Sequoia and Kings Canyon National Parks. 177 p.

Black, B. 1952. Erosion in the Roaring River District, Kings Canyon National Park: a pictorial review after ten years. National Park Service, Sequoia and Kings Canyon National Parks. 80 p.

Briggs, G. S. 1965. A report on backcountry conditions and resources with management recommendations: 1964–65. National Park Service, Yosemite National Park. 217 p.

DeBenedetti, S. H., and D. J. Parsons. 1979. Mountain meadow management and research in Sequoia and Kings Canyon National Parks: a review and update. P. 1305–1311 in R. M. Linn, ed. Proceedings of the 1st Conference on Scientific Research in the National Parks. National Park Service Transactions and Proceedings Series 5.

———. 1983. Protecting mountain meadows: a grazing management plan. Parks 8(3):11–13.

Graber, D. M. 1988. The role of research in wilderness. George Wright Forum 5(4): 55–59.

Harvey, H. T., R. J. Hartesveldt, and J. T. Stanley. 1972. Wilderness impact study report: human foot impact. Sierra Club, San Francisco. 87 p.

Holmes, D. O. 1972. Yosemite backcountry inventory, summer 1972. Final report. National Park Service, Yosemite National Park. 2295 p.

Kantola, W. 1975. A survey of backcountry visitors in Kings Canyon National Park. National Park Service, Sequoia and Kings Canyon National Parks. 41 p.

Keay, J. A., and J. W. van Wagtendonk. 1983. Effect of Yosemite backcountry use levels on incidents with black bears. International Conference on Bear Research and Management 5:307–311.

Lee, R. G. 1977. Alone with others: the paradox of privacy in wilderness. Leisure Science 1(1):3–20.

McClaren, M. P. 1989. Recreation pack stock management in Sequoia and Kings Canyon National Parks. Rangeland 11:3–8.

Muir, J. 1894. The mountains of California. Doubleday Inc., Garden City, New York. 300 p.

National Park Service. 1988. Management policies. Washington, D.C. 114 p.

Neumann, M. J. 1990. Past and present conditions of meadows in Sequoia and Kings Canyon National Parks. National Park Service, Sequoia and Kings Canyon National Parks. 723 p.

Parsons, D. J. 1983. Wilderness protection: an example from the southern Sierra Nevada, USA. Environmental Conservation 10(1):23–30.

———. 1986. Campsite impact data as a basis for determining wilderness use capacities. P. 449–455 in R. C. Lucas, ed. Proceedings of the National Wilderness Research Conference: Current Research. USDA, Forest Service Technical Report INT-212.

Parsons, D. J., and S. H. DeBenedetti. 1979. Wilderness protection in the high Sierra: effects of a 15-year closure. P. 1313–1317 in R. M. Linn, ed. Proceedings of the 1st Conference on Scientific Research in the National Parks. National Park Service Transactions and Proceedings Series 5.

Parsons, D. J., and D. M. Graber. 1991. Horses, helicopters and hi-tech: managing science in wilderness. P. 90–94 in Preparing to Manage Wilderness in the 21st Century. USDA, Forest Service General Technical Report SE-66.

Parsons, D. J., and S. A. McLeod. 1980. Measuring impacts of wilderness use. Parks 5(3):8–12.

Parsons, D. J., and T. J. Stohlgren. 1987. Impacts of visitor use on backcountry campsites in Sequoia and Kings Canyon National Parks. National Park Service Technical Report CPSU/UC-25. 79 p.

Parsons, D. J., T. J. Stohlgren, and P. A. Fodor. 1981. Establishing backcountry use quotas: an example from Mineral King, California. Environmental Management 5:335–340.

Parsons, D. J., T. J. Stohlgren, and D. M. Kraushaar. 1982. Wilderness permit accuracy: differences between reported and actual use. Environmental Management 6:329–335.

Ratliff, R. D. 1985. Meadows in the Sierra Nevada of California: state of knowledge. USDA, Forest Service General Technical Report PSW-84. 52 p.

Rutter J. A., and B. Black. 1953. Back country use report. National Park Service, Sequoia and Kings Canyon National Parks. 32 p.

Sharsmith, C. W. 1959. A report on the status, changes, and ecology of back country meadows in Sequoia and Kings Canyon National Parks. National Park Service, Sequoia and Kings Canyon National Parks. 122 p.

———. 1961. A report on the status, changes, and comparative ecology at selected back country meadows in Yosemite National Park that receive heavy visitor use. National Park Service, Yosemite National Park. 58 p.

Silverman, G., and D. C. Erman. 1979. Alpine lakes in Kings Canyon National Park, California: baseline conditions and possible effects of visitor use. Environmental Management 8:73–87.

Stohlgren, T. J. 1988. Analysis of campsite and trail impacts in Yosemite National Park, California. National Park Service Report CPSU/UC. 199 p.

Stohlgren, T. J., S. H. DeBenedetti, and D. J. Parsons. 1989. Effects of herbage re-

moval on productivity of selected high-sierra meadow community types. Environmental Management 13:485–491.
Stohlgren, T. J., and D. J. Parsons. 1986. Vegetation and soil recovery in wilderness campsites closed to visitor use. Environmental Management 10(3):375–380.
Sumner, E. L. 1936. Special report on a wildlife study of the High Sierra in Sequoia and Yosemite National Parks and adjacent territory. National Park Service, San Francisco. 61 p.
———. 1942. The biology of wilderness protection. Sierra Club Bulletin 27(8):14–22.
———. 1947. Erosion in the Roaring River District, Kings Canyon National Park: a checkup after six years. National Park Service, San Francisco. 46 p.
———. 1948. Tourist damage to mountain meadows in Sequoia-Kings Canyon National Parks: 1933 to 1948, a review with recommendations. National Park Service, San Francisco. 29 p.
Sydoriak, C. A. 1986. Yosemite wilderness trail and campsite impact monitoring system. National Park Service, Yosemite National Park. 25 p.
Taylor, T. P., and D. C. Erman. 1979. The response of benthic plants to past levels of human use in high mountain lakes in Kings Canyon National Park, California, USA. Environmental Management 9:2771–2782.
———. 1980. The littoral bottom fauna of high-elevation lakes in Kings Canyon National Park. California Fish and Game 66(2):112–119.
Thede, M., L. Sumner, and W. Briggle. 1963. Backcountry management plan for Sequoia and Kings Canyon National Parks. National Park Service, Washington, D.C. 106 p.
van Wagtendonk, J. W. 1978. Using wilderness permits to obtain route information. P. 197–203 in M. Shechter and R. C. Lucas, eds. Simulation of Recreational Use for Park and Wilderness Management. Johns Hopkins Press, Baltimore.
———. 1979a. A conceptual backcountry carrying capacity model. P. 1033–1038 in R. M. Linn, ed. Proceedings of the 1st Conference on Scientific Research in the National Parks. National Park Service Transactions and Proceedings Series 5.
———. 1979b. Use of a wilderness simulator for management decisions. P. 1039–1040 in R. M. Linn, ed. Proceedings of the 1st Conference on Scientific Research in the National Parks. National Park Service Transactions and Proceedings Series 5.
———. 1981. The effects of use limits on backcountry visitation trends in Yosemite National Park. Leisure Science 4(3)311–323.

———. 1986. The determination of carrying capacities for the Yosemite Wilderness. P. 456–461 in R. C. Lucas, ed. Proceedings of the National Wilderness Research Conference: Current Research. USDA, Forest Service General Technical Report INT-212.

van Wagtendonk, J. W., and J. M. Benedict. 1980a. Travel time variation on backcountry trails. Journal of Leisure Research 12(2):99–106.

———. 1980b. Wilderness permit compliance and validity. Journal of Forestry 78(7): 399–401.

van Wagtendonk, J. W., and P. R. Coho. 1986. Trailhead quotas: rationing use to keep wilderness wild. Journal of Forestry 84(11):22–24.

van Wagtendonk, J. W., E. P. DeGraff, J. M. Benedict, and N. Hunze. 1990. Interagency wilderness management: examples from California and Minnesota. P. 270–273 in D. L. Lime, ed. Proceedings of the Conference for Managing America's Enduring Wilderness Resource, 11–17 September 1989. University of Minnesota, St. Paul.

van Wagtendonk, J. W., D. J. Parsons, and E. P. DeGraff. 1992. Wilderness management in the Sierra Nevada, California: 23 years of interagency cooperation. P. 1–7 in E. E. Krumpe and P. D. Weingart, eds. Designation and Management of Park and Wilderness Preserves. Wilderness Research Center, University of Idaho, Moscow.

14

River Management at Ozark National Scenic Riverways

Kenneth Chilman, David Foster, and Thomas Aley

Ozark National Scenic Riverways consists of two exceptional free-flowing rivers: the Current and Jacks Fork. These rivers are fed by a system of large springs that maintain water clarity and a constant flow. Both rivers attract many recreation visitors, especially for canoe trips.

Large increases in recreational use in the 1970s prompted the initiation of a research program. Originally the program was directed toward recreational carrying capacities, i.e., the impacts of large numbers of visitors on water quality, soils, vegetation, and other visitors. Researchers soon recognized that water quality was also linked to other activities on the watersheds of the complex spring systems, and therefore they needed to expand the program in geographic and hydrological scope.

A hydrologist and social researcher saw the benefit of a long-term research program and began monitoring water quality and visitor use (Marnell et al. 1978). Today these form a systematic information base that is extremely helpful for management decision making.

Limited research funding was offset by vigorous National Park Service (NPS) recruitment of research assistance and development of low-cost methods for data collection on large areas. These enable managers to use current monitoring and research data to make daily decisions and address specific issues. Through vigilant review, this database of social, physical, and biological information is revised and improved so new research will ensure its relevance for current scientists and managers.

Setting and History of the Park

Ozark National Scenic Riverways extends along 216 km of the Current and Jacks Fork Rivers in the Ozark Highlands of southeastern Missouri (Fig. 14.1). This park is located within one of the largest karst geology regions in the nation, giving rise to a diverse landscape characterized by equally diverse ecosystems. Hills, steep valleys, intermittent and permanent streams, sinkholes, caves, vertical bluffs, and springs are especially numerous in this type of terrain. The riverways are believed to be the largest repository of caves (250+) and large springs in the national park system.

Although the river system is the primary resource within the park boundary, the watershed is the *élan vital* of the ecosystem. It covers 3,033 km^2 and is a complex system of both surface and groundwater sources

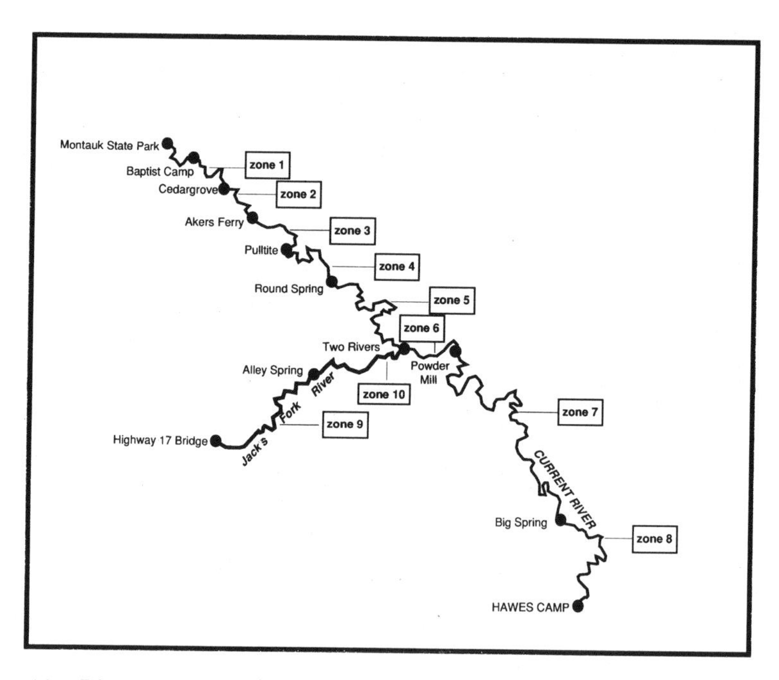

Figure 14.1 River zones, Ozark National Scenic Riverways.

typical of karst terrane. Groundwater supplies the major portion (60%) of the rivers' flow from 5 first-magnitude (>100 cubic feet per second [cfs]) springs and 53 others within the park and drainage area. Four springs within the park rank among the 10 largest in the state. The combined flow of the springs discharging into the two rivers exceeds 1.2 billion gallons per day.

The riverways landscape is dominated by mesic and dry-mesic bottomland forest in the riparian zone and dry or dry-mesic chert forest in the uplands (Nelson 1985). Interspersed among these are other diverse ecosystems representative of the region, as well as many smaller, specialized communities in the dolomite cliffs, igneous and dolomite glades, dolomite talus slopes, gravel washes, and wetland areas. The result is a high degree of biological diversity in the park: 70% of the natural communities in the region also thrive within the riverways, which comprise less than 2% of the land area (Nigh 1988).

Ozark National Scenic Riverways was the first national riverways system established in the country. It was authorized by Congress in 1964 (Public Law 88-492) to conserve and interpret unique scenic and other natural values and objects of historical interest, preserve portions of the Current and Jacks Fork Rivers as free-flowing streams, and preserve the springs and caves of the area, all for the use and enjoyment of the people of the United States. The act represented a milestone of nearly a half century of expanding state and federal government intervention in the recreational development of the Current River basin (Stevens 1991).

From 1880 to about 1910, the harvest of forest resources in this region reached its peak and was the foundation of the regional economy and culture. This unregulated rate of deforestation ended in a virtual collapse of that economic base and threatened what had been the traditional pattern of life along the Current River. The cultural habitat was gone and the need for a replacement was clearly evident by 1920. At this time, many regional and state leaders began to promote tourism to restore the economy of the Ozarks.

The early development of recreation was focused on sport hunting and fishing, which was short-lived because of the depletion of wildlife during and following the lumbering era. Logging drastically changed the river fisheries by causing massive amounts of rock and gravel to pour into the rivers. The next movement was directed toward attracting tourists to the area's scenic beauty. This effort was supported by the establishment of a state park system with three of the first parks—Big Spring, Alley Spring, and Round Spring—located on the Current and Jacks Fork Rivers. In spite of these developments and the concurrent expansion of railroad and highway travel, tourism was slow to develop in the more isolated portions of the Ozarks, such as the Courtois Hills (Current River) region.

The next major events occurred when the federal government embarked on several programs directed toward managing, developing, and preserving Ozark resources for a variety of uses. The Mark Twain National Forest was established in 1933 and introduced scientific management of timber resources. During the 1930s, Congress authorized the Corps of Engineers to develop 50 dams in the state. This decision precipitated action by conservationists and local opponents of the dams to devise strategies to preserve the Current as a free-flowing river. The creation of state parks at Big Spring, Alley Spring, and Round Spring (which were later incorporated into Ozark National Scenic Riverways) contributed to that "preservation ethic." By 1941 the Corps had constructed Wapappello Dam on the adjacent St. Francis River and turned its attention toward the Current River. World War II, however, delayed further planning by the Corps.

After the war, federal and state authorities revived plans to perpetuate the river as a free-flowing stream. In 1950, as opposition mounted against the proposed dams, the Corps of Engineers withdrew its plans for flood control and hydroelectric development of the Current River. After more than a decade of local and congressional discussion and consideration of several legislative proposals, the enabling act to establish Ozark National Scenic Riverways was finally passed in 1964.

Developing the Long-term Research Program

From discussions with persons involved with the 20-year history of research and monitoring programs at Ozark National Scenic Riverways, we identified the following four periods of activity: (1) initiating the research program, 1972–1976; (2) adding the watershed studies, 1972–1987; (3) resolving the court cases and doing management planning, 1976–1985; and (4) developing the monitoring systems, 1985–present.

Initiating the Research Program

After congressional authorization of Ozark National Scenic Riverways on 27 August 1964, land-acquisition activities were a major concern for several years. During this time, the area received an increasing number of visitors, as people became aware of the newly designated scenic riverways and the opportunities for float trips. Park managers estimated that use of the area in 1968 totaled about 40,000 floater days, including one-day float trips and some overnight camping along the river for more-than-one-day float trips. The increasing rate of use reached 142,850 floater days in 1972 (Marnell et al. 1978).

At the same time, news media began calling attention to problems associated with the increased use in articles published in newspapers and magazines. In 1970, during 4 hours, Jackson (1970) counted 476 persons and 214 canoes passing Cave Spring on the Current River. An NPS ranger estimated that the all-day tally, based on these numbers, would be 450 to 500 canoes. Ecological damage, too many entry points, and the use of firearms were also becoming increasing concerns.

Renken (1971) also documented the congestion of canoes: up to 700 per day on the 29.8-km stretch of the river between Akers Ferry and Round Spring. He noted that this segment of the Current River was heavily used before NPS began managing the park. Not only were large numbers of canoes a problem in that area, but also picnickers and campers, who gathered in groups of up to 100 persons on the more attractive gravel bars and destroyed vegetation. Visitors also complained about the crowding because of the floaters' behavior, including rowdiness and drunkenness.

Ironically, concerns about the Current River fishery finally led NPS to initiate a research program in 1972. Leo Marnell, a fisheries biologist who had done his Ph.D. research at Yellowstone National Park and had worked at Yosemite National Park, was transferred to Ozark National Scenic Riverways to investigate concerns about declines in fisheries. He soon recognized that the increasing number of canoes posed a more serious concern for river management. Consequently, he proposed a multifaceted, 5-year research program to investigate various aspects of recreational carrying capacity. During the research program, NPS called a moratorium on new canoe purchases by its concessionaires. Unfortunately, increased canoe rentals by unlicensed operators added many more canoes to the riverways.

The first priority of the research program was the development of a database on the recreational use of the Current and Jacks Fork Rivers (Marnell et al. 1978). To meet this goal, researchers focused on four areas: use monitoring, environmental impacts, sociological considerations, and water quality.

Counting users was the first concern because the increasing number of river users was the most rapidly changing element. Early counts of visitors at scattered points along the rivers indicated an increasing rate of 10–15% annually. Alleviating this situation formed the basis for river-use management.

Researchers selected designated river zones to survey through an innovative system based on time-lapse photography, which facilitated the task of counting canoes (Marnell et al. 1978). Ten zones were established on three sections of the river (Fig. 14.1). Electronic devices regulated and activated not only the cameras, but also the time intervals between exposures. Although logistical support for reloading the cameras was needed to carry out the operation, the costs of obtaining river-use data were considerably reduced through this method. Researchers verified the reliability of the sampling design by comparing the results to traffic counts made by an area resident and found an error of less than 4% (Marnell et al. 1978).

Observational data supported the time-lapse photography. Park rangers recorded daily the number of watercraft occupants and the ownership of the vessels. Researchers combined the number of occupants and vessels and the ownership categories to project the number of rented and privately owned watercraft and the total number of floaters. In recording observational data, researchers applied stratified sampling methods because the parameters varied between weekend days and weekdays (Marnell et al. 1978).

Environmental impacts were the second major concern of researchers. In 1973 researchers initiated a series of environmental impact studies on canoe landing sites and water quality to obtain baseline measurements of existing conditions. They conducted three categories of studies: (1) soils and vegetation; (2) aquatic communities; and (3) unique features, biota, and species. They also used two approaches to assess recreational impacts on the environment: (1) reconnaissance of the entire river system to identify sites affected by recreational use and (2) an examination of selected sites known to be altered by recreational use (Marnell et al. 1978). Results of these studies showed that the impacts of canoe traffic were minimal on gravel bars subject to periodic flooding. The impacts on areas outside of the floodplain, such as short trails to springs or caves, could be controlled by constructing well-drained and surfaced paths.

The third concern of researchers was sociological considerations, i.e., how visitors perceived conditions at Ozark National Scenic Riverways. Before the research program, Gisi (1971) conducted a study during the summer of 1970 to find out how first-time visitors learned about the area. Most of the respondents heard about the park from friends. Gisi discovered that 24% of the respondents came to the river to go canoeing and that 40% did not know that Ozark National Scenic Riverways was a unit of the national park system.

Because previous studies showed that crowding was a major complaint of visitors, Habermehl (1973) interviewed visitors to measure their perceptions of crowding. He found a strong relationship between expected and desired numbers of people encountered by floaters; i.e., floaters who encountered more people than expected tended to perceive "crowding." This view was held by more than 300 floaters. Habermehl also found that 23.5% of those interviewed perceived crowding as more than expected, 27.4% perceived crowding as more than desired, and 14.7% perceived crowding as a problem.

At about the time the Marnell et al. (1978) report was prepared, Andrews (1978) remeasured visitors' perceptions of crowding. He noted a dramatic increase: from 27.4% who perceived crowding as more than desired in 1972 to 51.4% in 1977. The number of canoeists who perceived other floaters as a problem also increased from 14.7% in 1972 to 34.3% in 1977. Estimated annual canoe use had risen during the same period from 121,000 canoeists (142,850 floater days) in 1972 to 243,000 in 1977. In 1979 perceptions of crowding were again measured and had again risen: 52.4% perceived crowding as more than desired, and 37.1% perceived other floaters as a problem (Chilman 1978). The estimated annual canoe use had also risen to 295,400 canoeists in 1979. The data from the 1972, 1977, and 1979 studies were to play a significant role in the federal court trial in 1982.

The fourth major concern of researchers was water quality. Assessing existing water-quality conditions and establishing a monitoring program became an important part of the initial research effort in 1972. Researchers planned studies based upon the knowledge that contaminants could enter the river system through two potential pathways: (1) from surface and groundwater watersheds outside the riverways' boundary, and (2) from sources on or near the rivers within the boundary. Because the spatial and temporal aspects of both pathways were largely unknown, researchers conducted studies to provide interpretive data for monitoring and management strategies.

In 1973 NPS asked the U.S. Geological Survey to conduct a water-quality study to determine existing physical, chemical, and biological conditions of the ground, spring, and surface waters of the riverways, with the emphasis on surface water. This information would constitute an initial status assessment and establish a baseline condition for future comparisons. From April 1973 to May 1975, researchers collected water samples from 19 wells, 7 large springs, 14 sites on the Current River, 7 sites on the Jacks Fork River, and 5 tributaries to the two rivers (Barks 1978). They resampled the areas and analyzed the data to see how common inorganic constituents, major nutrients, minor elements, pesticides, sediment, and aquatic biota varied with time, runoff, and water uses. Several other sites on the two rivers were sampled to determine any progressive changes in water quality, i.e., upstream and downstream of communities, park development areas, and areas of concentrated recreational activities. Results from this study indicated that the riverways' surface and groundwater was generally unaffected by pollution with the exception of temporary reductions in quality due to periodic runoff and some activities of man (Marnell et al. 1978).

Adding the Watershed Studies

The Current and Jacks Fork Rivers receive most of their flow from springs. Aley (1975) estimated that approximately 75% of the total runoff of the region passes through the groundwater system for at least some distance. This system results in less seasonal variation in river temperatures and flow rates than is typical of rivers that are not fed by major springs.

In 1966 the U.S. Forest Service (USFS) began the Hurricane Creek Barometer Watershed Study. Hurricane Creek, a tributary to the Eleven Point River, lies adjacent to the Current River and is west and southwest of Big Spring. The purpose of the Barometer Watershed Program, which included the

Hurricane Creek study, was to provide long-term monitoring to guide land-management decisions. Another program goal was to improve the integration of hydrological data into routine USFS land-management activities. Hurricane Creek was considered the type example of karst areas. The Hurricane Creek study was deemphasized in 1973 during a reorganization of USFS in Missouri. A consulting firm prepared a final report on the study in 1975 (Aley 1975). With the completion of this report, the Hurricane Creek study was essentially terminated.

A few months after the Hurricane Creek study began, USFS researchers realized that the creek yielded far less water to the Eleven Point River than they anticipated; they began a search for the "missing water." They concluded that Big Spring on the Current River was the likely discharge point for much of the missing water, although other possibilities existed.

In 1968 USFS conducted the first successful groundwater trace to Big Spring from Hurricane Creek using fluorescein dye. The straight-line distance between the point where the dye was introduced into the groundwater system and Big Spring was 27,370 m. The travel time for the first arrival of the dye was 10 days. Over the next 5 years, USFS conducted several other traces from points both inside and outside the topographic basin of Hurricane Creek; some of these were recovered from Big Spring and some from other springs in the Eleven Point River and Spring River basins. The farthest groundwater trace, 24.5 km away, was from the channel of the Middle Fork of the Eleven Point River to Big Spring. Although NPS was not involved in these tracing studies, the agencies cooperated well and openly shared study results.

The groundwater-tracing program that USFS conducted between 1968 and 1973 had three unique characteristics. First, researchers developed methodologies to successfully conduct long-distance groundwater traces to springs with large flow volumes. For example, the mean annual flow of Big Spring is about 435 cfs; one successful trace was conducted to this spring when the flow rate was about 820 cfs.

Second, researchers incorporated groundwater tracing into a program to delineate recharge areas for important springs. Rather than simply determining which spring received water from a particular point, researchers specifically designed the tracing program to define recharge areas. This goal required that the researchers locate many dye-injection sites near suspected groundwater divides. Such selection requires detailed fieldwork and the routine use of dye-injection sites that pose various difficulties.

Third, the groundwater-tracing program focused on interactions between land use and groundwater quality. Researchers injected dye at some sites with contaminated water. They developed tracing techniques using stained *Lycopodium* spores. These spores were used to replicate several groundwater traces; the recovery of the spores demonstrated the absence of effective natural filtration within the associated groundwater system.

Park managers in Ozark National Scenic Riverways recognized the long-term utility of the recharge studies for the Big Spring area. They knew that these approaches could (and should) be expanded to provide basic data on other important springs in the riverways. In 1972 NPS researchers started recharge-area delineation studies for Alley and Round Springs (Aley 1976). By 1976 they had expanded these studies to other springs on the west side of the Current River (Aley 1978), and by 1982 to springs on the east side of the Current River (Aley and Aley 1982). By 1987 they had conducted at least one successful groundwater trace to every major spring in the riverways and had prepared preliminary recharge-area delineation maps for all major springs (Aley and Aley 1987).

The USFS Hurricane Creek studies provided the impetus for recharge-area delineations of springs in the riverways. Subsequent studies funded by NPS resulted in improved techniques and important findings. The correlation between land use and water quality was greatly improved in 1976 with the development of a hazard-area mapping approach (Aley 1976). The hazard-area concept integrates the physical setting of sites with the use of these sites to qualitatively rank the risk posed to groundwater quality. Hazard-area maps have now been prepared for the recharge areas of most of the major springs in the riverways, except Big Spring.

Researchers' understanding of the complexity of spring systems in the riverways increased as analytical techniques improved and knowledge expanded. For example, the detection limit for tracer dyes during the 1987 studies was about 100 times smaller than during the studies of the late 1960s. The lower detection limits resulted in the discovery that some areas contribute water to two or more springs that may be several miles apart. Where recharge areas supply water to multiple springs, the relative contributions vary with flow conditions. Springs that may initially appear to be relatively minor may have important hydrological interactions with major springs. Another important discovery was that appreciable groundwater flow may pass beneath the major rivers of the region. For example, Round Spring and significant portions of its recharge area lie on opposite sides of the Current River.

Park managers now have a general understanding of the areas that recharge individual springs and hope to have more refined delineations for these areas in the future. They can now identify areas and land-use activities that will most likely adversely affect particular springs. They have a basic understanding of the functioning of the spring systems in the region, of groundwater transport rates (typically 1.6 km per day or more), and of other aquifer characteristics. They also have data that clearly demonstrate that the quality of water discharging from the springs is directly related to land-use activities in the related recharge areas.

The public has shown enormous interest in the groundwater-tracing and spring-system studies. The studies are a popular interpretive theme in the park. The results are widely known and have great credibility among resource managers, scientists, and area residents.

Resolving the Court Cases and Doing Management Planning

Long-term research projects provided important information for two major management issues at the park. The first issue involved a complex series of court cases that challenged NPS management of canoe use. The second issue is ongoing and addresses the possible impacts of proposed lead mining on the Big Spring recharge area and its water quality.

The question of whether NPS had the authority to require permits for commercial canoe-rental operations within the riverways was resolved through a series of court cases. NPS started issuing permits for canoe rentals in 1968 and continued to do so through 1971 with very little change and without effectively regulating the rapid increase of nonauthorized businesses. From 1971 to 1974, NPS placed limitations on the number of rental canoes owned by each permittee and issued no additional permits to protect the area's values that were threatened by the increasing floating activity. NPS also enforced the canoe "moratorium" (which was only partially effective) to stabilize use as much as possible so that research could be conducted under quasi-stable conditions.

NPS took civil and criminal actions against those who operated businesses within the park boundaries without legal authorization. One of the most important cases was filed against Irby C. Williams because of continued violations to the Code of Federal Regulations and NPS rules. On 4 February 1975, the NPS field solicitor advised Williams in writing that continued violations regarding the operation of nonauthorized businesses within the park would result in legal actions and referral to federal court (Sullivan 1985). This case was later tried in the U.S. District Court for the Eastern District of Missouri. The judge believed that the NPS regulations were not clearly stated, and therefore ruled that people did not have to obtain an NPS permit to operate their businesses.

In 1976 NPS issued a special regulation clearly defining its authority with respect to commercial canoe liveries and sent Williams new notices of violations. Williams was again brought before the U.S. District Court for the Eastern District of Missouri, and the result was again negative to NPS. Because of the NPS lack of authority and the creation of new nonpermitted businesses, NPS authorized 25 additional canoes to each permitted operator in 1977 and again in 1978, increasing dramatically the number of canoes on the river.

In 1980 the USFS won favorable decisions from the U.S. District Court for the Western District of Missouri and the Court of Appeals for the Eighth Circuit regarding the operation of commercial businesses on the Eleven Point River. NPS believed that the decision would directly affect future court decisions about commercial operations at Ozark National Scenic Riverways. Another decision made in Idaho, concerning law enforcement by federal agencies in similar cases, helped to strengthen the NPS position in new cases before the court.

With these precedents in hand, NPS lawyers sent letters to cease and desist, or face possible civil or criminal penalties, to all unlicensed operators. At that time, the operators were organized as the Free Enterprise Canoe Association of Missouri. In response to the NPS action, this association sought a court order prohibiting the NPS efforts. In 1982 the U.S. District Court for the Eastern District of Missouri (St. Louis) considered both cases simultaneously (the criminal suit against the operators and the civil suit against NPS).

The lawyer representing the unlicensed operators was a very experienced trial attorney from St. Louis. His law firm did a thorough job of researching the history of the decisions to issue permits and the data collection that formed the basis for those decisions. As part of the trial, the association's lawyer spent approximately 4 hours cross-examining researchers' testimony about the monitoring of water quality and visitors' perceptions. The questioning covered not only the research data, but also the methods used for gathering and analyzing the data.

After some delays, the district court judge ruled in favor of NPS in 1983. The judge's ruling essentially said that although funding for the research had been limited, NPS had conducted a thorough and systematic research program, which formed the basis for its decisions. The association appealed the ruling. The appeal was resolved in favor of NPS in August 1984.

In that same month, park managers organized a planning team. In September 1985, the team forwarded the Draft River Use Management Plan and Environmental Assessment to the NPS regional director for approval. This plan and assessment were prepared in accordance with the National Environmental Policy Act of 1969. In the draft, the team addressed river-use management separately from the General Management Plan for the riverways, which had been previously approved in December 1984. The team had two general purposes for this draft: (1) to list the actual and potential alternatives for river-use management and (2) to foresee probable environmental effects of management actions. The plan's strategy was to maintain different levels of canoe use in various river zones to offer choices to canoeists with preferences for different kinds of float experiences. The plan also established horsepower limits, by river zones, on the use of outboard motors. This part of the plan was very controversial and stimulated local opposition to the proposed regulations, which delayed approval of the plan until May 1989.

At about the same time as the river carrying-capacity issue was being resolved, another issue of serious environmental concern began to take shape. In November 1983, the U.S. Steel Corporation submitted two lease applications for mining lead, zinc, and associated metals in the Mark Twain National Forest. The applications led to an environmental impact statement prepared by USFS and the U.S. Bureau of Land Management on hardrock mineral leasing for a large area within and near the delineated recharge area for Big Spring (USDA, Forest Service and U.S. Bureau of Land Management, 1987).

The groundwater-tracing and spring-system studies were of critical importance in demonstrating that mining in the area would have hydrological impacts on Big Spring and the riverways. Although much of the area contributes water to Big Spring, most of the land lies outside the topographic basin for Ozark National Scenic Riverways. Adverse impacts on the riverways could result from the mining per se or from the disposal of mine tailings.

Tailings disposal is of particular concern because water passing through these wastes could become contaminated with significant concentrations of heavy metals. The routine method for disposing tailings from similar mines on national forest lands in Missouri is to build a dam and bury the tailings behind the dam in a landfill. However, in the area where the mining might occur, most of the valleys are important recharge areas for Big Spring and other springs (Aley 1975). Introducing wastes into these recharge areas is environmentally dangerous.

A detailed discussion of selected alternatives, appeals, and litigation is beyond the scope of this paper. No final resolution has been reached, and no mining is underway. However, the hydrological data developed in the studies that delineated the recharge areas and characterized the spring system continue to play a critically important role in impact assessments. The long-term research data has enabled NPS to play a major role in the issue; this involvement is essential to the protection of Big Spring.

NPS (Sullivan 1989) petitioned the U.S. Environmental Protection Agency (EPA) to designate the Big Spring Recharge Area as a Sole Source Aquifer under the provisions of Section 1424 (e) of the Safe Drinking Water Act of 1974. This section authorizes the EPA administrator to determine that an aquifer is the sole or principal source of drinking water for a region if more than 50% of the resident population rely on the aquifer for domestic purposes. The section also allows EPA to review federal financially assisted projects planned for the region to determine their potential for contaminating the aquifer. No federal funds may be committed to projects that EPA determines may contaminate such an aquifer (U.S. Environmental Protection Agency 1987).

NPS submitted the Sole Source Aquifer petition because the long-term data demonstrate that Big Spring and the associated aquifer system are particularly sensitive to contamination and pollution. NPS deems the designation essential because no other mechanism can currently provide NPS and the local residents with the necessary protection of the groundwater-recharge area. The petition could not have fulfilled the data requirements for Sole Source Aquifer designation without the data that NPS collected. A public hearing on the petition was held in November 1991. EPA ultimately denied the petition in December 1992, based upon the questionable conclusion that alternative sources of domestic water supplies are available.

During the development and execution of the river-recreation and water-quality research program, NPS recognized that several related resource issues needed attention. Because of the complexity of the resources and associated issues, NPS realized that development of a long-term systematic plan would be necessary to support such a research program. Three major resource categories were identified for planning purposes: water, riparian corridor, and terrestrial resources. Although interrelated to some degree, they are distinct enough for planning purposes; they are also related directly or indirectly to river-management efforts. NPS's next step was to plan an integrated, sequential research project to provide the information base for subsequent management decisions. The project emphasizes improving the basic knowledge of the structure and function of park ecosystems and their relationship to the larger system.

The water-resource research effort is a multidisciplinary integration of physical and biological sciences that provides baseline data about the aquatic ecosystem. Geomorphological and ecological projects are now underway to study the structure and function of the river ecosystem, stream morphometry, and fluvial dynamics associated with alterations of the physical characteristics of the river environment. Other ongoing projects include a study of the structural and functional ecology of a large spring system and the chemical and biological analysis of groundwater from shallow aquifers. Future studies are planned to refine the delineation of spring recharge areas and to interpret the complex relationships among several springs and their shared recharge areas.

Studies of the riparian corridor have led to an inventory and map of plant communities along the entire length of both rivers. Future research will emphasize ecological studies of these communities. Researchers are integrating studies of recreational visitor use with the resource studies to monitor use patterns and impacts on the riparian zone.

Terrestrial areas within the park have been inventoried, classified, and mapped according to the plant communities established by Nelson (1985). Monitoring has begun to determine patterns, types, and densities of visitor use of the backcountry areas. Future research will include ecological studies of both the major and rare plant communities.

Developing the Monitoring Systems

The 1985 River Use Management Plan recognized the need to "Monitor impacts on water quality, vegetation, and other riverine resources, and conduct surveys of visitors' reactions, preferences and use patterns to determine the need for adjustments in management strategies" (National Park Service 1985). Water-quality monitoring was established as an ongoing program at the riverways in 1976. Social monitoring (visitor counts and surveys) began as a continuing program in 1986 but was based on methods developed during previous research. The water-quality- and social-monitoring programs are expanded and improved as new needs are identified and low-cost methods of data collection are developed.

After the baseline water-quality studies were completed in 1975, NPS developed and implemented a monitoring plan to periodically remeasure selected constituents. The plan included collecting the physical, chemical, and biological data that could detect trends and changes in water quality and make the best use of limited support resources. Most of the physical and chemical constituents included in the baseline study have been sampled semiannually since 1977 at some river and spring stations. Seasonal and intermittent sampling of major nutrients (nitrogen and phosphorus) has also been continued at selected river and spring locations to detect any general or locality specific deviations from baseline levels.

For monitoring purposes, researchers significantly altered the methodology used in the collection of biological data because of the relatively high costs of collecting these data. They monitor lotic phytoplankton dynamics and primary production by repeatedly collecting water samples at river stations and analyzing them for chlorophyll "a" concentrations. Researchers sample benthic invertebrates triannually at selected river and tributary stations to detect any long-term changes in stream conditions. They also selectively collect bacterial data to monitor potential sources, e.g., communities and recreational areas, and to determine the effects of storm runoff and other variables on bacterial densities.

Social monitoring is of more recent origin than physical and biological monitoring in large parks and wilderness areas partly because the methods for survey research were still being developed in the 1970s. The Outdoor Recreation Resources Review Commission report (1962) called attention to the nationwide need for outdoor recreation management. Studies in the 1960s focused on physical and biological phenomena, such as site impacts. During the 1970s, research focused on measuring visitor numbers and characteristics associated with site impacts.

In the 1980s, researchers began to recognize the importance of monitoring systems for managing recreational carrying capacity (Chilman et al. 1981; Washburne 1982; Chilman 1983, 1985). Monitoring systems became an integral part of managing recreational carrying capacity (Stankey et al. 1985; Graefe et al. 1990). However, very few large parks have implemented recreational monitoring systems because of uncertainties about methodology and budget constraints.

Early research at Ozark National Scenic Riverways (Marnell et al. 1978)—especially the designation of 10 management zones along the riverways and the development of methods for doing visitor counts and surveys—was very useful for initiating a river-use monitoring system in 1986. Research initially focused on monitoring canoeists. The program was later expanded to add counts and interviews of boaters, tubers, and other river users.

In 1986 NPS began monitoring in 3-year cycles, covering one of the three ranger districts each year. This method allowed closer analysis of specific districts and zones and fit the limited budget. One researcher and one volunteer could do the counts and interviews at canoe access/takeout points at the ends of the designated river zones (Fig. 14.1) on sampling days during the heavy-use months of June, July, and August.

The river-use monitoring program began on the Upper Current District in 1986, moved to the Jacks Fork District in 1987 and then to the Lower Current District in 1988. Adjustments in sampling designs had to be made because of differences in zone conditions and the types and amounts of river traffic. Interviews of motorboaters and tubers were added on the Lower Current in 1988.

In the second monitoring cycle beginning in 1989, researchers added studies to develop and integrate methods for monitoring land-based recreation along the riverways. This recreation was primarily concentrated on primitive campsites and day-use areas on the riverbanks. Most of these sites are accessible to vehicles, but many of the access roads are down steep slopes and are difficult to maintain. Management concerns included a need to decide whether to close or relocate some of the roads and campsites where resources were damaged.

Because the primitive campsites were scattered, park managers tried to involve rangers in collecting monitoring data during their patrol activities. This effort has been only partially successful because of variations in district conditions and personnel, but research is continuing on this aspect. In the summer of 1991, the Lower Current District did commit one seasonal employee to measuring impacts at primitive sites. In this same district, volunteers in boats counted primitive-site users.

Current research is also addressing the use of computers to process, organize, and present the monitoring data. Most previous recreation-use studies had been one-time studies. Now the concept of monitoring as a flow of data over time calls for information systems to make the data more readily available and useful for decision making.

Social monitoring at the riverways began with a need to address the issue of canoe-carrying capacity, but the data were also useful for day-to-day decision making. The monitoring system is still evolving but provides current data on conditions in specific parts of the riverways. The low-cost data-collection methods are of particular interest to managers in other parks who are interested in initiating monitoring systems.

Success of the Program

The Ozark National Scenic Riverways research and monitoring program has integrated visitor research with natural resources research for 20 years. It has focused on two aspects central to management of a national scenic riverway: maintaining water quality and providing quality visitor experiences. From a continuing series of relatively small studies, a management information system has evolved that addresses specific issues and day-to-day management decision making. These studies have also led to several scientific discoveries that have significance beyond the immediate park situation.

The watershed program enabled researchers to improve methods (lower dye-detection limits), discover multidirectional subsurface flow systems, find a way to trace subsurface flows to springs that emerge in river channels, trace a subsurface flow under the Current River to Round Spring, and develop and implement hazard-area mapping.

The visitor research program enabled researchers to create low-cost, social-data collection methods for large wildland areas; develop a recreational carrying-capacity process as a way to improve recreational quality; develop and test integrated recreation-monitoring systems that embody the concepts of management information systems; and incorporate visitor perceptions into recreation management (Chilman et al. 1986, 1989, 1990).

The following factors have enabled the riverways research and monitoring program to be successfully continued for 20 years.

1. The fortunate circumstance of having two research biologists with extensive prior experience with park research programs. Leo Marnell had worked at both Yellowstone and Yosemite National Parks before arriving at the riverways in 1972. He was followed by David Foster in 1977. Foster had two decades of research experience with a variety of projects in the western United States.

2. The recognition that the recreational-carrying-capacity issue was complex and would require a multifaceted, multiyear research program. Leo Marnell was assigned to the riverways to address the political issue of declining fisheries but soon recognized that the rapidly increasing numbers of canoes posed a more serious problem. Marnell proposed a comprehensive, 5-year study program that was approved by superintendent Randy Pope, although political pressures were strong to deal only with the fisheries issues.

The same concept of a long-term research program guided the watershed studies. USFS began the Hurricane Creek Barometer Watershed Study in 1966 to demonstrate correlations between land-use activities and groundwater quality. Although the project was nationally recognized for producing important findings, it was deemphasized (and subsequently abandoned) after 7 years. Nevertheless, the studies had started with the concept of a long-term program.

3. Active recruitment of scientific assistance to implement the research program. The research biologist and superintendent recognized that funding for the program was very limited. Active recruitment of scientific cooperators would be needed to carry out the program. Marnell visited universities and government agencies and attended state and national scientific meetings. Fortunately, researchers viewed Ozark National Scenic Riverways as a unique park and important scientific opportunity. David Foster has continued this very active program of attending scientific meetings and recruiting cooperators.

4. Cooperating researchers with an interest in long-term research. Several researchers participated in projects at the riverways, but two had special long-term interests. Thomas Aley had worked with USFS on the Hurricane Creek project for several years before beginning spring recharge studies at the riverways. He recognized the importance of continuing and extending the watershed and spring studies. Kenneth Chilman had a particular interest in recreational carrying capacity, an issue that was receiving considerable attention in the 1970s and 1980s. He was especially interested in the need to develop and test low-cost, social-data collection methods for large wildland areas.

5. A series of court cases and appeals, which extended the data-collecting period and served as a prominent reminder of the need for research data for legal issues. The data-collecting period for recreational carrying capacity was extended from 5 to 10 years by the series of court cases that culminated in federal court in 1982. The data on visitors' perceptions of crowding from 1972, 1977, and 1979 were especially useful in the 1982 court case. Park managers also recognized that the data on carrying capacity were useful for answering day-to-day questions and for long-term decision making.

6. The willingness of researchers and other staff to discuss the research and its implications on a continuing basis. In many situations, research studies are seen as a contractual arrangement, i.e., just do the work and turn in the report. But at the riverways, considerable learning has taken place from intensive discussions of the research findings. These interactions may be

partly due to the immediate need for the data in court cases and other situations and to the need to collect the data on low budgets. However, the discussions go beyond that. The relationship of the long-term researchers with park managers is very cordial, like that of adjunct staff members. This cordiality extends the relationship: researchers frequently participate in riverways policy sessions and public information meetings.

7. The relative stability of key park staff. The superintendent and research biologist have both worked at the riverways since 1977, and other staff members have worked there for similar periods of time. This longevity enhances communication, understanding, and utilization of long-term research.

Literature Cited

Aley, T. 1975. A predictive hydrologic model for evaluating the effects of land use and management on the quantity and quality of water from Ozark springs. Ozark Underground Laboratory contract report to National Forests in Missouri. [Reprinted in Missouri Speleology, vol. 18, as the entire volume.]

———. 1976. Identification and preliminary evaluation of areas hazardous to the water quality of Alley, Round, and Pulltite Springs, Ozark National Scenic Riverways, Missouri. Ozark Underground Laboratory contract report to Ozark National Scenic Riverways. National Park Service, Van Buren, Missouri. 126 p.

———. 1978. Hydrologic studies of springs, Ozark National Scenic Riverways, Missouri. Ozark Underground Laboratory contract report to Ozark National Scenic Riverways. National Park Service, Van Buren, Missouri. 107 p.

Aley, T., and C. Aley. 1982. Hydrologic studies of springs draining areas east of the Current River in Missouri. Ozark Underground Laboratory contract report to Ozark National Scenic Riverways. National Park Service, Van Buren, Missouri. 108 p.

———. 1987. Groundwater study, Ozark National Scenic Riverways. Ozark Underground Laboratory contract report to Ozark National Scenic Riverways. National Park Service, Van Buren, Missouri. 176 p.

Andrews, M. 1978. Perceptions of crowding by canoe floaters in relation to the floating baseline concept. Unpubl. report, Ozark National Scenic Riverways. National Park Service, Van Buren, Missouri. 28 p.

Barks, J. H. 1978. Water quality in the Ozark National Scenic Riverways, Missouri. U.S. Geological Survey Water-Supply Paper 2048. 57 p.

Chilman, K. 1978. A remeasurement of perceptions of crowding at Ozark National Scenic Riverways. Paper presented at the Annual Meeting of the Rural Sociological Society, 3 September, San Francisco, California. 5 p.

———. 1983. Developing an information gathering system for large land areas. P. 203–215 in S. Lieber and D. Fesenmaier, eds. Recreation Planning and Management. Venture Publishing, State College, Pennsylvania.

———. 1985. Monitoring trends in recreation quality with a recreation resource inventory system. P. 327–336 in J. Wood, ed. Proceedings of the National Outdoor Recreation Trends Symposium II, February, Myrtle Beach, South Carolina.

Chilman, K., D. Foster, and A. Everson. 1986. A carrying capacity rationale and supporting data for river use planning, Ozark National Scenic Riverways. Paper presented at the National Conference on Science in the National Parks, 15 July, Fort Collins, Colorado. 14 p.

———. 1989. Updating the recreational carrying capacity process: recent refinements. P. 234–238 in D. Lime and D. Field, eds. Proceedings of the National Wilderness Management Conference, Minneapolis, Minnesota.

———. 1990. Designing recreation monitoring systems: some comments on the participant observer design. P. 163–172 in D. Hope, ed. Proceedings of the Southeastern Recreation Research Symposium, 14–16 February, Asheville, North Carolina.

Chilman, K., L. Marnell, and D. Foster. 1981. Putting river research to work: a carrying capacity strategy. P. 56–61 in D. Lime and D. Field, eds. Some Recent Products of River Recreation Research. USDA, Forest Service General Technical Report NC-63.

Gisi, D. B. 1971. Recreation expectations of first-time visitors to the Ozark National Scenic Riverways. Unpubl. M.S. thesis, University of Missouri, Columbia. 53 p.

Graefe, A., F. Kuss, and J. Vaske. 1990. Visitor impact management: the planning framework, vol. 2. National Parks and Conservation Association, Washington, D.C. 105 p.

Habermehl, J. 1973. Determining visitor perceptions of crowding on the Ozark National Scenic Riverways. Unpubl. M.S. thesis, University of Missouri, Columbia. 53 p.

Jackson, J. 1970. People problems on the riverways. National Parks and Conservation Magazine 44(275):24–27.

Marnell, L., D. Foster, and K. Chilman. 1978. River recreation research conducted at Ozark National Scenic Riverways, 1970–1977: a summary of research projects and findings. National Park Service, Van Buren, Missouri. 139 p.

National Park Service. 1985. River use management plan, Ozark National Scenic Riverways. Van Buren, Missouri. 46 p.

Nelson, P. W. 1985. The terrestrial natural communities of Missouri. Missouri Natural Areas Committee. 197 p.

Nigh, T. 1988. Missouri natural features inventory: Carter, Oregon, Ripley and Shannon Counties. Unpubl. report, Missouri Department of Conservation. 286 p.

Outdoor Recreation Resources Review Commission. 1962. Outdoor recreation for America. U.S. Government Printing Office, Washington, D.C. 246 p.

Renken, T. 1971. Canoes jamming, damming Current River. St. Louis Post-Dispatch, St. Louis, Missouri, 10 September. 2 p.

Stankey, G., D. Cole, R. Lucas, M. Petersen, and S. Frissell. 1985. The limits of acceptable change (LAC) system for wilderness planning. USDA, Forest Service General Technical Report INT-176. 37 p.

Stevens, D. L. 1991. A homeland and a hinterland, the Current and Jacks Fork riverways: historic resource study. National Park Service, Omaha, Nebraska. 248 p.

Sullivan, A. 1985. Paper prepared for River Use Management Plan meeting, Kansas City, Missouri. National Park Service. 16 p.

———. 1989. Big Spring recharge area sole source aquifer petition. Prepared by T. Aley and W. B. Creath for Ozark National Scenic Riverways. National Park Service, Van Buren, Missouri. 36 p.

USDA, Forest Service and U.S. Bureau of Land Management. 1987. Draft environmental impact statement; hardrock mineral leasing, Mark Twain National Forest, Rolla, Missouri. 110 p.

U.S. Environmental Protection Agency. 1987. Sole source aquifer designation; petitioner guidance. Office of Ground-Water Protection, Washington, D.C. 30 p.

Washburne, R. F. 1982. Wilderness recreational carrying capacity: are numbers necessary? Journal of Forestry 80:726–728.

5

Beyond Denial: Managing with Knowledge

15

Resource Issues Addressed by Case Studies of Sustained Research in National Parks

Gary E. Davis and William L. Halvorson

Has science done its job? Has research provided effective information to protect national parks unimpaired for future generations, or are we losing the parks and their irreplaceable values for science and society? Are national parks good sites in which to study long-term ecological processes, evaluate the effects of global change, and protect America's natural biological diversity? Can scientists and managers learn to protect the parks and apply that experience and knowledge to broader environmental issues for sustainable development?

The case studies in the preceding chapters addressed all of these issues. We hope they will encourage public-policy-makers, park superintendents, and other resource managers to invest in long-term science. We think they provide scientists with models of study designs, analyses, and applications of results that will be effective in communicating resource threats and remedial actions. We also trust that this helps the research community understand that the scientific value of national parks as natural areas is in serious jeopardy.

The 4,300 threats to park resources identified by the National Park Service (NPS) in 1980 can be summarized as six issues: ecosystem integrity and aesthetic degradation, polluted air, altered water quantity or quality, resource consumption, alien species invasions, and visitor impacts (including park operations). The following summary organizes the case studies by these issues and provides a brief synopsis of their findings.

Issue 1: Ecosystem Integrity and Aesthetic Degradation

Chapter 3. "Fire Research and Management in the Sierra Nevada National Parks"

Fire plays a major role in shaping landscapes of many national parks. People have long sought to control fire and use it to change their environment. When the Sierra Nevada national parks were established 100 years ago, protecting the forests from naturally occurring fires was believed essential for their management. Years of fire-suppression activities led to unnaturally high accumulations of fuels and shifts in composition from fire-tolerant to fire-intolerant species. Early observations and studies of the encroachment of pines into the meadows of Yosemite Valley and fires into the giant sequoia groves at Sequoia and Kings Canyon National Parks alerted park officials to these changes. Subsequent research firmly established the necessity of reintroducing fire into those ecosystems and the means for accomplishing that objective through prescribed burning.

As park personnel incorporated the results from these studies into active management programs, they established monitoring procedures to assess the effects of the prescribed burns. Monitoring data then indicated the need for refinements to the program, and additional research was initiated to address those needs. Incrementally, the fire program became more effective. Public challenges of the program and continued evaluation demonstrated the strength of scientifically derived knowledge versus belief-based consensus.

Chapter 5. "Wolf and Moose Populations in Isle Royale National Park"

Predator removal was one of the first resource-management actions taken in U.S. national parks. As with fire suppression, "protection" of desirable prey species was deemed necessary. Only by understanding predator-prey interactions and ecosystem dynamics developed by the scientific community over the past century can one begin to appreciate the profound changes wrought by such uninformed actions and begin restoration. The interaction of wolves (*Canis lupus*), moose (*Alces alces*), and the vegetation upon which moose feed in Isle Royale National Park is one of the longest studied predator-prey systems in terrestrial environments. Under the NPS aegis, various aspects of moose and wolf ecology have been studied for more than 30 years.

Because both moose and wolves are relative newcomers to Isle Royale, researchers have shown great interest—from both theoretical and applied perspectives—in the temporal stability of this predator-prey system. Indeed, much of the early work on the island was devoted to refining basic population-estimation techniques of both moose and wolves. Similarly, vegetation inventory and monitoring were integral parts of those early efforts.

Recently, the wolf population declined significantly to approximately one-half the long-term average. Intensive research identified three major hypotheses for decline: genetics (inbreeding), disease (canine parvo-virus or Lyme disease), and food availability. Scientific and public debate has focused on the causes of the decline, the uncertain prospects for the future, and management options in the event of further decline or extirpation. The future of the predator-prey system in Isle Royale is uncertain, highlighting the importance of the long-term data record of all three trophic levels. Future decisions regarding the management of this system require an evaluation of the utility of basic inventory and monitoring data that have been collected over the years and of species-specific management strategies. This case study clearly demonstrates how the belief that prey species must be selectively protected to sustain them has been replaced by management based on scientific study and an understanding of ecological dynamics.

Chapter 6. "Saguaro Cactus Dynamics"

Attempts to preserve the dynamic biological communities of the Sonoran Desert highlight the difficulties of managing systems without knowing their basic structure and function or without understanding the processes that sustain them. In 1933 Saguaro National Monument was established to protect the saguaro cactus, *Carnegiea gigantea*. The question of saguaro cactus survival is directly related to resource management and interpretative programs at the monument and has spawned crises, posed hypotheses, and provided answers as to how the saguaro survives. The saguaro population, especially in the Rincon Mountain District of the monument, has shown a marked decrease in population density and a dramatic shift in age-class structure. Several hypotheses have been postulated to explain the changes in the saguaro cactus population. These hypotheses include overgrazing and rodents, climate change, bacterial necrosis, and environmental limiting factors, such as freezing temperatures in winter, epidermal browning, and air pollution.

In the 1940s saguaros were cut down and removed from 130 ha of land because scientists believed that a bacterial disease was attacking the population. This approach, however, failed to solve the problem or prove that bacteria caused the saguaro decline. In the 1980s some investigators were still using this deductive approach and trying to prove that one cause was responsible, but other researchers were injecting more reason into the process by their inductive studies.

Fifty years after identifying the problem, researchers established long-term monitoring plots in 1987 to study changes in individual saguaros and the saguaro community. This study addressed such questions as how does browning develop, does epidermal browning occur in young plants at a rate higher than expected, and does mortality exceed recruitment?

Although researchers have conducted studies on the saguaro cactus throughout this century, there is still no consensus on its ecology and, therefore, no consensus on how best to manage the species or the ecosystem in which it resides. Scientists have made the transition, however, from attempting to prove preconceived causes to inductively exploring the broad range of possible causes.

Chapter 9. "Urban Encroachment at Saguaro National Monument"

National parks created in the late nineteenth and early twentieth centuries were isolated bits of wilderness, far from human population centers. During the past 75 years, however, agricultural and urban development has expanded virtually to park boundaries. Many parks are now islands of seminatural habitats engulfed by a sea of altered landscapes. In 1933, when Saguaro National Monument was set aside, the population of Tucson, Arizona, was approximately 35,000 and the city limits were more than 25 km from the monument boundary. Today, the population of greater Tucson is 670,000. The city limits are less than 2 km from the monument and residential construction is taking place right on the boundary.

Beginning in the mid-1980s, managers at the monument became increasingly concerned about the potential immediate and long-term impacts of adjacent community growth and development on monument resources and visitor enjoyment. This concern quickly translated into high levels of involvement with the scientific community, local authorities, and public interest groups. Park managers' interest in this issue coincided with community interest in the impacts of growth and the use of appropriate levels of sensitivity and restraint in development design and construction.

Scientific information provided vital assistance to park managers and the community as a whole in addressing concerns related to growth and development. The University of Arizona School of Renewable Natural Resources provided a scientific foundation for a wide variety of civic actions. The school published the proceedings of a national symposium, "Integrating Wildlife Conservation and New Residential Developments"; identified critical and sensitive habitats in eastern Pima County; developed a strategy for conservation adopted by Pima County and the city of Tucson; and prepared a map inventory of vegetation, riparian areas, land ownership, and zoning for all lands near the monument boundary. Scientists at the school also initiated studies on deer, collared peccary, and desert tortoise populations, on competition between native and nonnative cavity-nesting birds, and on patterns of domestic dog and cat predation. Social scientists conducted research on the sociology of neighboring residents' perceptions, experiences, and values. Local agencies used information from this research to develop land-use ordinances, and NPS used the data to negotiate its positions on adjacent development proposals.

A 1-year moratorium on rezoning close to the monument provided an opportunity to explore a variety of approaches that would allow more resource sensitivity in urban growth management. Resolution of conflicts along the boundary has involved workshops, conferences, attempts at referendum, formal mediation, buffer-overlay zoning ordinances, open-space and trail plans, and boundary adjustments to the monument.

The Tucson community recognized that managing Saguaro National Monument requires an in-depth ecological understanding by the monument staff and by local politicians and developers. Managing the monument has become a community effort. The belief that the monument could be protected simply by isolation has been replaced by an understanding of ecological processes and connectivity, based on continuing study and adjustment.

Issue 2: Polluted Air

Chapter 11. "Air Quality in Grand Canyon"

Clean air is essential to view the spectacular landforms and to conserve the forests and other sensitive plant communities protected in the national park system. NPS began monitoring visibility at Grand Canyon National Park shortly after Congress amended the Clean Air Act in 1977 to establish requirements for protecting visibility in national parks and wilderness. NPS researchers analyzed the monitoring data to characterize visibility conditions and to identify the pollutants and their sources, which were responsible for the degraded visibility. Simulation models showed that air transported from urban and industrialized areas in southern California and Arizona was the primary cause of impaired summer visibility. NPS suspected that local sources had a major impact in the winter.

In response to an inquiry from the Environmental Protection Agency (EPA), NPS identified the Navajo Power Generating Station as a possible source of visibility impairment at the Grand Canyon during the winter. In early 1987, NPS (in conjunction with numerous other parties) conducted the Winter Haze Intensive Tracer Experiment (WHITEX). A unique tracer was injected into the station's stacks, and a dozen monitoring sites were established throughout the Colorado Plateau. Data analysis, using numerous state-of-the-art techniques, showed that the plant was a major contributor to visibility impairment at the Grand Canyon. EPA has used the findings from the WHITEX report as the basis for a proposed regulation requiring substantial emission reductions at the plant to protect visibility at the Grand Canyon. In this case, deductive reasoning built a firm case to protect park resources.

Issue 3: Altered Water Quantity or Quality

Chapter 8. "Water Rights and Devil's Hole Pupfish at Death Valley National Monument"

Securing and protecting water rights are among the most significant water-resource issues that NPS faces in the western states. Of particular concern is the quantification of federal reserved water rights. Resolution of these matters typically occurs in a court of law and requires sound data as evidence.

United States v. Cappaert provides an excellent opportunity to discuss the issue of federal reserved water rights. This case illustrates how data collected through a monitoring program were used to protect NPS water rights and how an indicator species was used to evaluate ecosystem management.

Devil's Hole is a deep limestone cavern in Nevada, which contains a pool that is a remnant of the prehistoric Death Valley lake system. The pool contains the entire population of the federally listed, endangered Devil's Hole pupfish.

In 1968 Cappaert, an adjacent rancher, began pumping groundwater from a site about 2 km from Devil's Hole. The groundwater comes from an underground basin or aquifer that is also the source of water in Devil's Hole. With pumping, the summer water level of the pool in Devil's Hole began to drop. The U.S. Geological Survey has monitored the water level in Devil's Hole since 1962. Until 1969 the water level, with seasonable variations, had been stable.

When the water is at the lowest levels, a large portion of a rock shelf in Devil's Hole is above water. At a critical water level, most of the rock shelf is below water, enabling algae to grow on it. The algae and water, in turn, enable the desert pupfish to spawn. As the rock shelf becomes exposed, the spawning area is decreased, reducing the ability of the fish to spawn in sufficient quantities to prevent extinction.

In August 1971, a 6-year battle began in the courts over the rights to this essential water source. The U.S. district court ruled that pumping that lowers the water level below the critical level would not be allowed. The Court of Appeals for the Ninth Circuit and the U.S. Supreme Court have both affirmed this ruling. Again in a legal setting, deductive reasoning adequately showed how distant pumping from the aquifer affected park resources and changed transboundary water management.

Chapter 10. "Karst Hydrogeological Research at Mammoth Cave National Park"

Water quality in the Mammoth Cave National Park area of Kentucky has long been an issue. The park lies in a classic karst terrane with an adjacent sinkhole plain characterized by a lack of surface streams. Surface water and runoff quickly enters an underground conduit system and may flow several miles in only a few hours. Outside the park, researchers have shown that septic tanks and drain fields lead directly to underground water that enters the cave system.

Public tours began in Mammoth Cave in 1816. Authorized in 1926, the national park was finally established in 1946. It was recognized as a World Heritage Site in 1981 and as an International Biosphere Reserve in 1990. Because of the park's national and international significance and the ecological link between the surface and subsurface water, park managers are greatly concerned that Mammoth Cave be protected from pollution by sources beyond park boundaries.

In 1976 NPS began a program to delineate the movement of water in the park through dye-trace technology. One major element of the program was the development of methods to measure and monitor conduit flow characteristics of the underground system. Present monitoring builds on what has been learned and is developing more traditional measures of water-quality indices.

During this same period, EPA pressured two nearby communities to bring aging sewage-treatment plants into compliance with the Clean Water Act. Effluent from these plants was discharged directly into sinkholes, and thus, into the underground water system. A third community had no treatment facilities. Dye traces showed that pollutants from this community were potentially a major contributor to groundwater resources in Mammoth Cave.

Information derived from research conducted since the late 1970s influenced the development of facilities in the park and the communities adjacent to it. NPS also became a partner in regional planning efforts in 1984. Similar to conservation efforts in Saguaro National Monument, conservation of Mammoth Cave National Park has evolved from an isolationist strategy to one based on an understanding of the park's connectivity to adjacent lands—an understanding acquired through scientific knowledge.

Issue 4: Resource Consumption

Chapter 4. "Yellowstone Lake and Its Cutthroat Trout"

In spite of the inspired concept of leaving American national parks unimpaired for future generations—an idea more than 120 years old—submerged resources in parks are still treated with the axiom, "Out of sight, out of mind," and are very much considered expendable. Even catch-and-release programs, which are generally considered solutions to unsustainable harvests, present a stark contrast to broad national-park philosophy. (Sportsmen do not capture and release feathered and furry critters in parks; why fish?) Large quantities of aquatic plants and animals are regularly killed and removed from national parks in accordance with official policy. Management of these consumptive uses has only recently begun to incorporate ecological information that recognizes the interconnections among harvested species and other ecosystem components. Removing these species generates far-reaching problems that are virtually the same as those caused by predator removal and fire suppression.

Since the creation of Yellowstone National Park in 1872, the cutthroat trout fishery in Yellowstone Lake has been one of the park's main attractions. Reflecting a style of thinking that was common in the era, early beliefs concerning harvests and creel limits were very liberal, and even commercial fishing was authorized. By 1920, however, the glow of exploiting a virgin fishery began to wane, and trout stocks began a slow decline that resulted in a collapse of the sport fishery in the late 1960s. Trout populations and spawning stocks were low, and spawning populations were composed of few age classes instead of the multiple-age classes that thrived historically. The meager condition of the stocks was reflected in poor fishing success by pelicans, ospreys, and bears, as well as people. This knowledge, coupled with the increased awareness of the role and importance of these fish in park ecosystem processes, led to radical changes in management through the imposition of restrictive and often innovative angling regulations. The response of the cutthroat trout population was remarkable. The new angling regulations created some of the finest trout fishing in North America while restoring the native fish to historical densities. Carnivore populations dependent on fish for food also improved, partly because of the increased biomass and distribution of trout, dramatically demonstrating previously unrecognized fisheries impacts on wildlife.

The distinctive phases in the decline and recovery of the trout population are well documented. Spawning activity in key streams was measured periodically from the turn of the century and annually beginning in 1945. Researchers began estimating sport fishery harvests in 1938 and made annual measurements beginning in 1950. In addition to long-term monitoring, research efforts resulted in a body of knowledge on stock dynamics, life history, and ecology of cutthroat trout. This research, coupled with work on the limnology of the lake and dependent carnivores (e.g., grizzly bears, white pelicans, bald eagles, and ospreys) led to a holistic understanding of the ecological processes surrounding Yellowstone Lake. By replacing early beliefs that harvest did not significantly affect the lake and surrounding wildlife with a knowledge of ecosystem dynamics and connectiveness, park managers made possible the sustainable use of natural resources.

Issue 5: Alien Species Invasions

Chapter 7. "Alien Species in Hawaiian National Parks"

Interactions among species are always complex. The results are often devastating when species are taken out of ecological context and injected into new systems without the constraints of age-old "checks and balances" that evolved in their native systems. The damage is especially severe when the system into which alien species are introduced is a national park designed to protect the biodiversity of native ecosystems. Managing the impacts of alien (introduced) species in Hawaii has presented NPS with one of its greatest challenges. Ecosystems of the Hawaiian Islands, like those of other oceanic archipelagoes, are much more vulnerable to biological invasions than are continental ecosystems. Apparently, evolution in isolation from the continual challenge of forces that shape continental organisms—including foraging and trampling by herbivorous mammals, predation by ants and mammals, virulent disease, and frequent and intense fire—leaves island species less fit to compete. Foraging and trampling by feral goats and pigs have been the most destructive forces to the native biota of Hawaii, but numerous other introductions—vertebrates, invertebrates, and plants—contribute to the degradation of native ecosystems.

For more than 20 years, managers of Hawaiian parks have successfully used ecological knowledge to address these difficult problems. During that time, the belief that ecological interactions among alien and native species were not important was replaced with an understanding of the profound effects of aliens on the ecosystem.

Chapter 12. "Rare Plant Monitoring at Indiana Dunes National Lakeshore"

Indiana Dunes National Lakeshore supports a diverse and interesting community of unusual plant species. At present, about 1,315 plant species have been collected and are cataloged at the lakeshore herbarium. They represent not only typical species from the eastern deciduous forest, Great Lakes coniferous forest, and prairie biomes, but also remnant species from the boreal region and Atlantic Coast region. Many plant species in the lakeshore are state protected, one is federally protected, and several others are candidates for federal listing.

To make meaningful and ecologically justifiable decisions, park managers needed to understand the spatial and temporal distribution of rare, threatened, and endangered plant species. A monitoring program, which includes data that are displayed geographically, provides the information needed for the protection and management of plant species with respect to the lakeshore's overall management and development objectives.

Issue 6: Visitor Impacts

Chapter 13. "Wilderness Research and Management in the Sierra Nevada National Parks"

People—and the contrivances needed to house, feed, and transport them—were eagerly encouraged to enter national parks when the parks were first created. To manage hordes of adoring people and protect wilderness, park managers sacrificed some park lands to create developed zones, at the expense of limited wilderness. The challenge of providing wilderness access while maintaining wilderness values remains a major task in many parks. When the Sierra Nevada national parks were established 100 years ago, little thought was given to the problem of visitor impacts on backcountry areas. Years of use by large groups of visitors and stock led to accumulated impacts throughout the parks. Early observations and studies of backcountry impacts alerted park managers to the need for managing use. Subsequent impact inventories, studies, and accurate records of visitation patterns firmly established the need for controlling use and the means for accomplishing that objective through use limits.

Park managers incorporated the research results into their management programs and established monitoring procedures to assess the effects of the wilderness management system. As monitoring data indicated the need for refinements to the program, adjustments were made. Challenges to the system under the National Environmental Protection Act demonstrated the value of using scientifically developed knowledge instead of relying on belief-based consensus.

Chapter 14. "River Management at Ozark National Scenic Riverways"

The mandate to protect and manage the outstanding aquatic and related resources of the Current and Jacks Fork Rivers while meeting a large recreational demand required the development of a vigorous inventory and monitoring program. Park managers were faced with the perennial problem of limited funds to acquire large amounts of reliable information about visitors and resources. In this case, they turned outward to recruit as many scientists as possible from institutions and other agencies. This coordinated effort was designed to inductively study the impacts of visitor use—both types and patterns—on water quality, recharge areas, fisheries, aquatic invertebrates, caves, soils, vegetation, and the visitors themselves. What started out to be a rather simple visitor survey soon developed into a greater understanding of a very complex hydrological system of caves, springs, and rivers. Park managers used the information to develop a comprehensive management plan for visitor use that protects the river resources.

Conclusions

America's national parks are under siege, threatened by habitat fragmentation, invasions of alien species, development on their boundaries, and unsustainable use within the parks themselves. In the past, beliefs, such as those about fire and predators, sufficed to direct conservation strategies. Today, park managers need scientifically reliable information about park ecosystems and the threats to those resources. Without reliable knowledge, they cannot effectively protect and maintain these precious jewels or restore them when protection fails. Application of the scientific method and the results of long-term research in the 12 case studies presented here collectively identify five major themes.

First, ecosystems are dynamic. Long-term views of ecosystems yield dramatically different understandings than 5-year studies. These case studies show that park management needs to be viewed as an experimental, iterative process. The additional knowledge gained from sustained research complicates management options but allows treatment of causes through ecosystem management, not just treatment of symptoms, such as trying to fix system dysfunction one species at a time. Differences among short-term managerial views (which are often based on beliefs in static, isolated landforms) and long-term scientific knowledge of dynamic, interconnected ecosystems, frequently create conflict. Nevertheless, the understanding that comes from sustained research provides hope for undertaking such daunting tasks as ecological restoration and sustainable development and helps build public consensus. In the long-term, parks are clearly dynamic, not static, as once believed and portrayed.

Second, no park is an island. Transboundary forces influence park ecosystems and must be identified and addressed to adequately protect park resources. The myth of isolated parks, separate and apart from the rest of the world, has been deposed by sustained ecological research.

Third, knowledge is better than ignorance. Scientific processes can balance resource protection and visitor use better than belief-based consensus. These cases studies show that scientifically derived knowledge can resolve such issues as how much use is possible without ecological impairment, and how large parks must be to protect system function and avoid losses from habitat fragmentation. The answers obtained through such knowledge are more effective than those obtained through belief-based advocacy.

Fourth, sustained research reveals secrets that short studies never do. Without consistent, long-term data, understandings of the dynamics of ecosystems are flawed. Data are unavailable to park managers for many reasons, including lack of consistent funding, lack of management support for research, and lack of understanding the need for long-term data. These case studies illustrate the value of using parks for long-term research and show how long-term data can improve management decisions.

Fifth, research must be a cooperative effort. Many problems arise when research programs are isolated or when managers and scientists do not cooperate. To be meaningful, NPS research programs must be managed by both scientists and park managers, and those programs must be related to a broader, university-based community.

16

Lessons Learned from a Century of Applying Research Results to Management of National Parks

William L. Halvorson and Gary E. Davis

Americans are witnessing a paradigm shift in national park conservation and management. In the early years of national parks, conservation strategies and management actions were based on a belief that parks were static landscapes, isolated from human activities and adjacent lands (Sumner 1983; Sellars 1993a), and that they were solely for the pleasure of visitors (Sellars 1993b). Attempts to resolve system dysfunction—wrought by this erroneous view and the consequences of subsequent management policies, such as predator removal and fire suppression—were approached one species at a time (G. M. Wright et al. 1933; G. M. Wright and Thompson 1935; Leopold et al. 1963). Today, this paradigm is slowly shifting to a knowledge-based understanding of ecosystems as dynamic and interconnected (Robbins et al. 1963; Orians 1986). Conservation strategies now recognize the need to include people as part of the system and to address causes, rather than symptoms, of system dysfunction by managing whole ecosystems, not just single species (Agee and Johnson 1988).

We organized into five themes the lessons from the past century of research and management in national parks that these case studies teach: ecosystems are dynamic, no park is an island, knowledge is better than ignorance, sustained research reveals secrets that short studies never do, and research must be a cooperative effort.

Theme 1: Ecosystems Are Dynamic

National Parks Are Not the Static Entities That They Were Once Portrayed to Be

Through studies such as those presented here, National Park Service (NPS) managers have come to recognize a need to change attitudes toward resource management. The traditional view that "All we have to do is build a wall around the park and protect it" is changing. In the beginning, parks were perceived as being under siege from fire, predators, poachers, and eventually pollution, development, and too many visitors. This early approach was replaced in 1988 by a policy to "assemble baseline inventory data describing the natural resources under [NPS] stewardship and [to] monitor those resources at regular intervals to detect or predict changes. The resulting information will be analyzed to detect changes that may require intervention and to provide reference points for comparison with other, more altered environments" (National Park Service 1988). This modern policy clearly recognizes the dynamic nature of parks (Chapter 2). The shift is further demonstrated by recent NPS policy on fire (Chapter 3) and new conceptual designs for Yellowstone Lake (Chapter 4) and saguaro cactus ecology (Chapter 6).

Absence of Information Leads to False Conceptual Models and Costly Mistakes

The earliest conceptual models of NPS areas were of scenic places to be used by the public as pleasuring grounds. Managers were not too concerned about obtaining knowledge of what they were managing as long as it looked good (Chapter 2). This conceptual model was based mostly on belief and resulted in the practices of predator removals to assure deer in the meadows and total fire suppression to keep the forest a "politically correct" green. These practices, however, disrupted naturally functioning ecosystems. The short-term benefits were typically offset by longer term financial and ecological costs. Armed with better ecological understanding and information from a long-term view of the consequences of these actions, managers revised their conceptual models and set a new course—one that included a greater understanding of the complexity, structure, and function of biological systems as well as an appreciation of scenic beauty. This change to new, scientifically based models has been slow and cumbersome, being pushed from the outside and restrained by turmoil on the inside (Chapter 2; R. G. Wright 1992).

Fire management is one of the more publicized changes in policy that reflect a new understanding of ecosystem dynamics (Chapter 3). The early concept that fire disturbed stable "climax" communities led to the unfortunate decision that forests had to be protected by suppressing all fires. Only after years of gathering information on forest dynamics did scientists recognize fire as an important environmental force in many systems and identify fire suppression as the disturbance causing dramatic changes.

Predator-prey relationships provide classic examples of how much trouble false conceptual models can cause (Chapter 5). The Isle Royale story shows clearly how misguided were early actions to protect "game species" by removing predators. The litany of controversies over Yellowstone elk, wolves, coyotes, and grizzly bears and over Kaibab Plateau deer in Arizona exemplifies the cost of making decisions based on beliefs or without adequate knowledge of system functions (Leopold et al. 1963; R. G. Wright 1992).

The shifting paradigm of cutthroat trout management at Yellowstone National Park is another case in which sustained studies provided information to managers, who periodically revised their concept of how that complex system works (Chapter 4). This strategy is inductive—to add information until enough is known about the roles of trout, predators, and fishing to modify management of the lake, and then, to begin a new cycle and add again.

Theme 2: No Park Is an Island

NPS Areas Are Heavily Influenced by Transboundary Forces

Water rights at Devil's Hole (Chapter 8), air quality at Grand Canyon (Chapter 11), biological invasions in Hawaii (Chapter 7), water flow through Mammoth Cave (Chapter 10), and urban encroachment at Saguaro National Monument (Chapter 9) all show the influences of neighboring activities on national parks. Resolution of water-rights issues at Devil's Hole demonstrates a classic legal approach to resolving transboundary issues. Solving the air-quality issues at Grand Canyon National Park proved the power of combining state-of-the-art science with an advocate's deductive reasoning. The battles to protect Hawaiian parks from alien species and the efforts to protect the water quality in Mammoth Cave show the never-ending nature of trans-boundary issues. However, the cooperative management of Saguaro National Monument offers a promising solution to such threats by involving the neighboring community.

Theme 3: Knowledge Is Better than Ignorance

Verifiable Information Can Help Solve Use-Versus-Protection Dilemmas Better than Consensus-based Opinions

Instead of making decisions based on a belief system, park managers effectively used scientific knowledge in the Sierra Nevada parks (Chapter 13) and at the Ozark Riverways (Chapter 14). In both areas, visitor use increased incrementally. Only after they obtained sufficient knowledge of visitor effects on resources did park managers develop effective management strategies. At Indiana Dunes National Lakeshore, management of rare plants could not take place without detailed information on population biology and on human developments in and around the lakeshore (Chapter 12).

Theme 4: Sustained Research Reveals Secrets That Short Studies Never Do

NPS Areas Need Active Programs on Long-term Ecosystem Dynamics

These case studies are a stark reminder that no area in the national park system can do without information on its resources. Sharp and Appleton (1993) noted that NPS suffers from an embarrassing lack of knowledge about the natural resources in its parks. Furthermore, a 1992 study stated that informed resources management in NPS is impossible without science (National Research Council 1992).

Case studies show how information from sustained research averted the loss of endangered species in Death Valley National Monument (Chapter 8) and Indiana Dunes National Lakeshore (Chapter 12), reduced ecological damage from pollution in Mammoth Cave National Park (Chapter 10), and reduced the damage to visibility from air pollution in Grand Canyon National Park (Chapter 11). In fact, every case stresses that without resident research programs in the parks that make scientists regularly available to the park staff, false conceptual models continue to flourish, and consensus-based opinions continue to drive management decisions. Managers greatly benefit from such programs, which provide information on the dynamics of ecosystems—information that is vital to proper natural-resource management. Information gathered through research and long-term monitoring can be beneficial to all divisions of the park, including interpretation, protection, and maintenance.

Parks Are Good Places to Conduct Long-term Research

One of the most troubling problems in managing sustained research is that research sites are continually being lost to development as society marches across the landscape (Risser 1991). NPS areas are rapidly becoming "life rafts," where natural systems still survive in a sea of development. They can thus provide stable, protected sites for sustained studies. Agencies such as the National Science Foundation, however, have been reluctant to fund long-term research in any NPS areas because of a perception that NPS managers not only lacked an appreciation for sustained research but also opposed it (see NPS resource-management policy statements before 1988). As park managers change their views on monitoring and long-term resource management, a new relationship with scientists and the National Science Foundation could be forged (R. G. Wright 1992). With continuing NPS support through an operations-based program in resource monitoring, a more cooperative relationship with researchers will emerge in the national parks. This will benefit not only the park resources through increased information but also the understanding of landscape and ecosystem ecology.

Consistency Is Absolutely Necessary in Long-term Studies

Reasons for the paucity of long-term ecological studies in NPS areas include inconsistent support from park management and funding sources, lack of scientific leadership dedicated to sustained studies (Chapter 9), and lack of support to maintain a database management system in an environment of regularly changing personnel (Chapter 10; Likens 1989; Risser 1991). Before information that is useful for managers can be developed, all of these impediments need to be overcome. Short-term studies in general can lead to erroneous conclusions, either because they miss infrequent but important random events (Wiens 1977) or because they overestimate the importance of some unusual event, lacking the benefit of a long-term perspective (Weatherhead 1986).

Useful databases cannot be generated if there are gaps of years when data are not collected. Any resource-data collection program should be set up to be operational for a reasonably long time, but with built-in reviews at given intervals. Such programs should not continue unquestioned for years, but neither should it be possible for a new staff member to shut down a project because of personal preference. Likewise, a program should not be lost because a particular scientist moves to another area. The means must be found to institutionalize an active program of data collection, storage, summary, analysis, and reporting, despite changing personnel.

Long-term Data Sets Are Politically Powerful

Managers of natural areas and natural resources must regularly do battle in legal and political arenas. This need is increasing as parks become more and more affected by surrounding human developments. To effectively uphold the rights of natural resources, managers must have data and information that are sound enough to hold up (in both the public and legal sense) against the desires of those who want to use the resources for human activities or who will harm those resources by their activities. For instance, by monitoring air quality, researchers at Grand Canyon National Park were able to prove the effect of a nearby power-generating plant on visibility in the canyon (Chapter 11). The Devil's Hole pupfish was saved only because of research that showed the relationship between the habitat of this endangered species and regional groundwater use (Chapter 8). Managers at Saguaro National Monument must actively ameliorate the effects of urbanization on species within the monument boundary (Chapter 9). Understanding the interchanges of species between the monument and surrounding housing developments helped the superintendent to sensitize the public to the effects of developments near the natural area. This public awareness led to changes in zoning close to the monument that assist in protecting its biological resources, not just its scenic values.

Long-term Monitoring Is Cost-Effective for Protecting Resources

An active program to regularly assess the condition of resources helps managers identify problems and suggest solutions at an early stage. This program is analogous to regular physicals for individuals. In Hawaiian national parks, early detection of some alien species has allowed park managers to effectively remove them with minimum effort and cost (Chapter 7). By monitoring backcountry use, managers in the Sierra Nevada parks can adjust use to minimize damage, and thus, the need for large-scale restoration projects (Chapter 13). The use of monitoring protocols at Ozark National Scenic Riverways allowed park managers to effectively set carrying capacities on river use to protect resources (Chapter 14). This "stitch-in-time" policy saves the cost of restoring areas after they have been damaged.

Not All Long-term Studies Can Be Readily Applied to Short-term Management Issues

The dynamics of population interactions are extremely complicated, and the understanding of such interactions is still very shallow at best. The long-term studies of wolf and moose populations are a good example of this complexity (Chapter 5). Through them, scientists have understood that defining population interactions comes only with great diligence. Even after 15 or 20 years, the changes from year to year can still be puzzling (see also Wiens 1977 and Weatherhead 1986).

Managers tend to be uneasy when told that the information gained in long-term studies may be applied to management questions only with a great deal of uncertainty. Having spent time and money on gathering information, managers believe it is only logical that such information should be quite valuable. These long-term studies are important, both to science and management, but science does not guarantee certainty. The obstacles to making accurate predictions of the behavior of populations or communities are formidable. In most cases, it is best to view resource management projects as experiments. Every action should be followed by the collection of data to evaluate that action (National Research Council 1986). Ecosystems are characterized by combinations of stability and instability and by unexpected shifts in behavior from both internal and external forces. Ecosystem management, therefore, is a process of learning by doing, and documentation is critical to that process (Walters and Holling 1990).

Long-term data are essential to understanding many ecosystem processes and dynamics that ecosystem management must consider. They are fundamental to understanding such things as successional changes, fluctuations in weather patterns, and the effects on ecosystem function by weather patterns, storm regimes, and fire (Risser 1991).

Even with well-managed data sets from long-term ecological research, not all questions that managers deal with on a daily basis can be answered. Short-term studies must still augment the understanding that long-term monitoring and research bring. When patterns of change are identified, researchers often must conduct studies involving experimental manipulation to determine the cause of the change (Franklin 1989; National Park Service, n.d.).

Theme 5: Research Must Be a Cooperative Effort

The Relationship between Scientists and Park Staff Is Crucial

Recognition, support, and leadership in understanding the need for and value of long-term ecological research in our national parks must come from both scientists and NPS managers (Parsons 1989). The Cary Conference on long-term ecological research, held in May 1987 (Likens 1989), stated that because they have common long-term goals, scientists and resource managers must forge a new partnership. The conference suggested that this partnership include (1) an agreement by scientists to answer, to the best of their ability, the questions posed by managers, while making clear the level of uncertainty that exists and the additional research that needs to be done; and (2) an agreement by managers to consider these answers seriously and to support the continuing research toward better answers.

Data are important, but they must also be understandable; i.e., they must convey information to the managers who need it for day-to-day decisions. Scientists and managers must work cooperatively. Chapter 10 reveals the problems that arise when such teamwork is lacking. A partnership requires a change in attitude in both the manager and the scientist; each must understand and respect the values and needs of the other. Without such respect, there will be constant struggle in any NPS research program. Narrowly focused and obstinate scientists are often as troublesome to the search for truthful information as are superintendents who feel the need to impose their authority and direct research.

Research in National Parks Should Be Jointly Supervised by Local Superintendents, Regional Chief Scientists, and the Scientist's Research Supervisor

Many individuals, committees, commissions, and task forces have advised NPS to establish an independent research branch (Chapter 2). All of that advice has had little effect on Congress, the U.S. Department of the Interior, or the NPS research program. The relationship between research and management has become even more complicated because of the creation of the National Biological Service within the U.S. Department of the Interior.

However research is organized, these case studies clearly show that supervision of research must be done cooperatively at three levels. Scientists need input from superintendents to ensure that individual park needs are met and from the regional chief scientist to ensure that NPS research needs are fulfilled. In addition, the scientists' own need to maintain standing in the scientific community must be protected by proper supervision from a trained scientist.

Each Park's Research Program Should Be Related to a University Peer Group or the Larger Research Community

The case studies and program issues described in this volume indicate that each NPS research program should be strongly related to a larger research community. This tie could be made either through a university-based, cooperative-research unit or through an advisory group. A park's research program must be protected from becoming so ingrown that it is unrelated to the latest scientific knowledge or is insensitive to theories and concepts that are being developed by other researchers. In studies of the saguaro cactus (Chapter 6), interested superintendents got bad information and advice from researchers who failed to incorporate available information, failed to involve an adequate review process, and failed to use a holistic approach in planning their specific research projects. Many other park research programs have ignored the larger research community and thereby restricted their view of science, all to the detriment of the natural resources.

As scientists and managers, our primary concern should be that the resources of NPS areas are well cared for, i.e., preserved for the enjoyment of future generations. We must work more cooperatively to ensure this outcome. We must get beyond the personalities and personal agendas of researchers and managers and beyond the park boundaries. We must work with partners of many disciplines, looking at resources on a regional scale. The days of preserving the parks' scenery and popular game animals, and of providing good fishing and other visitor enjoyments in an isolated venue, are gone. It is time to embrace the challenge of managing ecosystems in an integrated landscape. To do that, NPS must adopt resource management programs that include long-term monitoring and research and that are an ongoing part of park operations.

Literature Cited

Agee, J. K., and D. R. Johnson. 1988. Ecosystem management for parks and wilderness. University of Washington Press, Seattle. 234 p.

Franklin, J. F. 1989. Importance and justification of long-term studies in ecology. P. 3–19 in G. E. Likens, ed. Long-term Studies in Ecology, Approaches and Alternatives. Springer-Verlag, New York.

Leopold, A. S., S. A. Cain, C. M. Cottam, I. M. Gabrielson, and T. L. Kimball. 1963. Report of the advisory board on wildlife management. National Parks Magazine, Insert 4-63, April:I–VI.

Likens, G. E., ed. 1989. Long-term studies in ecology, approaches and alternatives. Springer-Verlag, New York. 214 p.

National Park Service. 1988. Management policies. Washington, D.C. 4:4.

———. n.d. Planning for the future, a strategic plan for improving the natural resource program of the National Park Service. Washington, D.C. 156 p.

National Research Council, Commission on Life Sciences, Committee on Applications of Ecological Theory to Environmental Problems. 1986. Ecological knowledge and environmental problem solving. National Academy Press, Washington, D.C.

National Research Council. 1992. Science and the national parks. National Academy Press, Washington, D.C. 122 p.

Orians, G. H. 1986. Site characteristics favoring invasions. P. 133–148 in H. A. Mooney and J. A. Drake, eds. Ecology of Biological Invasions of North America and Hawaii. Springer-Verlag, New York.

Parsons, D. J. 1989. Additional views, evaluating national parks as sites for long-term studies. P. 171–173 in G. E. Likens, ed. Long-term Studies in Ecology, Approaches and Alternatives. Springer-Verlag, New York.

Risser, P. G., ed. 1991. Long-term ecological research, an international perspective. John Wiley & Sons, New York. 294 p.

Robbins, W. J., E. A. Ackerman, M. Bates, S. A. Cane, F. F. Darling, J. M. Fogg, Jr., T. Gill, J. M. Gillson, E. R. Hall, and C. L. Hubbs. 1963. A report by the advisory committee to the National Park Service on research. National Academy of Science, National Research Council, Washington, D.C. 156 p.

Sellars, R. W. 1993a. The rise and fall of ecological attitudes in national park management, 1929–1940. George Wright Forum 10(3):38–54.

———. 1993b. Manipulating nature's paradise. Montana: The Magazine of Western History 43(2):2–13.

Sharp, B., and E. Appleton. 1993. The information gap. National Parks 67(11–12): 33–37.

Sumner, L. 1983. Biological research and management in the National Park Service: a history. George Wright Forum 3(4):3–27.

Walters, C. J., and C. S. Holling. 1990. Large-scale management experiments and learning by doing. Ecology 71:2060–2068.

Weatherhead, P. J. 1986. How unusual are unusual events? American Naturalist 128: 150–154.

Wiens, J. A. 1977. On competitive and variable environments. American Scientist 65:590–597.

Wright, G. M., J. S. Dixon, and B. H. Thompson. 1933. A preliminary survey of faunal relations in national parks. National Park Service Fauna Series 1. U.S. Government Printing Office, Washington, D.C. iv + 157 p.

Wright, G. M., and B. H. Thompson. 1935. Wildlife management in the national parks. National Park Service Fauna Series 2. U.S. Government Printing Office, Washington, D.C. viii + 142 p.

Wright, R. G. 1992. Wildlife research and management in the national parks. University of Illinois Press, Urbana. 224 p.

List of Contributors

Chapter 1

William L. Halvorson
Unit Leader
National Biological Service
Cooperative Park Studies Unit
125 Biological Sciences East
University of Arizona
Tucson, AZ 85721

Gary E. Davis
Research Marine Biologist
National Biological Service
Channel Islands Field Station
1901 Spinnaker Drive
Ventura, CA 93001

Chapter 2

Ervin H. Zube
Professor
School of Renewable Natural Resources
University of Arizona
Tucson, AZ 85721

Chapter 3

David J. Parsons
Director
Aldo Leopold Wilderness Research Institute
P.O. Box 8089
Missoula, MT 59807

Jan W. van Wagtendonk
Research Biologist
National Biological Service
Yosemite Field Station
P.O. Box 577
Yosemite National Park, CA 95389

Chapter 4

John D. Varley
Paul Schullery
Research Biologists
Yellowstone National Park
P.O. Box 168
Yellowstone National Park, WY 82190

Chapter 5

R. Gerald Wright
National Biological Service
Cooperative Park Studies Unit
Department of Wildlife Resources
University of Idaho
Moscow, ID 83843

Chapter 6

Joseph R. McAuliffe
Research Director
Desert Botanical Garden
1201 N. Galvin Parkway
Phoenix, AZ 85008

Chapter 7

Charles P. Stone
Former Research Scientist (Retired)
Hawaii Volcanoes National Park
HC 53, Box 30D
Cass Lake, MN 56633

Lloyd L. Loope
National Biological Service
Haleakala Field Station
P.O. Box 369
Makawao, Maui, HI 96768

Chapter 8

Owen R. Williams
Jeffrey S. Albright
Paul K. Christensen
William R. Hansen
Jeffrey C. Hughes
Alice E. Johns
Daniel J. McGlothlin
Charles W. Pettee
Stanley L. Ponce
Hydrologists
National Park Service
Water Resources Division
1201 Oak Ridge Drive, Suite 250
Fort Collins, CO 80525

Chapter 9

William W. Shaw
Professor
School of Renewable Natural Resources
University of Arizona
Tucson, AZ 85721

Chapter 10

E. Calvin Alexander, Jr.
Professor
Department of Geology and Geophysics
University of Minnesota
Minneapolis, MN 55455

Chapter 11

Christine L. Shaver
Senior Attorney
Environmental Defense Fund
1405 Arapahoe Avenue
Boulder, CO 80302

William C. Malm
Research Physicist
Air Quality Division
National Park Service
P.O. Box 25287
Denver, CO 80225

Chapter 12

Noel B. Pavlovic
Research Biologist
Indiana Dunes National Lakeshore
1100 N. Mineral Springs Road
Porter, IN 46304

Marlin L. Bowles
Associate Ecologist
Morton Arboretum
Lisle, IL 60532

Chapter 13

Same authors as for Chapter 3

Chapter 14

Kenneth Chilman
Professor
Department of Forestry
Southern Illinois University
Carbondale, IL 62901

David Foster
Research Biologist
Ozark National Scenic Riverways
Van Buren, MO 63965

Thomas Aley
Director
Ozark Underground Laboratory
Protem, MO 65733

Chapter 15

Same authors as for Chapter 1

Chapter 16

Same authors as for Chapter 1

Index

About the Editors . . .

WILLIAM L. HALVORSON is a plant ecologist and research manager with the National Biological Service (NBS). He holds degrees in botany (plant ecology) from Arizona State University (B.S. and Ph.D.) and the University of Illinois (M.S.). He served 8 years on the faculty at the University of Rhode Island (1970–1978) before joining the National Park Service (NPS). After working 13 years as an NPS research biologist in Washington, D.C. and Channel Islands National Park, Halvorson returned to a university setting. He is both a faculty member and the leader of the NBS Cooperative Park Studies Unit (CPSU) at the University of Arizona. The CPSU is responsible for conducting and coordinating research for parks in southern Arizona and the deserts of eastern California.

Halvorson helped to develop an exemplary natural-resource monitoring program at Channel Islands National Park. Besides doing long-term monitoring, he is actively engaged in research related to the restoration and rehabilitation of natural ecosystems, rare and alien species biology, arid and semiarid plant community ecology, and management of natural areas. He serves on several national and regional government task forces and committees related to research and resources management in the U.S. Department of the Interior. He has organized a conference on research and resources management on and around the Channel Islands. Halvorson currently serves as treasurer of the Society of Ecological Restoration. For the Ecological Society of America (ESA), he served on the Public Affairs Committee during the development of the Washington Office (1983–1986) and, after helping to create the ESA Corporate Award, served on the Corporate Award Subcommittee of the Awards Committee (1987–

1993, the last 3 years as chair). He has also held the office of secretary of ESA's Applied Ecology Section.

GARY E. DAVIS is a research marine biologist with NBS. Assigned to the California Science Center, Davis, he conducts research on long-term ecosystem dynamics and explores ways to sustain marine fisheries in Channel Islands National Park and Cabrillo National Monument. He also consults on research and management of national parks and marine fisheries worldwide. A native of San Diego, he holds B.S. and M.S. degrees in biology from San Diego State University.

Davis began his career with NPS in 1964 as a ranger at Lassen Volcanic National Park. Following graduate school, he conducted research and administered scientific programs from 1968 to 1980 in the U.S. Virgin Islands and Florida at Virgin Islands, Everglades, Dry Tortugas, and Biscayne National Parks. He returned to California and assumed his present post when Channel Islands National Park was established in 1980, transferring from NPS to NBS with the latter's creation in 1993.

Davis has published more than 100 articles in professional journals, books, and technical reports on ecosystem management, spiny lobster and abalone fishery biology, sea turtle management, coral reef dynamics, and kelp forest ecology. He was certified as a fisheries scientist by the American Fisheries Society in 1980 and elected a fellow of the American Institute of Fishery Research Biologists in 1985. Active in several professional societies, he has served as president of the American Academy of Underwater Sciences, director of the Natural Areas Association, president of the George Wright Society, and district director of the American Institute of Fishery Research Biologists. Davis has received several professional awards, including U.S. Department of the Interior Superior Service Awards for his contributions as a TEKTITE I Aquanaut in 1969 and for his research on long-term ecological monitoring at Channel Islands National Park in 1986.